UNDER THE CANOPY

UNDER THE CANOPY

The Archaeology of Tropical Rain Forests

Edited by Julio Mercader

RUTGERS UNIVERSITY PRESS
New Brunswick, New Jersey, and London

Library of Congress Cataloging-in-Publication Data

Under the canopy : the archaeology of tropical rain forests / Julio Mercader (editor).
   p. cm.
   Includes bibliographical references.
   ISBN 0-8135-3142-X (alk. paper)
   1. Prehistoric peoples—Rain forests.   2. Hunting and gathering societies—Rain
forests.   3. Excavations (Archaeology)—Rain forests.   4. Rain forests—Antiquities.
I. Mercader, Julio, 1966–

GN890 .U53 2002
306′.08′09152—dc21

                       2002023717

British Cataloging-in-Publication data for this book is available from the British
Library.

Manufactured in the United States of America

Contents

PART III

The Last Frontier: Newcomers in a New World

Acknowledgments

The authors and the editor thank Stanley Ambrose, Richard Cooke, Michael Heckenberger, Charles Higham, Nicky Horsfall, Karl Hutterer, Renato Kipnis, Betty Meggers, John Miksic, Michael Morwood, Diogenes Patiño, Michael Petraglia, Scott Raymond, Martha Tappen, Pam Willoughby, and two anonymous readers for their help. The editor would like to thank Raquel Martí, Alison Brooks, Robert Bailey, and Richard Grinker for their support through the years.

UNDER THE CANOPY

Fig. I.1. Courtesy of Robert L. Humphrey.

Introduction
The Paleolithic Settlement of Rain Forests

Julio Mercader

Almost exactly in the middle [of the tropics] . . . lies . . . a vast expanse of
dense, damp and inhospitable-looking darkness. . . . Anyone who has stood in
the silent emptiness of a tropical rain forest must know how . . . [people] coming
. . . from an open country . . . of sunlight . . . [must feel]. Many people who . . .
have lived there, feel just the same, overpowered by the heaviness of everything—the damp air, the gigantic water-laden trees that are constantly dripping,
never drying out between the violent storms that come with monotonous regu-
larity, the very earth itself heavy and cloying after the slightest shower. And,
above all, such people feel overpowered by the seeming silence and the age-
old remoteness and loneliness of it all.

But these are feelings of outsiders, of those who do not belong to the forest.
If you are of the forest it is a very different place. What seems to other people to
be eternal and depressing gloom becomes a cool, restful, shady world with light
filtering lazily through the tree tops that meet high overhead and shut out the
direct sunlight—the sunlight that dries up the non-forest world of the outsiders
and makes it hot and dusty and dirty.

Even the silence is a myth. If you have ears for them, the forest is full of
sounds—exciting, mysterious, mournful, joyful. . . . And the most joyful sound
of all . . . is the sound of the voices of the forest people as they sing a lusty cho-
rus of praise to this wonderful world of theirs—a world that gives them everything they want. . . . But if you are an outsider from the non-forest world . . . this
glorious song would just be another noise to grate on your nerves.

—C. Turnbull, *The Forest People: A Study of the Pygmies of the Congo*

As Colin Turnbull noted in 1962, outsiders tend to depict rain forests as
impenetrable worlds of chaos, timeless "jungles" in which ancestral plants,
creatures, and humans are trapped. Romantic clichés portray counterfeit
descriptions of a pristine jungle, unchanged through time, in which animals
and plants are described with a plethora of aggrandizing superlatives. Con-
trarily, the human beings that inhabit this frozen homeland of botanical and
zoological wonders are referred to with degrading epithets allusive to ata-
vistic cultural features inherited from a timeless prehistoric past.

1

It is no surprise that archaeologists have shown little interest in discovering the prehistory behind the trees. Popular re-creations of early "primitive" life in the "jungle" draw on myths from nineteenth-century travel accounts, old ethnographic reports, and novels and perceive the forest as a barrier to "civilization." Thus, the early settlement of rain forests would be of little interest. Secondly, the tropical forest is often viewed as an extreme environment whose settlement requires great cognitive skills and technological endowment; that is, a high-risk and unhealthy ecosystem that was avoided by early hominids (Bar-Yosef and Belfer-Cohen, 2000), first colonized by anatomically modern humans (McBrearty and Brooks, 2000) or farmers (Bailey et al., 1989). Thirdly, there is the stereotype that the prehistory of the tropical forest is unknowable through archaeological research, given that organic remains decompose quickly in these environments, and potential archaeological materials would disintegrate and vanish in the acidic rain forest soils (see box 1). Yet, archaeological data to support or refute the above-described popular and academic assumptions have been totally lacking till recently.

Under the Canopy indicates that prehistoric foragers were fully capable of a long-term occupation of tropical forests and that, by late glacial times, the settlement of the world's rain forests was already well established. Uninterrupted human occupations for centuries must have inflicted a human signature on the makeup, structure, and geographical distribution of rain forests. Far from being pristine jungles, tropical forests today may be variable products of human and natural forces. And tropical forests in the past may be far from having modern analogues, as they occurred under Pleistocene climatic regimes not prevalent today. This is an important aspect to be observed when researching the many ways in which prehistoric groups responded to prehistoric ecosystems. The continuous and repeated inhabitation of the rain forest by prehistoric hunter-gatherers over hundreds of generations has brought about a tight interaction between human and biotic communities (Head, 1989; Piperno, 1994; Bush and Colinvaux, 1994), sometimes influencing tropical forest species composition through the use of fire (Piperno et al., 1991; Hopkins et al., 1993; Piperno, 1994, Bush and Colinvaux, 1994; Haberle, 1994; Kershaw, 1994; Hart et al., 1996).

Hunting and Gathering in Tropical Rain Forests: *Was* It Possible?

Perceived environmental and diet constraints in today's closed forests, especially lack of wild carbohydrates (Bailey et al., 1989), have sustained current anthropological depictions of the early prehistoric settlement of the tropical forest. It has been demonstrated that present forest dwellers do not live independently of farming. Therefore, the ability of prehistoric foragers to subsist in tropical forests on purely hunting and gathering grounds was questioned (Hart and Hart, 1986; Headland, 1987; Bailey and Peacock, 1988; Bailey et al., 1989; Bailey and Headland, 1991; Gamble, 1993; Headland, 1997). As a result, anthropological models portrayed dense tropical forests of the Holocene as unfriendly environments unable to support prehistoric foragers before the advent of farming (Headland, 1987; Bailey et al., 1989; Bailey and Headland, 1991; Headland, 1997). Prehistoric hunter-gatherers, thus, lived in tropical forests for the last few millennia, only after farmers colonized rain forests and enhanced a naturally low productivity by farming and subsequent environmental alteration of closed-canopy forests (Bailey et al., 1989: 73). The farming modification of the forest brought about a wider availability of game, which, in turn, made hunting and gathering feasible. This theory, referred to as the "null hypothesis," has very important implications for human evolution. These implications are (1) early humans lacked the capacity to settle extreme environments; (2) extensive population deserts existed throughout the wet tropical belt during the entire Pleistocene and most of the Holocene; (3) inherent human inability to colonize and live on rain forests was overcome during the late Holocene; (4) global colonization by archaic humans was highly differential and excluded tropical forests; and (5) hunter-gatherers were incapable of indirect or direct modification of the tropical forest's structure, composition, and productivity.

In spite of well-known ecological limitations for present-day humans dwelling in tropical forest environments (animal and plant food supplies are highly diverse, dispersed, and difficult to obtain, but see Townsend, 1990; Colinvaux and Bush, 1991; Bahuchet et al., 1991; Hladik and Dounias, 1993; Brosius, 1991; Dwyer and Minnegal, 1991; Endicott and Bellwood, 1991; Stearman, 1991), this volume presents archaeological evidence that the occupation of tropical forests has deep roots and much predates the horticulturalist colonization of these ecosystems. An early pre-farming settlement of tropical forests is the rule, not the exception. The archaeological sequences reported in this book, as well as those reported elsewhere (Horsfall, 1987; Gnecco and Mora, 1997; Roosevelt et al., 1996; Mercader et al., 2000), demonstrate that tropical forest environments supported a continuous settlement by hunter-gatherer groups for millennia.

But, Was There Any Forest at the Time of Pleistocene Occupation?

An assessment of the feasibility of tropical forest occupation by prehistoric hunter-gatherers relies on the available environmental data to demonstrate

Box 1. Preservation of archaeological materials and stratigraphic integrity in rain forests

Most workers would agree that overall preservation conditions in wet tropical environments are not as optimal as in cold and dry latitudes, and lead to a rapid decay of archaeological materials in the acidic environments of the rain forest. Bioturbation by plants and insects (Johnson, 1990; McBrearty, 1990) and geological processes such as "stone-line" formation (Mercader et al., 2002) undoubtedly contribute to the disturbance of the archaeological record in rain forests to variable degrees, ranging from slight to severe. The little direct evidence that we have, however, suggests that the processes and variables affecting site formation in lowland tropical rain forests vary widely even within the same region and environment.

Subaerial bone weathering as described by Behrensmeyer (1991) and studied by Kerbis and others (1993) and Tappen (1994) suggests that surface bones in rain forests do not weather as fast as bones in savanna environments. Diagenetic bone destruction resulting from burial in acidic soils is a different matter that, in the rain forest, remains poorly understood. Shell and plant remains such as carbonized pieces, charred parenchyma from tubers, opal silica from decomposed plant tissues, and starch grains (Balme and Beck, 2002) do not respond to acidity the same way bones do. A bibliographic survey demonstrates the existence of variably large bone and plant archaeological assemblages from wet tropical environments. Associations, age, and geochemical features of the matrix vary from one place to another. In East Africa, the fossil remains of *Ardipithecus ramidus kadabba* (~5.6 My) (Woldegabriel et al., 2001), as well as *Ardipithecus ramidus ramidus* (from ~5 My to 4.4 My) (White et al., 1994), among other hominid and forest primate species, are known to have been deposited in a forest context. Forest mammals from the Tragelaphini, Reduncini, and Bovini have been retrieved from the Omo valley (Bobe and Eck, 2001). Thousands of fossilized seeds of *Canthium* were found within the Aramis geological formation (Woldegabriel et al., 1994).

In the late Pleistocene and Holocene, the majority of the bone assemblages come from rockshelters. For example, in the lowland forests of Irian

the timing, geographical distribution, and nature of lowland forest formations in the distant past (see box 2). Until the late 1980s archaeological inquiry on the ability of humans to occupy lowland tropical rain forests was highly dependent on biogeographic models derived from the "refugia hypothesis" (box 2), geological indicators of perceived "aridity" (box 3),

Jaya, at the rockshelter site of Kria, Pasveer (1998) has reported large faunal samples from the last 8000 years consisting of burnt and unburnt bone, fishbone, and shell. Human bone from Niah cave (Borneo) (Harrison, 1972) and Tabon cave (Palawan) (Fox, 1970) have been dated to approximately 40,000 and 22,000 years respectively. Seven out of ten sites in the Ituri forest yielded animal bone from primates, bovids, and rodents dated to the last 800 years, along with shell (Mercader, chapter 3 in this book; Mercader et al., in press). Human remains were retrieved from the site of Matangai Turu in the Ituri forest of Congo (Mercader et al., 2001).

Botanical remains are more abundant than bone and form larger assemblages at both rockshelter and open-air sites. The Ituri rockshelters yielded charcoal, phytoliths, and endocarps of late Pleistocene to Holocene age (Mercader et al., in press). Open-air sites from Equatorial Guinea (Mercader et al., 2002) and southwest Cameroon (Mercader and Marti, chapter 3 in this book) have preserved charcoal for more than 30,000 years. Large macrobotanical assemblages were found in the Colombian and Brazilian Amazon (Roosevelt et al., 1996; Gnecco and Mora, 1997), some of which also contain faunal remains. Starch residues on stone artifacts were preserved in the open-air sites and rockshelters from the northern Solomon islands (Loy et al., 1992) and Panama (Piperno and Holst, 1998). Nutshells were recovered from Jiyer Cave, Northeast Australia (Horsfall, 1987).

Rockshelters create microenvironments with lower temperature and humidity. Sometimes rockshelters protect the sedimentary record well from subaerial processes; others do so to a limited extent. Archaeological materials may be preserved in rockshelters because of the protected nature of the environment (Balme and Beck, 2002; Mercader et al., in press), but given that pedogenesis occurs even inside rockshelters, showing intensive weathering in areas under the protection of the overhang (Mercader et al., in press), this may not be the only reason. Moreover, in the Ituri rain forest, Congo, the geochemical nature of the matrix in which archaeological remains were found, including water content and pH, was similar or identical both inside and outside rockshelters (Mercader et al., in press).

and paleoenvironmental records from forest-fringing sites and regions separated from the lowland rain forest by large geographical distances (box 4).

Current data suggest that during glacial episodes of the late Pleistocene the wet tropics may have sustained heterogeneous vegetational formations, including tropical forests that subsisted in a cooler, drier, and CO_2-starved

Box 2. Environmental change in rain forests

Ecologists define a tropical forest as a warmth-loving botanical formation dominated by broad-leaf angiosperms adapted to wet tropical climates, with little or no seasonality in rainfall and small daily temperature fluctuations (White, 1983). By definition, a tropical forest receives an annual precipitation in excess of 1400 mm, and often around 2000 mm or more (Grainger, 1996). However, low seasonality, not high rainfall, is the critical factor for a true rain forest to subsist.

Depending on how homogeneously distributed is the annual rainfall, the tropical forest will support variable proportions of evergreen and semi-evergreen taxa. Regions with rainfall below 1400 mm, pronounced seasonality, and dry periods of more than two months support seasonal forests with deciduous species, open forest structures, and mosaics. In Africa, the annual rainfall, seasonality, and altitude determine the existence of two types of rain forest (evergreen lowland tropical rainforest and semi-evergreen tropical wetland forest), an alpine formation known as montane forest, and two open ecosystems that include dry forests and woodlands (Grainger, 1996).

There is overall consensus about two climatic parameters and their effects on the forest paleogeography—temperature and CO_2 content in the paleoatmosphere—but very little with regard to rainfall. The first one of these climatic parameters is the decrease in temperature during glacial maxima. In equatorial Africa Bonnefille and others (1990) have estimated this temperature change to be around 4°C ± 2°C lower than the average temperature during glacial minima. In general, interglacials like the Holocene are believed to have been 6°C warmer than the coldest glacial phases (Colinvaux and Oliveira, 2001). Therefore, in the wet tropics, the temperature change from glacial to interglacial conditions was significant but not extreme. Some authors have suggested that temperature and humidity during the glacial phases of the middle Pleistocene could have been different from those of the late Pleistocene (Dupont et al., 2001) and possibly that the "aridity" seen during the last glacial maximum was not present during earlier glacial periods.

The second parameter under analysis is the amount of CO_2 present in the atmosphere. Interglacial levels of CO_2 were 50% higher than glacial ones (Cowling, 1999). Plants adapted their physiology to changing CO_2 conditions in the atmosphere by producing smaller leaf size during CO_2-starved glacial conditions and grew relatively more abundant during CO_2-enriched interglacial phases. These physiological changes may have contributed to changes in glacial forest structure with reduced-size leaves in the canopy, and a concomitant open forest structure that allowed more herbaceous taxa to grow in the forest floor.

Opinions about the degree of dryness during glacial episodes vary greatly (Hooghiemstra and Van der Hammen, 1998). On the one hand, those who defend the idea of "aridity" in the wet tropics and subsequent forest retreat to small forest havens argue for reductions in rainfall greater than 50%. On the opposite extreme, those who believe that the wet tropics remained wet, cool, and mostly forested support relatively minor reductions in precipitation. A majority of researchers posit a rainfall reduction of from 20 to 40% (Maley and Brenac, 1998; Thomas, 2000; Kutzback et al., 1998), inferred from an altitudinal shift in vegetation: a lowered treeline and montane species living in the lowlands during the last glacial maximum. The cooler temperatures moderated the evapo-transpiration cycle and reduced the frequency of convective rain, which in Central Africa accounts for most of the rainfall (Grainger, 1996). A smaller water budget triggered forest fragmentation, more open ecosystems, but only a *partial* savanna encroachment (Thomas, 2000; Farrera et al., 1999; Talbot and Johannessen, 1992).

The hypothesis that Pleistocene droughts fragmented the rain forest and drove it to small isolated pockets where it found haven from extinction is not new. This hypothesis is called the "refuge hypothesis" and draws on biogeographical patterns of endemic distribution in the Amazon and Congo basins. Tropical aridity was first proposed by Haffer for the Amazon in 1969; followed by Hamilton for Africa in 1972. In essence, this theory proposes that what during wet interglacials was a lush dense forest became a barren savanna during dry glacial periods, except for small havens. It was within these havens that the high species diversity that characterizes rain forests originally took place, to then spread to the rest of the rainforest region at times of massive expansion and coalescence.

The refuge hypothesis has wide support and has strongly influenced our perceptions of tropical forest prehistory. The main problem with it is that the glacial "aridity" that it assumes in many instances has not been directly confirmed in terrestrial sediments from sites inside the lowland rain forest, but inferred from marine records and marginal areas (well outside the lowland forest) in different climatic zones that are semi-dry and highly seasonal even at present time. Nobody disputes that the rain forest was fragmented during glacial phases of the Quaternary. Thus, the matter under discussion is the degree of deforestation, and if large blocks of glacial forest persisted and supported human occupation and rainforest adaptations.

Box 3. Geological indicators of "aridity"? The African example— "Kalahari sands" and "stone-lines"

For many years, archaeologists and geologists working in South Central Africa noticed the existence of an underlying basement consisting of wind-blown "Kalahari sands." These sands stretch 3500 km north to south, from Botswana to the southern Congo basin, and cover large regions across northern Angola, the Kinshasa region, Bandundu, Kasai, Shaba, and northern Zambia. In general, the aerial extent of this formation is always limited to latitudes south of the equator and, for the most part, they are scarce or not known north of the 5°S limit. The only known exception is the sands recorded in the Central Congo basin, around Lac Tumba (Fiedler and Preuss, 1985). The Kalahari sands from South Central Africa have been interpreted as proxy evidence for glacial aridity across the tropical belt, and archaeologists have reasonably concluded that sites in regions with Kalahari sands were savanna sites. Certainly, in a region that today supports semi-dry climates with marked seasonal precipitation this is not unexpected. In fact, an open paleoenvironment is predicted by current paleoclimatic models (Kutzback et al., 1998; Van der Hammen and Absy, 1994; Jolly et al., 1998) for regions that, like South Central Africa, yield less than 1500 mm of annual rainfall and receive seasonal precipitation.

In a recent review of sand formations in the Amazon, Colinvaux and others (2000) show that sandy deposits that could support the hypothesis of glacial "aridity" exist only in marginal nonforest locations outside the South American lowlands. These landforms have been repeatedly assigned to an arid phase in rainforest prehistory, without conclusive environmental evidence to confirm a glacial origin. The perceived arid landforms have not been

atmosphere and yielded distinct admixtures of highland and lowland species with many shrubs and herbaceous plants on the forest floor. Recent data worldwide suggests that, for the late Pleistocene at least,

1. Some tropical lowlands currently covered by evergreen forest were not severely deforested, as shown by the presence of arboreal taxa in the pollen and phytolith records older than 10,000 B.P. Botanical assemblages indicate a lowering of montane altitudinal vegetation belts and an admixture of lowland and highland species (Amazon: Colinvaux et al., 1996; Piperno, 1997; Haberle and Maslin, 1999; Africa: Talbot and Johannessen, 1992; Jahns et al., 1998; Maley and Brenac, 1998; Dupont et al., 2000; Mercader et al., 2000; Southeast Asia: Newsome

dated. Moreover, wind-blown sedimentation, as well as "stone-line" formation (see below), have been active intermittently for thousands of years before and after glacial phases, and some of them even date to the Holocene (Colinvaux, 2000).

"Stone-lines" are widespread geological features that appear in tropical and subtropical regions. They have a diverse nature and genesis. Stone-lines are a prevalent feature in the African tropics showing layers of "gravel" above bedrock that are capped by structureless clay or sand. Their genesis has been interpreted as either terrace deposits (Cornelissen, 1997), pedogenic mantles created by repeated termite disturbance (Johnson, 1990; McBrearty, 1990) or geogenic deposits that resulted from interruptions in sedimentary conditions leading to surface erosion with a partial loss of fines, consolidation by lowering of the surface, and subsequent lag (Thomas, 1994). The paleoenvironmental meaning of stone-lines is unknown. Mercader and others (2001) have reviewed the nature of stone-lines in the lowland forest of Equatorial Guinea and analyzed the energetics, sedimentary fabrics, taphonomic markers, and spatial distribution of artifacts in a stone-line context. The results support the geogenic nature of some stone-lines. Contrary to the prevailing paleoenvironmental assumption that stone-lines are desert pavements formed during the last glacial maximum, Mercader and others have shown chronological evidence that stone-line formation may be a discontinuous process that was active over long periods of time before, during, and after the last glacial maximum. Importantly, no evidence of varnish by deflation, aeolian agency, or physico-chemical modification caused by contrasting temperature and light conditions typical of savannas and deserts was found in the lowland forest of Equatorial Guinea.

and Flenley, 1988; Van der Kaars and Dam, 1995; Sahul: Hopkins et al., 1993; Hope and Tulip, 1994; Haberle, 1998).

2. There was an overall drop of temperature of approximately 5–7°C (Colinvaux et al., 1996). Therefore, cooling environments are expected during glacial periods, thereof called "hypothermals."

3. CO_2 content in the atmosphere could have been remarkably lower, causing significant changes in plant development, forest structure, and altitudinal distribution of plants (Jolly and Haxeltine, 1997; Cowling, 1999).

4. Rainfall may have decreased 20–30% (Newsome and Flenley, 1988; Colinvaux, 1996). Yet, sites in wet environments receiving approximately 2000 mm to more than 3000 mm of annual rainfall did not

**Box 4. Paleobotanical data indicating persistence
of rain forest during glacial phases**

Thomas (2000) has noticed that the paleobotanical and geological data often used to support the idea of aridity across the rainforest region during glacial episodes of the Pleistocene does not contain a single record from a site in the lowland forest. Moreover, models currently utilize records obtained from isolated forest-bordering sites that even in the Holocene have supported mosaics and savannas, such as Matupi cave in the Great African Western Rift (Van Noten, 1977) with a marginal geographical location (40 km away from the lowland forest border and 10 km away from the savanna) that has been part of a transitional altitudinal and vegetation zone for many millennia. In other instances, the data have come from seasonally arid regions such as Angola (Clark, 1963), the Southern Democratic Republic of Congo (Cahen, 1978), and Zambia (Barham, 2000; Clark and Brown, 2001) separated from the lowland forest by large distances (200/ > 500 km).

Based on terrestrial sediments from the lowland rainforest per se, the hard evidence conveys the following information:

One: In west Central Africa, Maley and Brenac (1998) have reconstructed the environments that existed in the lowlands of present-day southwest Cameroon based on direct palynological evidence from terrestrial records of the last 28,000 years. The entire sequence is characterized by high pollen frequencies from one of the most widely dispersed arboreal rainforest families in the African continent, the Caesalpiniaceae, as well as by pollen from montane taxa which, at times of glacial cooling, moved down to the lowlands. This palynological sequence leaves no doubt that the rainforest persisted in southwestern Cameroon for at least the last 30,000 years. Pollen cores from marine records off the coast of Gabon and Congo were used by Dupont and others to discern glacial and interglacial paleoenvironments in the Central Af-

undergo severe deforestation because even under a highly unlikely 50% reduction in rainfall, their water balance would be enough to support forests. The changes in the evergreen forest involved species reassembly resulting from downward displacement of montane elements, and/or invasions of trees characteristic of seasonal formations (Van der Hammen and Absy, 1994; Piperno, 1997), rather than simple cycles of contraction and expansion derived from the climatic aridity predicted by the refugia theory. On the other hand, the drier types of forests could have interdigitated with sclerophyll taxa to variable degrees (Hopkins et al., 1993; Kutzback et al., 1998; Jolly et al., 1998).

rican lowlands of the Atlantic seaboard during the last 130,000 years (Dupont et al., 2000). The reconstructed glacial paleovegetation from southwest Cameroon, Equatorial Guinea, and Gabon presents large regions covered by lowland rain forest. Similar observations were documented outside Central Africa, in the West African forest region, where pollen records taken from marine sediments off the Liberian coast indicate that rain forest persisted in Liberia and Western Ivory Coast over the last 400,000 years (Jahns et al., 1998).

Two: In the center of the continent, in the region that stretches from the Central Congo Basin to the Western Rift, marine records are not amenable to correlation, and lakes within the lowlands have not been cored. The amount of paleobotanical data is by far the sparsest in the continent and therefore is the least well known of any major area in the African wet tropics. The only evidence obtained from lowland terrestrial records comes from three phytolith sequences from the sites of Matangai Turu, Makubasi, and Lengbe in the Ituri rain forest, in the northeast Congo basin (Mercader et al., 2000). The oldest sequence was retrieved from Matangai Turu, covering the Pleistocene to Holocene transition, showing no indication of savanna encroachment in the last 20,000 years.

In sum, isolated records, by definition, do not represent regions of transcontinental size. Evidence from the lowland forest per se suggests that glacial forests were fragmented forests, but this is not to say that Central Africa was almost entirely deforested during the Pleistocene. It seems clear that any theory attempting to understand rainforest paleogeography needs to appreciate the huge size of the area under discussion, and the effects over time that geographical, ecological, and climatological variations have had in past climates and ecosystems.

5. During cooler and drier periods of the Pleistocene, the open forest structures that appeared in some regions yielded a large supply of economically exploitable taxa in the form of undergrowth, improving food supply and overall suitability for occupancy by early humans.

The Early Settlement of Tropical Forests

Recent research shows that some sort of tropical forests, dry forests, woodlands, and montane forests from terrestrial, riverine, and lacustrine settings were the cradle of the earliest biped hominids during the Miocene and

Pliocene, from 6 million to 3 million years ago (Bonnefille, 1994, 1995; Wesselman, 1995; Bobe and Eck, 2001; Leakey et al., 2001; Woldegabriel et al., 1994; Bonnefille and Letouzey, 1976; Boesch-Achermann and Boesch, 1994). Yet no hominid species has been shown to be exclusively adapted to a closed rain-forest environment of Central African type—rather than a combination of Guineo-Congolian taxa in cohort with species from alpine regions and variably dry woodlands and savannas. In a recent review of the archeological evidence supportive of an early and middle Pleistocene occupation of the African forest, Mercader indicates that there is an almost complete lack of industrial, chronological, and environmental evidence from the lower Paleolithic or Acheulian period in the Congo Basin or West Africa due to lack of research (Mercader, 2002). About five Acheulian sites characterized by large bifaces and cleavers have been reported in the High Sangha, at the western corner of the Central African Republic, in areas today covered by dense lowland forest (Bayle des Hermens, 1975). In total, the archaeological evidence consists of more than 70 bifacial hand axes and cleavers retrieved during road construction and mining. Some sites pose stratigraphic problems regarding their taphonomic integrity or suggest extensive rounding and redeposition. But, regardless of the dubious validity of this type of data, it is unquestionable that the geographical range of early Pleistocene hominids had expanded beyond East Africa and South Africa, at times considerably greater than 200,000 B.P., to include the Western Rift (Boaz, 1990), Central Africa (Bayle des Hermens, 1975; Cahen and Moeyersons, 1969–71), Chad (Brunet et al., 1995), and North Africa (Sahnouni et al., 2002).

Paleoanthropological, archaeological, and palynological evidence from Sangiran and Mojokerto, eastern Java, Indonesia (Swisher et al., 1994; Sémah et al., chapter 5 in this book), indicates that *Homo erectus* was present in wet equatorial latitudes at the gates of Sahul between 1.8 and 1.6 million years ago. Further support for hominid presence in equatorial environments comes from the debated site of Mata Mengue, Flores, Indonesia, in which lower Paleolithic industries have been dated at circa 0.7 to 0.5 million years (Sondaar et al., 1994; Van der Bergh et al., 1996; but see Keates, 1998). In addition, lower Paleolithic industries have been securely dated to about 0.8 million years at Lampang, in northern Thailand (Pope and Keates, 1994), and Bose, in southern China (Yamei et al., 2000). The latter has yielded Acheulian-like industries associated with a tektite fall episode and wood burning, and Yamei et al. (2000) have inferred a behavioral adaptation to deforested ecosystems. Yet, the available data do not constitute demonstration that hominids avoided rainforest environments. Also, Bose falls outside the lowland forest climate zone. Paleoecological data from the Sangiran dome (Sémah et al., chapter 5 in this book) indicates that the Javanese specimens of *Homo erectus* lived in diverse ecological settings, some of which were characterized by patchy botanical formations that included mangroves; swamp forest; closed tropical forest with typical Southeast Asian families, such as the very tall canopy trees of the *Dipterocarpaceae;* seasonal forests; and open savanna environments. So far there is no clear proof that *Homo ergaster* and *Homo erectus* focused on the exploitation of forest ecosystems, but the paleoenvironmental data and the geographical dispersion of lower Paleolithic sites throughout the equatorial and tropical belts suggest that tropical forests may have been part of the landscape potentially available to African and Southeast Asian hominids.

Indeed, early hominids are considered fully capable of major behavioral and adaptive accomplishments such as the earliest colonization of the Old World (Gamble, 1993), moving out of Africa and traversing the great distances and many environments that existed across the Middle East, India, China, Indonesia, and southern Europe. Yet, many scholars consider premodern hominids as incapable of settling the tropical forest zones that existed in their homelands. Did hominids settle the open environments of the tropical landscapes of Africa and Asia, shunning and circumventing the ubiquitous forest blocks that existed across the landscape? Were early humans so drastically limited in their capacity to inhabit rain forests that they constantly avoided the forest barrier?

In this line, some have suggested that the initial colonization of tropical forests is part of the behavioral package that comes with the emergence of anatomically modern humans (McBrearty and Brooks, 2000). The earliest behavioral evidence for modernity comes from Ethiopia (Wendorf et al., 1994), Zambia (Barham and Smart, 1996; Barham, 2000), and Kenya (McBrearty et al., 1996) 300,000 to 200,000 years ago. McBrearty and Brooks see the rain forest as a marginal and extreme environment, one that was not colonized by premodern hominids. These authors interpret the late Pleistocene settlement of tropical forests as a sign of behavioral modernity, since the colonization of marginal regions implies increased geographic range by our species and the utilization of superior technologies in that quest.

We do not know when it was that our ancestors first came into the rain forest, if they ever left it, but, whether *Homo sapiens* was the first colonizer of rain forests or not, current archaeological evidence indicates that before 40,000 B.P. modern humans had already encountered, crossed, and settled tropical forests of all types during their expansion across the vast expanses of land that go from Africa to Australia.

The earliest African industries presumably made by modern humans are those of the Middle Stone Age (MSA), which includes several technocomplexes characterized by radial technologies, flake tools, and, sometimes, lanceolate points. Outside the forest region, on stratigraphic grounds, the majority of these industries postdate the end of the Acheulian techno-complex around 200,000,000 B.P. (McBrearty and Brooks, 2000). In southern Central Africa, also outside the forest, Middle Stone Age industries have been dated to the middle Pleistocene, starting as early as 270,000 B.P. (Barham, 2000) through 95,000 B.P. and later (Clark and Brown, 2001). Early ages into the middle Pleistocene have also been published for the Sangoan industries from the rain-forest region in Côte d'Ivoire, dated to 255,000 years (Lioubine and Guede, 2000). In neighboring western East Africa, McBrearty (1987, 1988) obtained dates between 100,000 B.P. and 65,000 B.P. Casey also reports 35,000-year-old sites in northern Ghana (see chapter 1 in this book). Within the lowland rain forest, sites considerably older than 35,000 B.P. have been reported in southwestern Cameroon (Mercader and Martí, chapter 2 in this book) and Equatorial Guinea (Mercader et al., 2002; Mercader and Martí, chapter 2 in this book). Additional evidence for a long-term occupation of the African forest comes from Later Stone Age (LSA) sites in a demonstrated paleoenvironmental context of lowland tropical forest, at the time of occupation (Mercader et al., 2000), in which foraging communities of the northeastern Congo Basin settled variably dense forest environments during 20,000 years (Mercader, chapter 3 in this book). A 40,000-year-old LSA site has recently been discovered in the lowland forest of Gabon (Oslisly et al., 2001), and early LSA sites are also known from forest-bordering locations with a paleobotanical context of gallery forest and mosaics at 40,000 B.P. (Van Noten, 1977), 30,000 B.P. (Lavachery et al., 1996; Moeyersons, 1997), and 25,000 B.P. (Mercader and Brooks, 2001).

Humans were present in the Far East coastal lowlands of Borneo (Harrison, 1972; Krigbaum, in press), Malaysia (Bulbeck, chapter 4 in this book), and the Philippines (Fox, 1970) before, during, and after the last glacial maximum. Moreover, the Borneo groups seem to have been well adapted to a broad spectrum exploitation of forest plants for their diets (Krigbaum, 2000). To judge by the early dates in which modern humans had settled the variably forested coastal lowlands of the Sunda Islands near Irian Jaya (Bellwood et al., 1998; Pasveer, 1998), the Solomon group (Loy et al., 1992), western Melanesia (Pavlides and Gosden, 1994; Gosden, 1995; Pavlides, 1999), the Australian forest-savanna domain (Asmussen, chapter 6 in this book; Morwood and Hobbs, 1996: 751; Gosden, 1995: 810; Kershaw, 1986; Cosgrove, 1996: 911), and New Guinea (Groube et al., 1986), it is beyond doubt that modern humans traversed and perhaps exploited all types of tropical forest before or at the same time as neighboring open terrains, that is, prior to 40,000 B.P.

A similar picture emerges from the New World. Archaeological evidence from Panama (Ranere and Cooke, chapter 7 in this book), Colombia (Gnecco and Mora, 1997; Mora and Gnecco, chapter 9 in this book), Brazil

(Roosevelt et al., 1996; Roosevelt, 1998; Meggers and Miller, chapter 10 in this book), and Venezuela (Barse, chapter 8 in this book) suggests that by about the late Pleistocene to Holocene all environments of the Neotropics, comprising evergreen lowlands, alpine highlands, seasonal forests, and open woodlands, had been settled by Paleoindians (Piperno et al., 1991; Cooke, 1998).

On Technology and Early Subsistence

Clive Gamble sees the expansion of Holocene agricultural populations into the challenging rain forest habitat as a sign of the higher capacity of recent humans to colonize high-risk landscapes by means of improved technology (1993: 197–200), but it is unclear that the early stone technologies employed by tropical forest occupants were in any way superior to or more effective than contemporaneous savanna tool-kits. The authors of this book put a question mark on the assumption that sophisticated technologies are needed for migration to or exploitation of rain forests. If modern humans were the first people to enter and settle tropical forests, one would expect the earliest rain forest technologies, especially those with the oldest sequence of occupation in the African continent, to show an ecological imprint and adaptive response to new environments. However, the earliest Middle and Later Stone Age techno-complexes known in the tropical forest and its vicinity do not show any clear signature of adaptation to a new landscape or the distinct environment of the rain forest. Stone technologies from forest sites are not particularly adapted to a tropical forest environment (Mercader and Brooks, 2001; cf. Asmussen, chapter 6 in this book). Moreover, these industries persist without major technological transformations through climatic and environmental boundaries (Mercader and Martí, chapter 2 in this book; Mercader, chapter 3 in this book). This lack of ecological distinctiveness seems to be the rule everywhere, with the exception of recent Australian Aboriginal adaptations (Dixon, 1976; Harris, 1978; Asmussen, chapter 6 in this book).

Late Pleistocene to Holocene stone tool kits consist of relatively unstandardized stone tools in which simple percussion with little preparation of core platforms dominates. In some instances, tropical forest technologies include spears and arrowheads produced by careful bifacial percussion (Mercader and Martí, chapter 2 in this book).

In other cases, the Pleistocene to Holocene boundary comes with major technological changes. Thus, in Southeast Asia, the onset of the Holocene sees a switch from flake-dominated industries to unifacial and bifacial pebble-based kits (Bulbeck, chapter 4 in this book). But, in wet tropical Africa, the trend toward microlithism, which characterizes the Pleistocene to

Holocene boundary in other places, appears much prior to the Holocene, before the last glacial maximum (Van Noten, 1977; Lavachery et al., 1996; Mercader and Brooks, 2001; Oslisly et al., 2001).

Economic practices remain largely conjectural. In relative terms, two diet restrictions seem to be inherent to the early settlement of the tropical forest: wild starch procurement and fat calorie extraction. The African forest is home to an extensive number of plant species, which provide oils, starches, vegetables, nuts, fruits, spices, and stimulants (table I.1) potentially available for consumption and manipulation by prehistoric foragers. Although, present-day foragers throughout the world obtain their dietary starch from their agriculturalist neighbors (Bailey et al., 1989; Headland, 1997), several ecological studies indicate that wild tubers played an important role (Hladik and Dounias, 1993) during the early settlement of tropical forests (Loy et al., 1992; Therin, 1994; Piperno and Holst, 1998; Piperno, 1998). The consumption of oily fruits is widely attested in the archaeological record. A demonstrated late Pleistocene to early Holocene arboricultural use of edible varieties of *Canarium* and palm fruits is available from Southeast Asia/Sahul (Maloney, 1998; Gosden, 1995), Africa (Mercader, chapter 3 in this book), and the New World (Roosevelt et al., 1996; Mora and Gnecco, chapter 9 in this book). Early arboriculture may have favored the spread of economically important forest trees that, to an extent, relied on human manipulation and consumption for a wider dissemination through the forest by inadvertent disposal of their seeds after consumption. *Canarium* trees, for instance, are known to abound in forest gaps created around human settlements such as camps, garbage areas, trails, etc. (Laden, 1992). Thus, although inadvertently, early foragers practiced some type of forest management that expanded the available pool of oil resources during pre-farming periods.

Diverse economic strategies were certainly present in South America. For instance, forager groups of the lower Amazon used plants, small game, and fish (Roosevelt et al., 1996). Arboriculturalists lived off the early Holocene forests of the Colombian Amazon (Gnecco and Mora, 1997; Mora and Gnecco, chapter 9 in this book). Mixed generalist-specialist hunters existed in the seasonal forests of central and southern Amazonia (Meggers and Miller, chapter 10 in this book) and in the cyclically forest-invaded savannas of the Orinoco (Barse, chapter 8 in this book). Open-terrain foragers lived in the dry forests and woodlands of the Greater Amazon region (Kipnis, 1998).

Under the Canopy: Contents and Future Prospects

This volume is organized around a central theme: the testing of the provocative question originally posed by Bailey et al. (1989), addressing the archaeological evidence to refute the categorical hypothesis that prehistoric hunting and gathering in tropical forests was impossible. Were the energies invested to disprove this hypothesis justified? Yes, since the debate will now focus on more narrowly focused questions and new hypotheses, some of which have already generated initial answers in this volume. How and why did prehistoric foragers use the resources available in different rain forests over the world before there was domesticated food? Were stone technologies crucial to tame the rain forest? Or, were other less archaeologically visible technologies, such as fire, detoxification, or inadvertent dissemination of economically important species, the actual key technologies to promote a long-term occupation of rain forests? If archaic and modern humans inhabited the rain forest, for how long a time in the annual cycle did they stay inside the forest? Were they purely forest foragers, or did they exploit a number of ecosystems and therefore were they able to subsist and exploit multiple environments? What type of hunter-gatherer society was required in order to stay in the tropical forest for prolonged periods of time?

Early hominids may or not have settled tropical rain forests. Absence of evidence derived from differential research should not be conflated with evidence of absence, for no modern archaeologist has excavated yet any Early Stone Age site within the present-day lowland tropical rain forest. Archaeological sequences point out that modern humans lived in both tropical forests and neighboring nonforest ecosystems since early ages. The close match between the initial colonization dates for woodlands, savannas, and deserts dispersed through Africa, Sahul, and the New World and the dates for the early colonization of the wet tropics suggests that early humans did not perceive tropical forests, closed or open, seasonal or not, as badlands or barriers. In Africa, Southeast Asia, Australia, and the Neotropics, Pleistocene to Holocene foragers colonized tropical forests, near the coast or inland, and remained in these regions for many millennia. Densely forested or not, the environments settled by Stone Age hunter-gatherers continued to be used at times of both forest expansion and retreat, sometimes over 40,000 years. The prolonged settlement of the same tropical forest regions for many millennia, prior to the last glacial maximum in some cases, shows that cyclical ecological changes and constraints in local economic resources were overcome early on. These groups seem to have been able to rapidly explore and take advantage of local forest resources. This ready disposition to exploit tropical forests and persist through major landscape adjustments may tell us that the behavioral characteristics that allowed early humans to adapt to tropical forests had been in place for a long time, probably much prior to what the earliest archaeological evidence so far known suggests today. This idea is reinforced by the fact that the stone tool technology from rain forests is like that of savannas.

By documenting when and where humans first learned to occupy rain

Table I.1. Edible Plants from the African Wet Tropics

Ethnobotany	Species	Family	Currently under partial cultivation	Currently collected	Currently collected/eventually cultivated
Fat calories/oil extraction	Canarium schweinfurthii	Burseraceae		X	
	Elaeis guineensis	Palmae	X	X	
	Baillonella toxisperma	Sapotaceae		X	
	Panda oleosa	Pandaceae		X	
	Poga oleosa	Rhizophoraceae		X	
	Dacryoides edulis	Burseraceae		X	
	Sesamum radiatum	Pedaliaceae		X	
	Pentadesma butyracea	Clusiaceae		X	
	Chrysobalanus icaco atac.	Chrysobalanaceae		X	
	Klainedoxa gabonensis	Irvingiaceae		X	
	Parinari excelsa	Chrysobalanaceae		X	
	Phyllobotryon spathulatum	Flacourtiaceae		X	
	Telfairia occidentalis	Cucurbitaceae	X	X	
	Cucumerops sp.	Cucurbitaceae	X	X	
	Luffa cylindrica	Cucurbitaceae		X	
	Ricinodendron heudelotti	Euphorbiaceae			X
	Irvingia gabonensis	Irvingiaceae			X
	Coula edulis	Olacaceae			X
Starch consumption	Dioscorea cayenensis	Dioscoraceae	X		
	Dioscorea rotundata	Dioscoraceae	X		
	Dioscorea dumetorum	Dioscoraceae			X
	Dioscorea praehensilis	Dioscoraceae			X

Category	Species	Family			
	Dioscorea abyssinica	Dioscoraceae		X	
	Dioscorea sansibarensis	Dioscoraceae			X
	Dioscorea burkilliana	Dioscoraceae		X	
	Dioscorea colocasfoliai	Dioscoraceae		X	
	Dioscorea bulbifera	Dioscoraceae			X
	Dioscorea Sagittifolia	Dioscoraceae	X		
	Dioscorea smilacifolia	Dioscoraceae			X
	Dioscorea macroura	Dioscoraceae			X
	Dioscorea minutiflora	Dioscoraceae			X
	Magnistipula tessmanii	Chrysobalanaceae		X	
Vegetable	*Talinum triangulate*	Portulacaceae	X	X	
	Justicia insularis	Acanthaceae		X	
	Amaranthus spinosus	Amaranthaceae	X	X	
	Vernonia amygdalina	Compositae	X	X	
	Acanthus montanus	Acanthaceae		X	
	Althernanthera sessilis	Amaranthaceae		X	
	Anchomanes difformis	Araceae		X	
	Ancistrocarpus densispinosus	Tiliaceae		X	
	Asplenium sp.	Pteridophyte		X	
	Cochorus olitorius	Tiliaceae		X	
	Cyathea camerooniana	Pteridophyte		X	
	Gnetum buchholzianum	Gnetaceae		X	
	Gnetum africanum	Gnetaceae		X	
	Heinsia crinita	Rubiaceae		X	
	Impatiens macroptera	Balsaminaceae		X	
	Laportea ovalifolia	Urticaceae		X	
	Lasianthera africana	Icacinaceae		X	

(continued)

Table I.1. *Continued*

Ethnobotany	Species	Family	Currently under partial cultivation	Currently collected	Currently collected/ eventually cultivated
	Phytolacca dodecandra	Phytolaccaceae		X	
	Plumbago zeylandicum	Plumbaginaceae		X	
	Portulaca oleracea	Portulacaceae		X	
	Pteridium aquilinum	Pteridophyte		X	
	Scaphopetalum sp.	Sterculiaceae		X	
	Solanum nigrum	Solanaceae		X	
Seeds and fruits	*Vigna ungiculata*	Leguminosae	X		
	Hibiscus sp.	Malvaceae	X		
	Pentaclethra macrophylla	Mimosaceae		X	
	Klainedoxa gabonensis	Irvingiaceae		X	
	Oubanguia alata	Scytopetalaceae		X	
	Treculia africana	Moraceae		X	
	Tetracarpidium conophorum	Euphrobiaceae		X	
	Garcinia kola	Clusiaceae		X	
	Antrocaryon klaineanum	Anacardiaceae		X	
	Raphia gentiliana	Palmae		X	
	Coelocaryon presusii	Myristicaceae		X	
	Blighia sapida	Sapindaceae		X	
	Vitex grandiflora	Verbenaceae		X	

Category	Species	Family		
	Angylocalyx talbotii	Papillionaceae	X	
	Carpolobia lutea	Polygalaceae	X	
	Cola lepidota	Sterculiaceae	X	
	Myrianthus arboreus	Moraceae	X	
	Pseudospondias microcarpa	Anacardiaceae	X	
	Trichoscypha acuminata	Anacardiaceae	X	
	Sacoglottis gabonensis	Humiriaceae	X	
Spices	*Piper guineensis*	Piperaceae		X
	Tetrapleura tetraptera	Mimosaceae	X	
	Afromomun sp.	Zingiberaceae	X	
	Xylopia aethiopica	Annonaceae	X	
	Raphia hoockeri	Palmae	X	
	Afrostyrax lepidophyllus	Styracaceae	X	
	Beilschmiedia sp.	Lauraceae	X	
	Dennettia tripetala	Annonaceae	X	
	Eribroma oblonga	Sterculiaceae	X	
	Monodora brevipes	Annonaceae	X	
	Tetracarpidium conophorum	Euphorbiaceae	X	
Stimulants	*Cola acuminata*	Sterculiaceae	X	
	Coffea sp.	Rubiaceae		X
	Guibourtia tessmanii	Leguminosae	X	

SOURCE: Data supplied by Raquel Martí.

The World's Lowland Tropical Forests

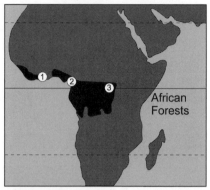

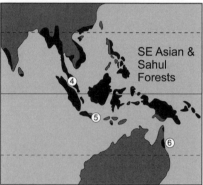

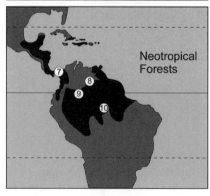

Fig. I.2. Tropical rain forests in the world, with regions covered in this volume and chapter numbers: *1,* Ghana; *2,* Cameroon–Equatorial Guinea; *3,* Democratic Republic of Congo; *4,* Malaysia; *5,* Indonesia; *6,* Australia; *7,* Panama; *8,* Venezuela; *9,* Colombia; *10,* Brazil.

forests, we stand to learn about fundamental clues to human origins, the sociocultural and adaptive capabilities of our species and ancestors to life in closed environments, expanding territorial breadth, residential stability, and progressive ecological control of tropical forests, among other topics. In this sense, the African record plays an important role in understanding some of the most crucial dynamics behind the early prehistoric settlement of rain forests worldwide. Thus, by looking at when and where humans first began to solve the problems inherent to long-term rain forest settlement without using farming techniques, and the paths followed by African forest groups, we can start to understand some of the trends and behavioral variables associated with the earliest settlement of rain forests.

Although tropical forests have long been viewed as too harsh for human occupation and cultural development, the research presented in this book indicates that forests could harbor important clues to human evolution, origins of modern human behavior, cultural diversity, and the fundamental role that humans may have played in the distribution, structure, and species composition in present and future tropical forests. The questions raised in this volume and the body of archaeological data used for hypotheses are of great appeal to anyone interested in tropical archaeology, historical ecology, hunter-gatherer studies, and management or conservation of tropical rain forests presently occupied by humans.

References

Bahuchet, S., D. McKey, and I. de Garine (1991): Wild yams revisited: Is independence from agriculture possible for rainforest hunter-gatherers? *Human Ecology* 19(2):57–65.

Bailey, R., and N. Peacock (1988): Efe pygmies of northeast Zaire: Subsistence strategies in the Ituri forest. In *Coping with uncertainty in food supply,* edited by I. de Garine and G. A. Harrison, 88–117. Oxford and New York: Oxford University Press.

Bailey, R., and T. Headland (1991): The tropical rain forest: Is it a productive environment for human foragers? *Human Ecology* 19(2):261–285.

Bailey, R., G. Head, M. Jenike, B. Owen, R. Rechtman, and E. Zechenter (1989): Hunting and gathering in tropical rainforest: Is it possible? *American Anthropologist* 91(1):59–82.

Balme, J., and W. Beck (2002): Starch and charcoal: Useful measures of activity areas in archaeological rockshelters. *Journal of Archaeological Science* 29:157–166.

Barham, L. S. (2000): *The Middle Stone Age of Zambia, South Central Africa.* Bristol: Western Academic & Specialist Press.

Barham, L., and P. Smart (1996): Early date for the Middle Stone Age of central Zambia. *Journal of Human Evolution* 30:287–290.

Bar-Yosef, O., and A. Belfer-Cohen (2000): The healthier choice: Early coloniza-
tions of Eurasia. *Journal of Human Evolution* 38(3):A6.

Bayle des Hermens, R. (1975): *Recherches préhistoriques en Republique Centrafricaine.*
Paris: Labethno.

Behrensmeyer, A. (1991): Terrestrial vertebrate accumulations. *Taphonomy: Releas-
ing the data locked in the fossil record,* edited by P. Allison and E. Briggs, 291–335.
New York: Plenum Press.

Bellwood, P., G. Nitihaminoto, G. Irwin, Gunadi, A. Waluyo, and D. Tanudjiro
(1998): 35,000 years of prehistory in the northern Moluccas. In *Bird's Head
approaches: Irian Jaya studies—a programme for interdisciplinary research,* edited by
G. Bartstra, 233–268. Rotterdam: A. A. Balkema.

Boaz, N., ed. (1990): *Evolution of environments and Hominidae in the African Western
Rift valley.* Martinsville: Virginia Museum of Natural History.

Bobe, R., and G. Eck (2001): Responses of African bovids to Pliocene climatic
change. *Paleobiology* 27:1–47.

Boesch-Achermann, H., and C. Boesch (1994): Hominization in the rainforest:
The chimpanzee's piece of the puzzle. *Evolutionary Anthropology* 3:9–16.

Bonnefille, R. (1994): Palynology and paleoenvironment of East African hominid
sites. In *Integrative paths to the past: Paleoanthropological advances in honor of F. Clark
Howell,* edited by R. Corrucini and R. Ciochon, 415–427. Englewood Cliffs, N.J.:
Prentice-Hall.

Bonnefille, R. (1995): A reassessment of the Plio-Pleistocene pollen record of
East Africa. In *Paleoclimate and evolution, with emphasis on human origins,* edited
by E. Vrba, G. Denton, T. Partridge, and L. Burckle, 299–310. New Haven: Yale
University Press.

Bonnefille, R., and R. Letouzey (1976): Fruits fossiles d'Antrocaryon dans la vallé
de l'Omo (Ethiopie). *Andasonia* 16:65–82.

Bonnefille, R., J. C. Roeland, and J. Guiot (1990): Temperature and rainfall estimates for the past 40,000 years in equatorial Africa. *Nature* 246:347–349.

Brosius, P. (1991): Foraging in tropical rain forests: The case of the Penan of Sarawak, East Malaysia. *Human Ecology* 19(2):123–150.

Brunet, M., A. Beauvilain, Y. Coppens, E. Heintz, A. Moutaye, and D. Pilbeam
(1995): The first australopithecine 2,500 kilometres west of the rift valley
(Chad). *Nature* 378:273–275.

Bush, M., and P. Colinvaux (1994): Tropical forest disturbance: Paleoecological
records from Darien, Panama. *Ecology* 75(6):1761–1768.

Cahen, D. (1975): *Le site archéologique de la Kamoa (Région du Shaba, République
du Zaïre): De L'Age de la Pierre Ancien à L'Age du Fer.* Tervuren: Musée Royal de
L'Afrique Centrale.

Cahen, D. (1976): Nouvelles fouilles à la Pointe de la Gombe (Ex-Pointe de Ka-
lina), Kinshasa, Zaire. *L'Anthropologie* 80:573–602.

Cahen, D., and J. Moeyersons (1969–71): Le site prehistorique de la Kamoa, Katanga, Zaire. *Palaeoecology of Africa* 6:237–238.

Clark, D., and K. Brown (2001): The Twin Rivers Kopje, Zambia: Stratigraphy, fauna, and artefact assemblages from the 1954 and 1956 excavations. *Journal of Archaeological Science* 28:305–330.

Clark, J. D. (1963): *Prehistoric cultures of northeast Angola and their significance in tropical Africa.* Lisbon: Companhia de Diamantes de Angola (DIAMANG).

Colinvaux, P. (1996): Quaternary environmental history and forest diversity in the Neotropics. In *Evolution and environment in tropical America,* edited by J. Jackson, A. Budd, and A. Coates, 359–405. Chicago: University of Chicago Press.

Colinvaux, P. (2000): Paradigm lost: Holocene environments of the Amazon basin (physical evidence and paleoclimates, part 1). *Review of Archaeology* 21:11–17.

Colinvaux, P., and M. Bush (1991): The rain forest ecosystem as a resource for hunting and gathering. *American Anthropologist* 93(1):153–160.

Colinvaux, P., and P. Oliveira (2001): Amazon plant diversity and climate through the Cenozoic. *Palaeogeography, Palaeoclimatology, Palaeoecology* 166:51–63.

Colinvaux, P., P. Oliveira, and M. Bush (2000): Amazonian and neotropical plant communities on glacial time-scales: The failure of the aridity and refuge hypotheses. *Quaternary Science Reviews* 19:141–169.

Colinvaux, P., P. Oliveira, E. Moreno, C. Miller, and M. Bush (1996): A long pollen record from lowland Amazonia: Forest and cooling in glacial times. *Science* 274: 85–88.

Cooke, R. (1998): Human settlement of Central America and northern South America, 14,000–8,000 B.P. *Quaternary International* 49–50: 177–190.

Cornelissen, E. (1997): Central African transitional cultures. In *Encyclopedia of precolonial Africa,* edited by J. C. Vogel, 312–321. Walnut Creek, Calif.: Altamira Press.

Cosgrove, R. (1996): Origin and development of Australian aboriginal tropical rainforest culture: A reconsideration. *Antiquity* 70:900–912.

Cowling, S. (1999): Simulated effects of low atmospheric CO_2 on structure and composition of North American vegetation at the last glacial maximum. *Global Ecology and Biogeography* 8:81–93.

Dixon, R. (1976): Tribes, languages, and other boundaries in northeast Queensland. In *Tribes and boundaries in Australia,* edited by N. Peterson, 207–238. Canberra: Australian Institute for Aboriginal Studies.

Dupont, L., B. Donner, R. Schneider, and G. Wefer (2001): Mid-Pleistocene environmental change in tropical Africa began as early as 1.05 Ma. *Geology* 29: 195–198.

Dupont, L., S. Jahns, F. Marret, S. Ning (2000): Vegetation change in equatorial West Africa: Time slices for the 150ka. *Palaeogeography, Palaeoclimatology, Palaeoecology* 155:95–122.

Dwyer, A., and M. Minnegal (1991): Hunting in lowland tropical rainforest: Towards a model of non-agricultural subsistence. *Human Ecology* 19(2):187–212.

Endicott, K., and P. Bellwood (1991): The possibility of independent foraging in the rain forest of peninsular Malaysia. *Human Ecology* 19(2):151–186.

Farrera, I. S. Harrison, I. Prentice, G. Ramstein, J. Guiot, P. Bartlein, R. Bonnefille, M. Bush, W. Cramer, U. von Grafenstein, K. Holmgren, H. Hooghiemstra,

G. Hope, D. Jolly, S. Lauritzen, Y. Ono, S. Pinot, M. Stute, and G. Yu (1999): Tropical climates at the Last Glacial Maximum: A new synthesis of terrestrial paleoclimate data. I. Vegetation, lake levels, and geochemistry. *Climate Dynamics* 15:823–856.

Fiedler, L., and J. Preuss (1985): Stone tools from the Inner Zaïre Basin (Région de l'Equateur, Zaïre). *African Archaeological Review* 3:179–187.

Fox, R. (1970): *The Tabon caves 1.* Manila: National Museum Monographs.

Gamble, C. (1993): *Timewalkers: The prehistory of global colonization.* Stroud: Alan Sutton.

Gnecco, C., and S. Mora (1997): Late Pleistocene/early Holocene tropical forest occupations at San Isidro and Peña Roja, Colombia. *Antiquity* 71:683–690.

Gosden, C. (1995): Arboriculture and agriculture in coastal Papua New Guinea. *Antiquity* 69:807–817.

Grainger, A. (1996): Forest environments. In *The physical geography of Africa*, edited by W. Adams, A. Goudie, and A. Orme, 173–195. Oxford: Oxford University Press.

Groube, L., J. Chappell, J. Muke, and D. Price (1986): A 40,000-year-old human occupation site at Huon Peninsula, Papua New Guinea. *Nature* 324:453–455.

Haberle, S. (1994): Anthropogenic indicators in pollen diagrams: Problems and prospects for late Quaternary palynology in New Guinea. In *Tropical archaeobotany: Applications and new developments*, edited by J. Hather, 172–201. London: Routledge.

———. (1998): Late Quaternary vegetation change in the Tarin Basin, Papua New Guinea. *Palaeogeography, Palaeoclimatology, Palaeoecology* 137:1–24.

Haberle, S., and M. Maslin (1999): Late Quaternary vegetation and climate change in the Amazon Basin based on a 50,000-year pollen record from the Amazon fan, odp site 932. *Quaternary Research* 51:27–38.

Haffer, J. (1969): Speciation in Amazonian forest birds. *Science* 165:131–136.

Hamilton, A. (1972): The significance of patterns of distribution shown by forest plants and animals in tropical Africa for the reconstruction of upper Pleistocene palaeoenvironments. *Palaeoecology of Africa* 9:63–97.

Harris, D. (1978): Adaptation to a tropical rain-forest environment: Aboriginal subsistence in Northeast Queensland. In *Human behaviour and adaptation*, edited by N. Blurton-Jones and V. Reynolds, 113–134. London: Taylor and Francis.

Harrison, T. (1972): The Prehistory of Borneo. *Asian Perspectives* 13:17–45.

Hart, T., and J. Hart (1986): The ecological basis of hunter-gatherer subsistence in African rain forests: The Mbuti of eastern Zaire. *Human Ecology* 14(1):29–55.

Hart, T., J. Hart, M. Dechamps, M. Fournier, and M. Ataholo (1996): Changes in forest composition over the last 4000 years in the Ituri Basin, Zaire. In *The biodiversity of African plants*, edited by L. Van der Maesen, 545–563. Dordrecht: Kluwer.

Head, L. (1989): Prehistoric Aboriginal impacts on Australian vegetation: An assessment of the evidence. *Australian Geographer* 20(1):37–46.

Headland, T. (1987): The wild yam question: How well could independent hunter-gatherers live in a tropical rain forest ecosystem? *Human Ecology* 15(4):463–491.

————. (1997): Revisionism in ecological anthropology. *Current Anthropology* 38(4): 605–630.

Hladik, A., and E. Dounias (1993): Wild yams of the African forest as potential food resources. In *Tropical forests, people, and food: Biocultural interactions and applications to development*, edited by C. Hladik, vol. 1, 163–176. Paris: UNESCO.

Hooghiemstra, H., and T. Van der Hammen (1998): Neogene and quaternary developments of the Neotropical rain forest: The forest refugia hypothesis, and a literature overview. *Earth-Science Reviews* 44:147–183.

Hope, G., and J. Tulip (1994): A long vegetation history from lowland Irian Jaya, Indonesia. *Palaeogeography, Palaeoclimatology, Palaeoecology* 109:385–398.

Hopkins, M., J. Ash, A. Graham, J. Head, and R. Hewett (1993): Charcoal evidence of the spatial extent of the Eucalyptus woodland expansions and rainforest contractions in North Queensland during the late Pleistocene. *Journal of Biogeography* 20:59–74.

Horsfall, N. (1987): Living in rainforests: The prehistoric occupation of North Queensland's humid tropics. Ph.D. diss., James Cook University of North Queensland.

Jahns, S., M. Huls, and M. Sarnthein (1998): Vegetation and climate history of west equatorial Africa based on a marine pollen record off Liberia (site GIK 16776) covering the last 400,000 years. *Review of Palaeobotany and Palynology* 102:277–288.

Johnson, D. (1990): Biomantle evolution and the redistribution of earth materials and artifacts. *Soil Science* 149:84–101.

Jolly, D., and A. Haxeltine (1997): Effect of low glacial atmospheric CO_2 on tropical African montane vegetation. *Science* 276:786–787.

Jolly, D., S. Harrison, B. Damnati, and B. Bonnefille (1998): Simulated climate and biomes of Africa during the late Quaternary: Comparison with pollen and lake status data. *Quaternary Science Reviews* 17:629–658.

Keates, S. (1998): A discussion of the evidence for early hominids on Java and Flores. In *Bird's Head approaches: Irian Jaya studies—A programme for interdisciplinary research*, edited by G. Bartstra, 179–191. Rotterdam: Balkema.

Kerbis, J., R. Wrangham, M. Carter, and M. Hauser (1993): A contribution to tropical rain forest taphonomy: Retrieval and documentation of chimpanzee remains from Kibale forest, Uganda. *Journal of Human Evolution* 25:485–514.

Kershaw, A. (1986): Climatic change and aboriginal burning in north-east Australia during the last two glacial/interglacial cycles. *Nature* 322(6074):47–49.

————. (1994): Pleistocene vegetation of the humid tropics of northeastern Queensland, Australia. *Palaeogeography, Palaeoclimatology, Palaeoecology* 109:399–412.

Khan, K. (1993): *Catalogue of the Roth collection of Aboriginal artefacts from North Queensland: Technical reports of the Australian Museum.* Sydney: Australian Museum.

Kipnis, R. (1998): Early hunter-gatherers in the Americas: Perspectives from central Brazil. *Antiquity* 72:581–592.

Krigbaum, J. (2000): Human paleodiet in tropical Southeast Asia: Isotopic evidence from Niah Cave and Gua Cha. New York: New York University.

————. (in press): Unstable and stable isotopes of carbon from Niah Cave (Sarawak, East Malaysia): Recent work on charcoal and tooth enamel apatite. *Bulletin of the Indo-Pacific Prehistory Association.*

Kutzback, J., R. Gallimore, S. Harrison, P. Behling, R. Selin, and F. Laarif (1998): Climate and biome simulations for the past 21,000 years. *Quaternary Science Reviews* 17:473–506.

Laden, G. (1992): Ethnoarchaeology and land-use ecology of the Efe (Pygmies) of the Ituri rain forest, Zaire: A behavioral ecological study of land-use patterns and foraging behavior. Ph.D. diss., Harvard.

Lavachery, P., E. Cornelissen, J. Moeyersons, and P. de Maret (1996): 30,000 ans d'occupation, 6 mois de fouilles: Shum Laka, un site exceptionnel en Afrique centrale. *Anthropologie et Prehistoire* 107:197–211.

Leakey, M. G., F. Spoor, F. H. Brown, P. Gathogo, C. Kiarie, L. N. Leakey, and I. McDougall (2001): New hominin genus from eastern Africa shows diverse middle Pliocene lineages. *Nature* 410:433–440.

Lioubine, V., and F. Guede (2000): *The Paleolithic of Republic Côte d'Ivoire (West Africa).* In Russian with an English summary. St. Petersburg: Russian Academy of Sciences.

Loy, T., M. Spriggs, and S. Wickler (1992): Direct evidence for human use of plants 28,000 years ago: Starch residues on stone artefacts from the northern Solomon islands. *Antiquity* 66:898–912.

Maley, J., and P. Brenac (1998): Vegetation dynamics, palaeoenvironments, and climatic changes in the forests of western Cameroon during the last 28,000 years B.P. *Review of Palaeobotany and Palynology* 99:157–187.

Maloney, B. (1998): The long-term history of human activity and rain forest development. In *Human activities and the tropical rainforest,* edited by B. Maloney, 65–85. Netherlands: Kluwer Academic Publishers.

McBrearty, S. (1987): Une evaluation du Sangoen: Son age, son environement et son rapport avec l'origine de l'Homo sapiens. *L'Anthropologie* 91:127–140.

————. (1988): The Sangoan-Lupemban and Middle Stone Age sequence at the Muguruk site, Western Kenya. *World Archaeology* 19:379–420.

————. (1990): Consider the humble termite: Termites as agents of post-depositional disturbance at African archaeological sites. *Journal of Archaeological Science* 17:111–143.

McBrearty, S., and A. Brooks (2000): The revolution that wasn't: A new interpretation of the origin of modern human behavior. *Journal of Human Evolution* 39(5):453–563.

McBrearty, S., L. Bishop, and J. Kingston (1996): Variability in traces of middle Pleistocene hominid behavior in the Kapthurin formation, Baringo, Kenya. *Journal of Human Evolution* 30:563–580.

Mercader, J. (1997): Bajo el techo forestal: La evolución del poblamiento en el bosque ecuatorial del Ituri, Zaire. Ph.D. diss., Universidad Complutense, Madrid.

————. (2002): Forest people: The role of African rain forests in human evolution and dispersal. *Evolutionary Anthropology* 11:117–124.

Mercader, J., and A. Brooks (2001): Across forests and savannas: A comparison of later stone assemblages from Ituri and Semliki, Northeast Democratic Republic of Congo. *Journal of Anthropological Research* 57(2):197–217.

Mercader, J., M. D. Garralda, P. Pearson, and R. Bailey (2001): Eight hundred-year-old human remains from the Ituri rain forest, Democratic Republic of Congo: The rock shelter site of Matangai Turu NW. *American Journal of Physical Anthropology* 115:24–37.

Mercader, J., R. Martí, I. González, A. Sánchez, and P. García (in press): Archaeological site formation in tropical forests: Insights from the Ituri rock shelters, Congo. *Journal of Archaeological Science.*

Mercader, J., F. Runge, L. Vrydaghs, H. Doutrelepont, E. Corneille, and J. Juan-Tresseras (2000): Phytoliths from archaeological sites in the tropical forest of Ituri, Democratic Republic of Congo. *Quaternary Research* 54:102–112.

Mercader, J., R. Martí, J. Martinez, and A. Brooks (2002): The nature of "stone-lines" in the African Quaternary record: Archaeological resolution at the rainforest site of Mosumu, Equatorial Guinea. *Quaternary International* 89: 71–96.

Moeyersons, J. (1997): Geomorphological processes and their palaeoenvironmental significance at the Shum Laka cave (Bamenda, western Cameroon). *Palaeogeography, Palaeoclimatology, Palaeoecology* 133:103–116.

Morwood, M., and D. Hobbs (1996): Themes in the prehistory of tropical Australia. *Antiquity* 69:747–768.

Newsome, J., and R. Flenley (1988): Late Quaternary vegetational history of the central highlands of Sumatra; II: Palaeopalynology and vegetational history. *Journal of Biogeography* 15:555–578.

Oslisly, R., H. Doutrelepont, M. Fontugne, H. Forestier , P. Giresse, C. Hatte, and L. White (2001): Le site de Maboué 5 dans la réserve de la Lopé au Gabon: Premier résultats pluridisciplinaires d'une stratigraphie vielle de plus de 40,000 ans. Paper presented at the Congress of the Panafrican Association for Prehistory and Related Studies.

Pasveer, J. (1998): Kria Cave: An 8000-year occupation sequence from the Bird's Head of Irian Jaya. In *Bird's Head approaches: Irian Jaya studies—a programme for interdisciplinary research*, edited by G. Bartstra, 67–89. Rotterdam: A. A. Balkema.

Pavlides, C. (1999): The story of Imlo: The organisation of flaked-stone technologies from the lowland tropical rainforest of West New Britain, Papua New Guinea. Ph.D. diss., La Trobe University.

Pavlides, C., and C. Gosden (1994): 35,000-year-old sites in the rain forests of West New Britain, Papua New Guinea. *Antiquity* 68:604–610.

Piperno, D. (1994): Phytolith and charcoal evidence for prehistoric slash-and-burn agriculture in the Darien forest of Panama. *The Holocene* 4(3):321–325.

———. (1997): Phytoliths and microscopic charcoal from leg 155: A vegetational and fire history of the Amazon basin during the last 75 K.y. In *Proceedings of the ocean drilling program, scientific results*, edited by R. Flood, D. Piper, A. Klaus, and L. Peterson, vol. 155, 411–418.

———. (1998): Paleoethnobotany in the Neotropics from microfossils: New insights into ancient plant use and agricultural origins in the tropical forest. *Journal of World Prehistory* 12(4):393–449.

Piperno, D., and I. Holst (1998): The presence of starch grains on prehistoric stone tools from the humid Neotropics: Indications of early tuber use and agriculture in Panama. *Journal of Archaeological Science* 25:765–776.

Piperno, D., M. Bush, and P. Colinvaux (1991): Paleoecological perspectives on human adaptation in central Panama; I: The Pleistocene. *Geoarchaeology* 6(3): 201–226.

Pope, G., and S. Keates (1994): The evolution of human cognition and cultural capacity: A view from the Far East. In *Integrative paths to the past: Paleoanthropological advances in honor of F. Clark Howell,* edited by R. Corrucini and R. Ciochon, 531–567. Prentice Hall.

Roosevelt, A. (1998): Ancient and modern hunter-gatherers of lowland South America: An evolutionary problem. In *Advances in historical ecology,* edited by W. Balee, 190–212. New York: Columbia University Press.

Roosevelt, A., M. Lima, C. Lopes, M. Michab, N. Mercier, H. Valladas, J. Feathers, W. Barnett, M. Imazio, A. Henderson, J. Silva, B. Chernoff, D. Reese, J. Holman, N. Toth, and K. Schick (1996): Palaeoindian cave dwellers in the Amazon: The peopling of the Americas. *Science* 272:373–384.

Sahnouni, M., D. Hadjouis, S. Abdesselam, A. Olle, J. Verges, A. Derradji, H. Belarech, and M. Medig (2002): El-Kherba: A lower Pleistocene butchery site in northeastern Algeria. *Journal of Human Evolution* 42:A31–A32.

Sondaar, P., G. Van der Bergh, B. Mubroto, F. Aziz, J. de Vos, and U. Batu (1994): Middle Pleistocene faunal turnover and colonization of Flores (Indonesia) by Homo erectus. *C. R. Academy of Science* 319:1255–1262.

Stearman, A. (1991): Making a living in the tropical forest: Yuquí foragers in the Bolivian Amazon. *Human Ecology* 19(2):245–260.

Swisher, C., G. Curtis, T. Jacob, A. Getty, and A. Suprijo (1994): Age of the earliest known hominids in Java, Indonesia. *Science* 263:1118–1121.

Tappen, M. (1994): Bone weathering in the tropical rain forest. *Journal of Archaeological Science* 21:667–673.

Talbot, M., and T. Johannessen (1992): A high-resolution palaeoclimatic record for the last 27,500 years in tropical West Africa from the carbon and nitrogen isotopic composition of lacustrine organic matter. *Earth and Planetary Science Letters* 110:23–37.

Therin, M. (1994): Subsistence through starch: The examination of subsistence changes on Garua island, West New Britain, Papua New Guinea, through the extraction and identification of starch from sediments. BA thesis, University of Sydney.

Thomas, M. (1994): *Geomorphology in the tropics: A study of weathering and denudation in low latitudes.* Chichester: University of Stirling; John Wiley.

———. (2000): Late Quaternary environmental changes and the alluvial record in humid tropical environments. *Quaternary International* 72:23–36.

Townsend, P. (1990): On the possibility/impossibility of tropical forest hunting and gathering. *American Anthropologist* 92(3):745–747.

Turnbull, C. (1962): *The forest people: A study of the Pygmies of the Congo.* New York: Museum of Natural History.

Van der Bergh, G., B. Mubroto, F. Aziz, P. Sondaar, and J. de Vos (1996): Did *Homo erectus* reach the island of Flores? *Bulletin of the Indo-Pacific Prehistory Association* 14:27–36.

Van der Hammen, T., and M. Absy (1994): Amazonia during the last glacial. *Palaeogeography, Palaeoclimatology, Palaeoecology,* 109:247–261.

Van der Kaars, W., and M. Dam (1995): A 135,000-year record of vegetational and climatic change from the Bandung area, West Java, Indonesia. *Palaeogeography, Palaeoclimatology, Palaeoecology* 117:55–72.

Van Noten, F. (1977): Excavations at Matupi Cave. *Antiquity* 51:35–40.

Wendorf, F., A. Close, and R. Schild, (1994): Africa in the period of *Homo sapiens neaderthalensis* and contemporaries. In *History of humanity; vol. 1: Prehistory and the beginnings of civilization,* edited by S. de Laet, A. Dani, J. Lorenzo, and R. Nunoo, 117–135.

Wesselman, H. (1995): Of mice and almost-men: Regional paleoecology and human evolution in the Turkana basin. In *Paleoclimate and evolution, with emphasis on human origins,* edited by E. Vrba, G. Denton, T. Partridge, and L. Burckle, 356–368. New Haven: Yale University Press.

White, F, ed. (1983): *The vegetation of Africa: A descriptive memoir to accompany the Unesco/Aetfat/Unso vegetation map of Africa.* Paris: Unesco.

White, T., G. Suwa, and B. Asfaw (1994): *Australopithecus ramidus,* a new species of early hominid from Aramis, Ethiopia. *Nature* 371:306–312.

Woldegabriel, G., T. White, G. Suwa, P. Renne, J. de Heinzelin, W. Hart, and G. Heiken (1994): Ecological and temporal placement of early Pliocene hominids at Aramis, Ethiopia. *Nature* 371:330–333.

Woldegabriel, G, Y. Haile-Selassie, P. Renne, W. Hart, S. Ambrose, B. Asfaw, G. Heiken, and T. White (2001): Geology and paleontology of the late Miocene Middle Awash valley, Afar rift, Ethiopia. *Nature* 412:175–178.

Yamei, H., R. Potts, Y. Baoyin, G. Zhengtang, A. Deino, W. Wei, J. Clark, X. Guangmao, and H. Weiwen (2000): Mid-Pleistocene Acheulian-like stone technology of the Bose Basin, South China. *Science* 287:1622–1626.

PART I

AFRICAN PIONEERS

CHAPTER 1

The Archaeology of West Africa from the Pleistocene to the Mid-Holocene

Joanna Casey

The degree to which the forested areas of West Africa were occupied prior to the advent of farming and metal technology is difficult to address because the archaeology of the West African forests is not well known. Not only is West Africa a vast region that has received comparatively little archaeological attention, but investigations into its earlier prehistory have been hampered by a perception that people could not have lived south of the dry savannas prior to the advent of farming and/or metal working (cf. Davies, 1967; Shaw, 1978). Climatic changes throughout the Pleistocene dramatically shifted the boundaries of West Africa's forests and savannas, so sites that occur today in the forested parts of the region may not necessarily have been forested at the time they were occupied. Unfortunately, at present neither paleoenvironmental information nor chronological resolution at most sites are adequate to address the issue.

This chapter assesses what is known about the archaeology of the rain forests in West Africa. Most of the information comes from the work of Oliver Davies, who undertook extensive archaeological research in Ghana in the 1950s and 1960s. Davies worked in an era when the hominid finds in East and southern Africa dominated the discipline (Kense, 1990). In West Africa he expected to find, and subsequently thought he had found, the full sequence of African prehistory from the Oldowan to the present. Later researchers who reanalyzed Davies's material dismissed his earliest sites and had grave misgivings about the validity of many others. After Davies's overzealous designation of early sites, West African prehistory entered an era where researchers appear to have been reluctant to find any early material at all. Increasingly, however, evidence is emerging from across West Africa that humans were present well before the Later Stone Age (LSA). Furthermore, information from other parts of Africa is changing our ideas about the timing of prehistoric events and what this implies for the evolution, migration, and cultural development of *Homo sapiens*. However flawed, Davies's work remains the most comprehensive survey of Stone Age materials in West Africa, and it is therefore useful to review his data in light of more recent evidence from around the continent.

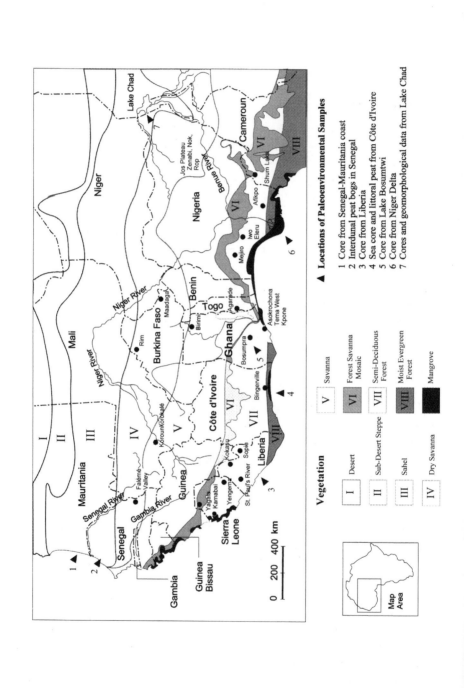

Vegetation

I	Desert
II	Sub-Desert Steppe
III	Sahel
IV	Dry Savanna
V	Savanna
VI	Forest Savanna Mosaic
VII	Semi-Deciduous Forest
VIII	Moist Evergreen Forest
	Mangrove

▲ **Locations of Paleoenvironmental Samples**

1 Core from Senegal-Mauritania coast
2 Interdunal peat bogs in Senegal
3 Core from Liberia
4 Sea core and littoral peat from Côte d'Ivoire
5 Core from Lake Bosumtwi
6 Core from Niger Delta
7 Cores and geomorphological data from Lake Chad

Map Area

0   200   400 km

The Region of Study

Sub-Saharan West Africa (fig. 1.1) stretches from the Atlantic coast of Senegal in the west to Lake Chad (approximately 14 degrees east) and from the Guinea Coast in the south to the edge of the Sahara Desert at approximately 20 degrees north. It is a region of generally low-lying plains and plateaux with altitudes rarely exceeding 500 meters above sea level. Exceptions are the Fouta Djallon and associated mountains in Guinea, which are almost 2000 meters in elevation, and the Cameroon Mountains, which reach a maximum of 4070 meters. Two major air masses that shift their positions throughout the year bring about a seasonal regime that has little annual fluctuation in temperature, but large differences in rainfall. The length and intensity of the dry season increases from south to north, and this results in vegetation zones that appear as broad east-west bands across West Africa, with dense rain forests occurring in the extreme south, followed by progressively drier forests, woodlands, and savannas up to the Sahara Desert (Lawson, 1986). (See fig. 1.1.) Locally, the vegetation zones are complicated by variations in soils and topography (Ahn, 1970; Cole, 1986; Morton, 1986; Moss and Morgan, 1970).

West Africa has two types of forest. The moist evergreen forest occurs only below 6 degrees north and receives annual rainfall of 1600–2000 mm with no months that are completely dry. The most typical tree species of this forest are *Cynometra ananta, Lophira alata,* and *Tarrietia utilis,* and within this area are numerous low-lying swamps containing climbing raffia and other palms (Lane, 1962). Semi-deciduous forest occurs between 6 and 8 degrees north in a region where the annual precipitation is 1200–1600 mm and there is a dry season of about three months (Flenley, 1979; Lane, 1962; Lawson, 1966; Steentoft, 1988). Characteristic tree species are *Celtis zenkeri, Triplochiton schleroxylon, Antiaris africana,* and *Chlorophora excelsa.* Progressively northward, the effects of the dry season are more keenly felt, the proportions of the primary trees shift, the upper story becomes more broken, and the difference between the emergent and upper canopy becomes less pronounced (Lane, 1962). Cultivation and logging have altered the forests to such a degree that the original structure is barely apparent (Baker, 1962). Montane vegetation is present in the mountains of West Africa and on some of the higher escarpments and hills.

The savanna stretches north from approximately 8 to 15 degrees north and receives between 1200 and 500 mm of annual rainfall. Savanna vegetation is composed of short, widely spaced trees adapted to withstand seasonal

Fig. 1.1. West African vegetation zones, sites for paleoenvironmental samples, and archaeological sites mentioned in the text

burning and long periods of aridity. Typical savanna species are *Daniella oliveri, Borassus aethiopum, Entada africana, Terminalia glaucescens,* and *Terminalia macroptera.* The more heavily wooded Guinea savanna in the south gives way to the sparser northern Sudan savanna in response to the diminution in rainfall and lengthening of the dry season. Drier, northern savannas contain progressively more *Acacia, Combretum, Ficus,* and *Gardenia.* The savanna has been severely impacted by grazing, burning, and farming that have altered the species composition of the region. Among the results is a predominance of species that are of economic importance, species such as *Ceiba pentandra, Adansonia digitata, Parkia biglobosa, Acacia albida, Tamarindus indica,* and *Vitellaria paradoxa.* North of the savanna, the Sahel is a region of sparse grass cover, thorn scrub, and hardy trees dominated by *Acacia senegal, Acacia radian, Leptadena pyrotechnica, Salvadora persica,* and *Grewia sp.* The Sahel receives only 150–500 mm of rainfall per year and has a dry season that can last up to 11 months. (Gritzner, 1988). Human activity and, especially, grazing are thought to have had a severe impact on this region.

The Dahomey Gap, an anomalously dry region of savanna vegetation, extends through Togo, Benin, and the eastern coast of Ghana and effectively separates the forests of West Africa from Central Africa. Forests on either side of the gap are similar, with 380 out of 389 tree species being common to both, suggesting that the two forests were joined during wetter climatic periods (Dupont et al., 2000; Hall and Swaine, 1981; Maley, 1987, 1989). At the river mouths along the coast, *Rhizophora* dominates mangrove swamps.

The Paleoenvironment

Information on the paleoenvironment of West Africa comes from deep-sea cores taken off the coast (Assemien et al., 1970; deMenocal et al., 2000; Dupont et al., 1998, 2000; Hébrard et al., 1971; Huang et al., 2000; Jahns et al., 1998; Martin, 1972; Zhao et al., 2000); from cores taken from terrestrial sources at the delta of the Niger River (Sowunmi, 1981, 1986), the interdunal peat bogs in Senegal (Lézine, 1991), Lake Bosumtwi in Ghana (Hall et al., 1978; Talbot, 1983; Talbot and Hall, 1981; Talbot et al., 1984), and Lake Chad (Maley, 1981; Servant and Servant-Vildary, 1980); from geomorphological evidence left by sea-level changes (Allen, 1964, 1965a,b, 1970; Andah, 1980b; Barusseau and Giresse, 1987; Bruckner and Anderson, 1957; Gregory, 1962; Maron, 1969; McMaster et al., 1970a,b); and from other geomorphological features found at random throughout the region (Fritsch, 1969; Hervieu, 1970a,b; Moeyersons, 1997). (See fig. 1.1). These widely scattered loci can provide glimpses of the paleoenvironmental record from which large-scale trends can be reconstructed. However, in a large and diverse area such as West Africa, small-scale environmental changes that are not always detectable within the geological time frame (Rognon, 1980; Rossignol-Strick et al., 1998; Sarnthein, 1978) may have been of enormous significance to plant and animal communities and consequently to humans.

The global climatic changes that brought the glaciations to Europe brought equally dramatic climate change to West Africa. Arid phases that generally coincided with glacial maxima contracted the vegetation belts and expanded the deserts. These phases were followed by humid conditions that prompted a rise in sea levels and forest expansion and caused what is now the Sahara desert to became a lush environment with abundant surface water. Recent evidence suggests that these climatic changes may have been extremely rapid, taking only a few hundred years to become fully realized (deMenocal et al., 2000).

Information is sketchy for the earliest time periods, but it appears that between 1 million and 700,000 years ago there was a wet phase and period of rising sea levels that was followed by several cycles where sea levels rose and fell. The evidence is primarily geological: raised beaches, clay deposits, and leaf impressions preserved in volcanic ash (Sowunmi, 1986). Much more is known about the paleoenvironment during the last glacial cycle. Dupont et al. (2000) provide a series of "time slices" showing the vegetational changes in West Africa during the last 150,000 years by drawing on marine cores and terrestrial sources. The earliest part of the sequence (the latter part of Oxygen Isotope Stage 6) shows that just before deglaciation, mangrove swamps along the coast were very much reduced as compared with the situation today, and savanna and open dry forest predominated to the north. Many changes took place during Oxygen Isotope Stage 5, so it is divided into five substages. There are few undisturbed records from the earliest part of the sequence (substage 5e), but those that do exist show rain forests and mangrove swamps expanding while dry forests and savannas move to higher latitudes. After about 115,000 B.P. (substage 5d), the extent of the rain forest reduced while montane forest expanded. Mangroves became scarce along the central and eastern coasts but were still abundant toward the west. Between 105,000 B.P. and 100,000 B.P. (substage 5c), montane forests declined and savanna and dry forests shifted south. Rain forests and mangroves recovered along the eastern coast at the end of this stage. Substage 5b again brought the expansion of montane vegetation, but an increase in grasses, suggesting arid conditions in the Niger Delta; and a contraction of the limits of the forest occurred off the coast of Senegal. Substage 5a is the last interglacial period. There was a general expansion of montane vegetation throughout the mountainous areas as well as an expansion of the rain forests in the lower elevations. Not much is known about Oxygen Isotope Stages 3 and 4 because records are so poor, but there is evidence for arid phases during this time (Lézine, 1991; Pokras and Mix, 1987;

Hooghiemstra, 1989). Much more is known about Stage 2, which coincides with the glacial maximum. There was a reduction of mangrove in all areas and a drastic increase in grasses, indicating expansion of the dry savanna. Forests disappeared inland but remained in patches along the coast, although the composition of the forests had changed. *Canthium cf. subcordatum* and *Holoptelea grandis* were among the few forest species represented, while the dry savanna species *Borreria caetocephala* and *Huphaenae thebaica* and an abundance of grasses suggest an environment that was similar to the northern parts of the present Guinea savanna (Sowunmi, 1986). Montane vegetation was extremely rare. At this point the Sahara desert extended to about 14 degrees north (Dupont and Hooghiemstra, 1989; Hooghiemstra, 1989; Rossignol-Strick and Duzer, 1979).

The African Humid Period starts at the glacial-interglacial transition, but it begins at different times in different parts of West Africa. The earliest evidence is seen in the proliferation of ferns on the coasts of Nigeria and Senegal as early as 17,000 B.P. and 15,000 B.P., respectively, and in the expansion of rain forests in Nigeria at 14,000 B.P. During the early Holocene the northern limits of the forests extended to 10 to 12 degrees north (Lézine, 1991; Rossignol-Strick and Duzer, 1979), and it is likely that at this time the forest was continuous across the Dahomey Gap (Dupont et al., 2000). The highest lake levels, indicating heavy precipitation, are recorded right across the Sahara between 12,500 B.P. and 8000 B.P. (Rossignol-Strick, 1983; Street and Grove, 1976; Talbot, 1980). A short arid event interrupted the humid phase at around 8000 B.P. (deMenocal et al., 2000). Once the humid phase had resumed, the climatic optimum in the central Sahara occurred at about 5500 B.P. (Hébrard et al., 1971; Lauer and Frankenberg, 1980; Sowunmi, 1981, 1986; Talbot et al., 1984), after which the climate rapidly deteriorated (deMenocal et al., 2000; Holmes et al., 1997; Petit-Maire, 1991; Servant and Servant-Vildary, 1980; Sowunmi, 1986; Talbot 1980). By 3000 B.P., evidence from both Lake Bosumtwi and coastal Nigeria indicates a dramatic increase in the pollen of oil palm *(Elaeis guineensis),* which has been interpreted as indicating intentional clearing of the forest by human beings (Sowunmi, 1981, 1986; Talbot et al., 1984). In Nigeria a coincident increase in the pollen of weeds associated with cultivated land and waste places would seem to support this idea, as does archaeological evidence for the earliest farming communities in West Africa at around this time (Sowunmi, 1986).

The Archaeology

Throughout West Africa, archaeology began with sporadic collections and reports done by Europeans working with the colonial administrations. None of this work was undertaken by professional archaeologists, but by interested officials at various levels and, most importantly, by geologists and other technicians who published descriptions of antiquities, undertook minor excavations, and speculated on the origins of the West African people (e.g., Kitson, 1913, 1916; Braunholtz, 1936; Wild, 1934, 1935a,b; Migeod 1926; Todd, 1903; Hamy, 1904; Laforgue, 1925, 1936; Delafosse, 1900, 1922). After the creation of the Institut Français d'Afrique Noire (IFAN) in 1938, archaeology in Francophone and Anglophone West Africa took different trajectories. IFAN created an archaeology-prehistory section in 1941 and installed two professional archaeologists who undertook excavations and synthesized the archaeology of the French colonies in West Africa (e.g., Mauny, 1949, 1950, 1957; Szuomowski, 1957). Meanwhile, in Anglophone West Africa the first professional archaeologists conducted their research when they could get breaks from their other duties (Shaw, 1944; Fagg, 1945). The establishment of universities in West Africa during the 1950s and 1960s brought about the proliferation of European archaeologists in West Africa and the training of the first professional African archaeologists. In Francophone West Africa, archaeology initially focused on the arid regions and was primarily oriented toward locating the directions from which African civilizations had come under a diffusionist explanatory paradigm (DeBarros, 1990; Mauny, 1947, 1952; Vaufrey, 1953). The idea that Africa had no history led other archaeologists, primarily in Anglophone West Africa, to concentrate on the Stone Age (Posnansky, 1982). This interest in the Stone Age continued through the 1960s, fueled by the hominid discoveries in East and southern Africa (Kense, 1990). Oliver Davies, who did extensive surveys and excavations throughout Ghana and the neighboring countries (Davies, 1957a,b,c, 1961a,b, 1964, 1967, 1968, 1971, 1973, n.d.), undertook much of this work. Since the late 1970s archaeology in West Africa has been consciously aimed at investigating the more recent time periods that have more direct ties to the present (DeBarros, 1990; Kense, 1990; McIntosh and McIntosh, 1983; Stahl, 1994), and research has primarily been focused well north of the rain forests (McIntosh, 1994).

There appear to have been few geographic impediments to the early occupation of West Africa. The Sahara, which is often perceived as the greatest barrier to modern human movement, was fully capable of supporting human occupants throughout much of the Pleistocene. The recovery of australopithecine and *Homo ergaster* remains in Chad (Brunet et al., 1995; Coppens, 1961, 1965, 1966) would appear to suggest that neither the environment nor the capabilities of early hominids restricted their movement into the area. There is a substantial Acheulian presence in the central and western Sahara (Pasty, 1999; Petit-Maire, 1988, 1991; Tillet, 1983, 1985), on the Jos Plateau in Nigeria (Bond, 1956; Fagg, 1956; Soper, 1965), in the Falémé Valley in Senegal (Camara and DuBoscq, 1990); and there is emerging evidence for an Early Stone Age occupation in Mali (Huysecom and Sow, 2001; Robert and Soriano, 2000). However, evidence for the early human occupation of West Africa south of the dry savanna has been scarce.

Oliver Davies has found what he believed to be a significant number of early sites in West Africa (see figs. 1.2 and 1.3). Davies classified most of his artifacts on the basis of overall morphology, weathering, and their association with geological formations. He was able to date his finds by coordinating them with raised beaches and river terraces that were thought to relate to the sea-level changes that took place throughout the Pleistocene. Archaeologists and geologists have criticized Davies's use of geology because his assumptions about the timing and signatures of geological events were proven to be incorrect. Davies's "Oldowan" ("Pebble Tool" and "Chellean") and "Acheulian" artifacts have not withstood reevaluation and have all been reclassified or discarded by later researchers (Allsworth-Jones, in Sutton, 1981; Nygaard and Talbot, 1984; Swartz, 1972; Wai-Ogosu, 1973). Despite many reports from other West African countries, there is no good evidence for Oldowan tools in West Africa (Isaac, 1982; but see Huysecom and Sow, 2001; Robert and Soriano, 2000), and Acheulian artifacts have never been found in lowland West Africa (Swartz, 1980).

The earliest compelling evidence for human presence in sub-Sahelian West Africa is a macrolithic complex, characterized by heavy-duty picks and bifaces, that has been found at numerous sites in Ghana (Davies, 1964, 1967; Nygaard and Talbot, 1976a,b, 1977, 1984), Nigeria (Davies, 1957b; Soper, 1965; Allsworth-Jones, 1987), Guinea (Delcroix and Vaufrey, 1939), Mali and Mauritania (Vaufrey, 1947), Liberia (Gabel, 1976), and Côte d'Ivoire (Chenorkian, 1983; Chenorkian and Pradis, 1982; Lioubine and Guede, 2000). Davies (1957a,b,c) attributed this complex to the Sangoan because of its perceived similarity to the Early Stone Age industry, in equatorial and East Africa, that is characterized by heavy-duty bifacial tools such as core axes, picks, core scrapers, and, in some places, spheroids and choppers, along with a light-duty component of scrapers and denticulates (Kleindienst, 1962; McBrearty, 1988). Davies designated sites as "Sangoan" by the presence of crude bifaces, chopping tools, and especially picks, and by the occurrence of these and other artifacts on stone lines, river terraces, and raised beaches that were thought to be contemporaneous with the known dates for the Sangoan elsewhere in Africa. Davies did not provide a technological analysis and description of the Sangoan, making it impossible to compare the sites to each other or to other Sangoan sites beyond superficial similarities of the larger formal tool types. Elsewhere in Africa, Sangoan sites remain few in number; and despite intensive investigations at some of them (Clark, 1969, 1974; Cornelissen, 1995; McBrearty, 1987, 1988; Sheppard and Kleindienst, 1996), the complex remains poorly understood

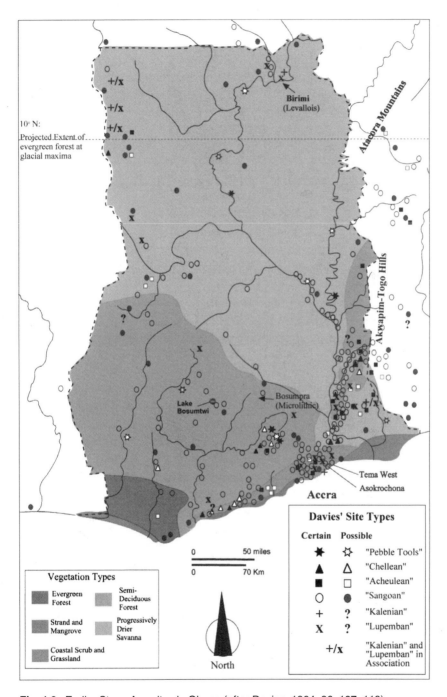

Fig. 1.2. Earlier Stone Age sites in Ghana (after Davies, 1964: 96, 107, 119)

(Cornelissen, 1997). Davies's estimation of the age of the Sangoan at about 50,000 B.P. has also turned out to be false. Earlier age determinations for the Sangoan elsewhere in Africa were characteristically much younger than they should have been because of the limits of the dating methods in use at the time (cf. Clark and Brown, 2001; Sheppard and Kleindienst, 1996; McBrearty and Brooks, 2000). Where it is found in context, the Sangoan overlies the Acheulian and underlies the Middle Stone Age (MSA), giving it a likely date of somewhere between 300,000 B.P. to 250,000 B.P. (McBrearty and Brooks, 2000).

In equatorial Africa, the Sangoan is followed by the early Middle Stone Age Lupemban industrial complex, which is characterized by light-duty tools such as small, retouched tools and scrapers, heavy-duty tools such as picks and core-axes, and large cutting tools such as hand axes and lanceolate points that become more refined through time (Clark and Brown, 2001). Like the Sangoan, the Lupemban is not well understood. While many early Lupemban artifacts are similar to those found in the Sangoan, it is the presence of lanceolate points that distinguishes the earliest Lupemban industries. McBrearty and Brooks (2000) have suggested that the Early Stone Age–Middle Stone Age transition is not just a shift in technology, but marks the appearance of fully modern *Homo sapiens*. Points are a defining characteristic of the MSA. They indicate the use of composite tools and advanced hunting methods, and they show regional variation—all traits that are considered to be associated with "modern" behaviors (McBrearty and Brooks, 2000). The transition between Sangoan and Lupemban industries is therefore potentially very significant. In West Africa, Davies had difficulty distinguishing between Sangoan and Lupemban industries in the absence of lanceolates, and tended to lump them. Subsequent researchers who have reevaluated Davies's work have found little justification for attributing the Ghanaian material to the Sangoan or Lupemban (Andah, 1979a; Sutton, 1981; Wai-Ogosu, 1973). At the suggestion of Wai-Ogosu (1973), Davies proposed a local name, the "Awudome Industry" for what he had earlier referred to as "Sangoan" (Davies, 1976). Allsworth-Jones (in Sutton, 1981) suggested relegating most of Davies's "Sangoan" and "Lupemban" material to the Middle Stone Age.

Although there have been many attempts to abandon the term "Sangoan" and its associated implications in West Africa, the term persists primarily because of the evidence from Asokrochona and Tema West on the coast of Ghana. The sites were found by Davies in 1958 (Davies, 1964) and excavated in the 1970s (Andah, 1979a; Nygaard and Talbot, 1976a,b, 1977, 1984). At these sites an industry that Davies defined as "Sangoan" was found in stratigraphic sequence on top of a stone line—a layer of lateritic gravel interspersed with quartz pebbles and resting on a highly irregular eroded surface

(Nygaard and Talbot, 1984). The lateritic layer is similar to gravel formations present in Benin and Côte d'Ivoire that are thought to date to the Mio-Pliocene (Houessou and Lang, 1978). At Asokrochona the layer underlies a late Pleistocene formation which probably accumulated during the glacial maximum between 25,000 B.P. and 13,000 B.P. (Nygaard and Talbot, 1984). Elsewhere in Africa, Sangoan materials are characteristically found on top of or in the upper part of stone lines (Cornelissen, 1997: 314). A full description of the material has still not been published, but Nygaard and Talbot (1984) place the lower levels of Asokrochona within the larger Sangoan industrial complex due to the presence of picks, core-axes, core-scrapers, spheroids, and choppers. They suggest the name "Asokrochona Industry" to denote the local manifestation. Differences in the size and morphology of artifacts from the Ghanaian sites compared to those further east are thought to be due to differences in raw material and to the fact that the Ghanaian sites are lithic workshops (Nygaard and Talbot, 1984).

Other Asokrochona Industry and Sangoan sites have been found in similar geological contexts elsewhere in the Accra plains (Nygaard and Talbot, 1984; Davies, 1959, 1961b, 1968) and in more ambiguous contexts in other parts of the country (Davies, 1964, 1967; Sordinas, 1971, cited in Swartz, 1974, 1980; Swartz, 1974). Davies found about 280 "Sangoan" sites in Ghana and about 50 others in Togo and Benin (Davies, 1967: 108). (See fig. 1.2.) These numbers contrast sharply with the thin scatter of known sites from elsewhere in West Africa, but Davies felt that the distribution of sites was a real one and that the Sangoan people had entered West Africa west of the Niger River through the Atacora Mountains in northern Togo and eastern Ghana. The problem with most of Davies's designations is that they are based on the overall morphology and weathering of "diagnostic" tools collected from undated, secondary contexts. Many of the tools Davies illustrates (e.g., Davies, 1967: 112–115) are generalized flaked pieces that are not particularly diagnostic.

A hiatus separates the Sangoan from some Middle Stone Age occurrences. In general, Davies's MSA sites (fig. 1.3) seem to be a collection of miscellaneous aggregates that do not fall into any coherent pattern (Nygaard and Talbot, 1984). The "Guinea Aterian" is found in the savanna region of Ghana and Togo and is defined on the basis of presumed similarities to the Aterian, a MSA industry from North Africa and the Sahara that dates from around 90,000 B.P. to some time before the last glacial maximum (Mc-Brearty and Brooks, 2000). True Aterian sites have not been found below 19 degrees north at the southern extent of the lakes that were present in the Sahara during the Pleistocene humid periods (Pasty, 1999: 12). Davies describes the Guinea Aterian as a "degenerate" industry made primarily on pebbles of quartz, quartzite, and chert, with very few formal tools and some

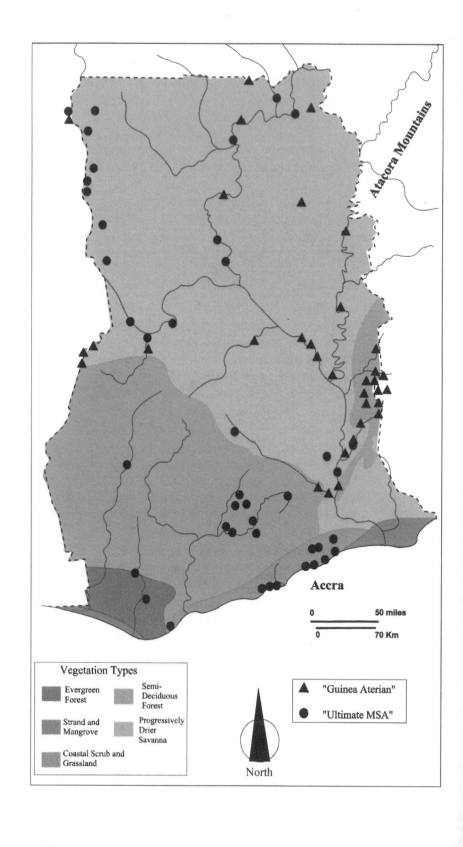

knowledge of Levallois technique (Davies, 1967: 135). Tanged implements are the *fossile directeur,* or type tool, for the Aterian, but those illustrated by Davies (e.g., Davies, 1964: 131) are unconvincing, and he provides no detailed technological description that would enable comparison with the Aterian proper. Allsworth-Jones (in Sutton, 1981), among others, feels that the term "Aterian" is inappropriate in this context.

The "Ultimate MSA" (fig. 1.3) is a similarly ambiguous industry that is characterized by a variety of small, crudely made picks, push-planes, Levallois flakes, tanged implements, flaked tools, and the occasional microlith. Davies also refers to "undifferentiated MSA," which appears to mean generalized macroflake industries with no diagnostic elements: "From Ghana, Dahomey and the forest and bush-savannah to the south there is a good deal of undifferentiated MSA. It is not, however worth discussion, because I am sure that it is either Lupemban or Guinea Aterian, from sites which have yielded nothing recognizable. Some have been identified as MSA, principally from their position stratified on the upper stone-line or in the basal gravels" (Davies, 1964: 128).

Regardless of how one ultimately decides to regard Davies's classifications, the fact remains that, at some point after the Sangoan and prior to the Later Stone Age, archaeological lithic aggregates that included prepared-core techniques and/or Levallois elements existed. MacDonald and Allsworth-Jones (1994: 84) have recommended that the term "Levallois" be dropped in favor of the term "flakes from prepared cores" because West African examples are not typical of the Levallois as it is known in Europe and North Africa. The more recent definitions of Levallois technology (cf. Chazen, 1997) give considerable flexibility in defining Levallois and recognize that it encompasses a variety of techniques.

Dates for the Levallois and/or prepared-core industries in West Africa are inconsistent. It has been difficult to date MSA or earlier materials in West Africa because the contexts are beyond the reach of conventional radiocarbon dating; and West Africa has no recent volcanic activity that would permit the kind of chronometric dating that is possible in East Africa. Many of Davies's sites were not in primary context, so only limiting dates could be obtained in any case. At Tema West and Asokrochona the MSA appears in the upper part of the formation that dates to the dry period recorded between 20,000 b.p. and 13,000 b.p. Similar sites containing Levallois, disk core, or other prepared-core techniques are found elsewhere in the Accra plains in southern Ghana (Davies, 1961a, 1968; Sordinas, 1971, cited in Swartz, 1980) but none of these sites have been dated.

Fig. 1.3. Middle Stone Age sites in Ghana (after Davies, 1964: 130, 137)

Most recently, a flake industry with a strong Levallois component has been found stratified under a sterile layer below a Later Stone Age component at the Birimi site in northeastern Ghana (Casey et al., 1997; Hawkins et al., 1996). An Optically Stimulated Luminescence (OSL) date obtained from Birimi provides a minimum date of 35,000–30,000 B.P. (Casey et al., 1997). OSL offers the possibility for dating early West African materials because it dates quartz-rich sediments that are abundant in West Africa, and its range extends beyond the limits of conventional radiocarbon dating. Elsewhere in West Africa, variations of similar lithic aggregates have been found in Senegal, where several sites containing Levallois components date to between 27,000–13,500 B.P. and 16,900–5660 B.P. (Allsworth-Jones, 1987). In Cameroon and Nigeria similar sites have been estimated to date to the late Pleistocene dry period around 20,000 B.P. (Allsworth-Jones, 1986). Strangely, the few radiocarbon-dated sites in Nigeria have produced even later dates: 5440 ± 100 B.P. at Zenabi (Allsworth-Jones, 1979: 244) and 5490 ± 85 B.P. and 5660 ± 90 B.P. at Nok (Allsworth-Jones, 1987: 119). Allsworth-Jones (1986: 166) suggested that the mid-Holocene dates at these sites are due to the process of re-deposition through erosion that took place 5000 years ago during the Holocene climatic optimum. At Shum Laka in Cameroon, triangular points made with a prepared-core method that distinctly resembles Levallois are located in levels dated to around 7000 B.P. and are possibly associated with ceramics (Lavachery et al., 1996: 206). Such evidence needs to be treated with caution, but it does lend support to the late dates for Levallois-influenced materials elsewhere in West Africa and also suggests that the presence of Levallois technique is not, in itself, an indicator of the MSA, if that is regarded as being a chronostratigraphic term. Undated sites are found in northern Nigeria and on the Jos Plateau (Soper, 1965), in Cameroon (Marliac, 1974, 1975, 1978; Marliac and Gavaud, 1975) and Guinea (Boriskovsky and Soloviev, 1978).

The appearance of microlithic industries is thought to signal the Later Stone Age, but the designation of "microlithic industries" is also problematic. The issue is whether individual researchers are referring to true, geometric microliths, which are a specific class of purposely formed tools, or to small-scale flake-tool industries (Casey, 1993; Nygaard and Talbot, 1984). The earliest geometric microliths appear in Central Africa at around 40,000 B.P. (Van Noten, 1977) and at Shum Laka at around 30,000 B.P. (Cornelissen, 1996; Lavachery et al., 1996). Industries that consist of prolific flaked-stone aggregates of unmodified, retouched, and utilized flakes, frequently in quartz, are well represented all over West Africa prior to the advent of iron-working technology. These industries are sometimes associated with geometric microliths, which appear to have a somewhat more restricted distribution, although this is difficult to gauge from the literature.

Elsewhere in Africa, MSA and LSA technologies appear to have coexisted from about 50,000 B.P. until the end of the Pleistocene (McBrearty and Brooks, 2000; Mercader and Brooks, 2001; Mercader and Martí, 1999). In West Africa these small-flake industries occur at 13,000 B.P. to 6000 B.P., although at Shum Laka they appear as early as 30,000 B.P. (Cornelissen, 1996; Lavachery et al., 1996). These industries are often overlain by similar industries that also contain ground-stone tools and/or pottery. The two types of industries have not been found in stratigraphic sequence in Ghana. Davies does not distinguish between them but refers to the whole microlithic industry as the "Mesoneolithic." The use of geometric microliths and small chipped-stone tools continued in West Africa after the appearance of settled village life and domestic plants and animals. It is therefore possible that among the sites that Davies classifies as Mesoneolithic are many that are specialized activity areas belonging to larger and later complexes.

A preceramic/ground-stone microlithic industry is found at 13,050 ± 230 B.P. at Bingerville, near Abidjan in Côte d'Ivoire (Chenorkian, 1983: 130), at several sites in Burkina Faso, including Rim, which has been dated at 12,000 B.P. to 5000 B.P. (Andah, 1979b, 1980a,b; Wai-Ogosu, 1971), and Maagada, which dates to 7590 ± 90 B.P. and 7000 ± 100 B.P. (Breunig and Wotzka, 1993). Similar stratification is noted at several sites in Nigeria, including Rop (Eyo, 1972; Fagg, 1945, 1972; Rosenfeld, 1972), Afikpo (Andah and Anozie, 1980; Hartle, 1980), Dutzen Kongba (York, 1978), Mejiro Cave (Willett, 1962), and Iwo Eleru, the lower level of which dates to 11,200 ± 200 B.P. (Shaw and Daniels, 1984). Kourounkorokale in Mali shows similar stratification, with the preceramic layer underlying at a date of 5185 ± 95 B.P. (MacDonald, 1997).

The later small flaked-stone industries are found all over West Africa. In Ghana the earliest and best-investigated site of this period in is Bosumpra, in central Ghana, and has a lithic industry consisting of geometric microliths in quartz and ceramics dated to 5330 ± 100 B.P. (Smith, 1975; Shaw, 1944). Similar sites appear in Sierra Leone at Yengema, Yagala, and Kamabai (Atherton, 1972; Coon, 1968), though no true microliths are recorded here, and in Liberia at Sopie and Kokasu (Gabel, 1976), where ground-stone tools are absent. Dates at these sites range from around 4200 B.P. to 1500 B.P. (Atherton, 1972; Coon, 1968). In Cameroon, Sumpa is undated but closely resembles the upper layer at Rop (David, 1980: 149), and in central Togo, Agarade rock shelter has a prolific quartz industry with pottery with radiocarbon dates ranging from around 4200 B.P. to 2650 ± 60 B.P. (DeBarros, 1992). Several shell middens on the coast of West Africa, most notably Kpone in Ghana, also date from this time period (Dombrowski, 1977; Nygaard and Talbot, 1976a, 1977, 1984).

By 3500 B.P., the Kintampo complex was well established in Ghana, bringing the first settled communities and domestic plants and animals (D'Andrea et al., 2001). There can be no doubt that the changes brought about by permanent settlements and land clearing enhanced humans' ability to survive in the rain forest by attracting light-dependent edible plant species, commensal faunas, and animals attracted to new shoots growing on cleared fields. This change is noted at the few Kintampo sites where organic remains have been preserved (Stahl, 1985).

Discussion

Figures 1.2 and 1.3 show the locations of the sites that Oliver Davies designated as Early and Middle Stone Age, superimposed on maps showing the current limits of the modern vegetation zones and the limit of the forest during the glacial maxima. In these figures Lupemban and Kalenian are lumped with the Early Stone Age because Davies was unable to distinguish between them and the Sangoan in the absence of lanceolates. These are Davies's designations, and for the reasons already discussed his terminology is not necessarily valid. Some of these sites can also be ruled out as not being the result of human activity (e.g., Swartz, 1972), and still others are likely due to later human activities that produced macrolithic debitage. MacDonald and Allsworth-Jones (1994) have proposed that all the materials that Davies would have classified as Sangoan through Mesoneolithic be included as components of the Later Stone Age. While typical LSA sites are primarily microlithic in nature, the Sangoan would be part of a macrolithic component of the LSA, a component that would also include macroflake industries and Levallois technology. The scheme does have merit because it manages to circumvent the problem of fitting the often contradictory West African data into a unilinear framework. On the other hand, by lumping everything into the LSA, the scheme itself may suggest that West Africa was uninhabited until relatively recently.

If all of Davies's earlier sites are disregarded, then it appears as though there was little or no occupation of the southern part of West Africa until relatively recently. Shaw (1977, 1984) has suggested that disease was a barrier to the early occupation of sub-Saharan West Africa, and that it was only with the development of immunities and cultural practices that reduced the incidence or effects of disease that populations had the opportunity to grow. It is also possible that early peoples relied heavily on perishable materials for their technology, and consequently their sites cannot be found except under exceptional preservational circumstances. Equally likely is the possibility that there is a cognitive problem with identifying early sites. It is as though, with the dismissal of much of Davies's material, archaeologists are now reluctant to designate any material as possibly being early. With so many archaeological projects aimed at investigating later time periods, it is also likely that the expertise needed to identify early material is also lacking.

Regardless of the actual affiliation of the sites in figures 1.2 and 1.3, they nevertheless represent an impressive collection of macrolithic occurrences that contain neither pottery nor ground stone. The consistency and quantity of the sites argue for there having been a significant human presence in Ghana, and consequently in West Africa, prior to the advent of farming and iron technology. In the absence of secure dates, the possibility remains that some of these typologically ambiguous assemblages really are of considerable antiquity.

The sites seem to cluster in the Akwapim Hills, along the coast, to the northwest of Accra, and along the northwest border of Ghana. While Davies was, by all accounts, a tireless searcher of archaeological sites, the preponderance of sites located near the University of Ghana and along the roads linking major centers would seem to suggest that his survey methods were opportunistic rather than systematic and that the distribution of sites reflects modern accessibility rather than prehistoric preference. A paucity of sites in the extreme southwest of Ghana, the only area that is today in evergreen rain forest, likely reflects a conscious decision on Davies's part because he did not believe that occupation of the rain forests was possible prior to advanced technology: "It is a mistake in terminology to associate [the Sangoan] with the equatorial forest. This was never penetrable by primitive man, and where the Sangoan is found in areas now forest, one is justified in assuming that forest did not exist" (Davies, 1964: 98).

The early Equatorial complexes, such as the Sangoan, that the macrolithic material is thought to superficially resemble were originally thought to be an adaptation to heavily forested environments (Clark, 1970, 1972, 1982), but this has been challenged by paleoenvironmental evidence from some sites that indicate more open environments (McBrearty, 1987, 1988). It is likely that these pick and heavy-duty bifacial complexes represent versatile human adaptations. Undoubtedly, in the past as now, differences in soils, topography, and other ecological variables created numerous microenvironments in the forested zones of West Africa. It is also likely that in the past, as now, people sought ecotones where they would have had access to a variety of resources. Figure 1.1 shows the locations of the excavated sites mentioned in the text. Of those located in the present-day forest, some, such as Shum Laka, Asokrochona, Tema, Kpone, Bingerville, and Bosumpra, are in microenvironments that are considerably more arid and less heavily wooded than the surrounding area. Few of the sites contained plant remains that would enable paleoenvironmental reconstruction, but at Sopie and Kokasu the presence of oil palm remains would indicate a more open and less heavily wooded environment.

Davies saw all new innovations after the Sangoan as being due to invasions from the north, but MacDonald (1997) suggests a southern origin for the early microlithic complex on the basis of the early dates from southern sites such as Shum Laka and Iwo Eleru and from the Ivory Coast. During the last arid period those zones that are now in forest supported a Guinea savanna vegetation, and, as the forests advanced with the climatic amelioration, people migrated north with the open forest and wooded savannas. This scenario suggests that during the arid periods the dry forests in the south were the likely routes of migration throughout equatorial and West Africa. The picture is complicated by the early dates on microliths at more northern sites such as Rim, but if the climatic changes happened as quickly as deMenocal et al. (2000) have suggested, perhaps we should not expect to see much difference in the ages of the sites. Conceivably, people also followed the forests earlier in prehistory, lending logistical support to Davies's idea that the macrolithic pick and biface industries of West Africa share affinities with the Sangoan and Lupemban in Central Africa. This scenario is also much more suggestive of a prehistoric preference for more open environments.

Conclusions

Considerably more sites need to be found and verified before the antiquity of human occupation of the forested regions of West Africa can be established. While there are real, practical problems with finding sites in the forests, perhaps the biggest problem to be overcome is a cognitive one, which makes archaeologists reluctant to designate possible early sites as such. The data we do have seem to suggest the possibility of a substantial early presence in Ghana, and consequently West Africa; however, most of these sites remain ambiguous due to a lack of context, dates, and truly diagnostic artifacts. The dramatic and often quick changes in climatic conditions and consequently in environments make it difficult, if not impossible, to determine the actual environmental conditions in which early prehistoric peoples found themselves. The little evidence we do have seems to indicate a preference for more open and less heavily wooded environments, but this is based on very little evidence and does not preclude the possibility that early peoples utilized a variety of ecological zones within their territory.

Acknowledgments

I would like to thank the Department of Anthropology at the University of South Carolina and the Social Sciences and Humanities Research Council of Canada for providing funds that enabled me to undertake library research at the University of Toronto and the Institute of Archaeology at University College London, respectively. Thanks to Kevin MacDonald and Jay Woodhouse for providing access to Thurstan Shaw's collection of offprints at the University College London and to Heather Henderson and Clive and Valerie Waterer for hosting me during those trips. The manuscript was improved considerably by Maxine Kleindienst, Alicia Hawkins, Julio Mercader, and the anonymous reviewers who read it and made comments.

References

Ahn, P. M. (1970): *West African soils.* Oxford: Oxford University Press.

Allen, J.R.L. (1964): The Nigerian continental margin: Bottom sediments, submarine morphology, and geological evolution. *Marine Geology* 1:289–332.

———. (1965a): Late quaternary of the Niger Delta. *Bulletin of the American Association of Petroleum Geologists* 49:587–600.

———. (1965b): Coastal geomorphology of eastern Nigeria: Beach ridge barrier islands and vegetated tidal flats. *Geologie en Minmbouw* 44:1–21.

———. (1970): Sediments of the modern Niger delta: A summary and review, deltaic sedimentation (edited by J. P. Morgan). *Society of Economic Paleontologists and Mineralogists,* Special Publication 15:138–157.

Allsworth-Jones, P. (1979): The Middle Stone Age industry from Zenabi, northern Nigeria. In *Proceedings of the 8th Panafrican Congress of Prehistory and Quaternary Studies, Nairobi, 1977,* edited by R. E. Leakey and B. A. Ogot, 244–247. Nairobi: International Louis Leakey Memorial Institute for African Prehistory.

———. (1986): Middle Stone Age and middle Palaeolithic: The evidence from Nigeria and Cameroun. In *Stone Age prehistory: Studies in memory of Charles McBurney,* edited by G. N. Bailey and P. Callow, 153–168. London: Cambridge University Press.

———. (1987): The earliest human settlement in West Africa and the Sahara. *West African Journal of Archaeology* 17:87–128.

Andah, B. W. (1979a): The early Paleolithic in West Africa: The case of Asokrochona. *West African Journal of Archaeology* 9:47–85.

———. (1979b): The Later Stone Age and Neolithic of Upper Volta reviewed in a West African context. *West African Journal of Archaeology* 9:87–110.

———. (1980a): Excavations at Sindou and Kawara (Upper Volta). *West African Journal of Archaeology* 10:1–59.

———. (1980b): Excavations at Rim, north-central Upper Volta: A paleoecological study. In *West African culture dynamics,* edited by B. K. Swartz and R. A. Dumett, 41–65. The Hague: Mouton.

Andah, B., and F. Anozie (1980): Preliminary report on the prehistoric site of Afikpo (Nigeria). *West African Journal of Archaeology* 10:83–102.

Assemien, P., J. C. Filleron, L. Martin, and J. Taset (1970): Le Quaternaire de la zone littorale de Côte d'Ivoire. *Bull. Assoc. Sénég. Et. Quatern. Ouest Afr. Dakar* 25:65–78.

Atherton, J. H. (1972): Excavations at Kamabai and Yagala rockshelters, Sierra Leone. *West African Journal of Archaeology* 2:39–74.

Baker, H. G. (1962): The ecological study of vegetation in Ghana. In *Agriculture and land use in Ghana*, edited by J. B. Wills, 151–159. London: Oxford University Press.

Barusseau, J. P., and P. Giresse (1987): Evolution of the Atlantic coastal zone of Africa in the Quaternary. In *Quaternary coastal geology of West Africa and South America*, 1–27. UNESCO Reports in Marine Science 43. Paris, France: UNESCO.

Biberson, P. (1968): Review of *The Quaternary in the coastlands of Guinea*, by Oliver Davies. *West African Archaeological Newsletter* 9:20–36.

Böeda, E. (1994): *Le concept Levallois: Variabilité des méthodes*. Paris: Monographie du CRA 9.

Bond, G. (1956): A preliminary account of the Pleistocene geology of the Plateau Tin Fields region of northern Nigeria. In *Proceedings of the III International West African Conference, Ibadan, Nigeria, 1949*, 187–202. Lagos: Nigerian Museum.

Boriskovsky, P. I., and V.V. Soloviev (1978): New data on the Stone Age of Guinée. *West African Journal of Archaeology* 8:51–74.

Braunholtz, H. J. (1936): Archaeology in the Gold Coast. *Antiquity* 10:469–74.

Breunig, P., and H.-P. Wotzka (1993): Archäologische forschungen im südosten Burkina Fasos 1989/90: Vorbericht über die erste grabungskampagne des Frankfurter Sonderforschungsbereiches 268 "westafrikanische savanne." *Beiträge sur Allgemeinen und Vergleichenden Archäologie* 11:145–187.

Bruckner, W. D., and J. P. Anderson (1957): Note on raised shorelines of the Gold Coast. In *Proceedings of the Third Panafrican Congress on Prehistory*, edited by J. D. Clark, 86–92. London: Chatto and Windus.

Brunet, M., A. Beauvilain, Y. Coppens, E. Heintz, A.H.E. Moutaye, and D. Pilbeam (1995): The first australopithecine 2,500 kilometres west of the Rift Valley (Chad). *Nature* 378:273–275.

Camara, A., and B. DuBoscq (1990): La fouille d'un site Acheuléen à Djita (basse vallée de la Falémé, Sénégal). *L'Anthropologie* 94(2):293–304.

Casey, J. (1993): Geometric microliths from northern Ghana and notes for a tentative morphological typology. *Nyame Akuma* 40:22–29.

——. (2000): *The Kintampo Complex: The Late Holocene on the Gambaga Escarpment, Northern Ghana*. Cambridge Monographs in African Archaeology 51, British Archaeological Reports International 906. Oxford: Archaeopress.

Casey, J., R. Sawatzky, D. R. Godfrey-Smith, N. Quickert, A. C. D'Andrea, M. Wollstonecroft, and A. Hawkins (1997): Report of investigations at the Birimi site in northern Ghana. *Nyame Akuma* 48:32–38.

Chazan, M. (1997): Redefining Levallois. *Journal of Human Evolution* 33:719–735.

Chenorkian, R. (1983): Ivory Coast prehistory: Recent developments. *African Archaeological Review* 1:127–142.

Chenorkian, R., and G. Pradis (1982): Un industrie paléolithique découverte dans la "Terre de Barre" d'une terasse proche d'Anyama (Région d'Abidjan). *Nyame Akuma* 21:18–27.

Clark, J. D. (1968): Review of *The Quaternary in the coastlands of Guinea,* by Oliver Davies. *West African Archaeological Newsletter* 9:37–40.

———. (1969): *The Kalambo Falls prehistoric site.* Vol. 1. London: Cambridge University Press.

———. (1970): *The prehistory of Africa.* New York: Praeger.

———. (1972): Problems of archaeological nomenclature and definition in the Congo Basin. In *Actes du 6ᵉ Session du Congrés Panafricain de Prehistoire, Dakar, 1967,* edited by H. Hugot, 584–593. Imprimeries Réunies de Chambéry.

———. (1974): *The Kalambo Falls prehistoric site.* Vol. 2. London: Cambridge University Press.

———. (1982): The cultures of the middle Palaeolithic/Middle Stone Age. In *The Cambridge History of Africa,* vol. 1, edited by J. D. Clark, 248–341. Cambridge: Cambridge University Press.

Clark, J. D., and K. S. Brown (2001): The Twin Rivers Kopje, Zambia: Stratigraphy, fauna, and artefact assemblages from 1954 and 1956 excavations. *Journal of Archaeological Science* 28:305–330.

Cole, M. (1986): *The savannas.* New York: Academic Press.

Coon, C. S. (1968): *Yengema cave report.* Philadelphia: University of Pennsylvania Museum.

Coppens, Y. (1961): Découverte d'un Australopitheciné dans le Villafranchien du Tchad. *Comptes Rendus de l'Academie des Sciences* (Paris) 252:3851–3852.

———. (1965): *L'hominien du Tchad.* Actes du Vᵉ Congres Panafricain de Prehistoire et de l'Etude du Quaternaire. Museo Arqueológico de Tenerife Publicaciones del S.I.A. del Excmo. Cabildo insular Nᵒ 5. Santa Cruz de Tenerife, Islas Canarias.

———. (1966): Le Tchadanthropus. *L'Anthropologie* 70(1–2):5–16.

Cornelissen, E. (1995): Indications du post-Acheuléen (Sangoen) dans la Formation Kapthurin, Baringo, Kenya. *L'Anthropologie* 99:55–73.

———. (1996): Shum Laka (Cameroon): Late Pleistocene and early Holocene deposits. In *Aspects of African Archaeology,* edited by G. Pwiti and R. Soper, 257–63. Harare: University of Zimbabwe Publications.

———. (1997): Central African transitional cultures. In *Encyclopedia of Precolonial Africa,* edited by J. O. Vogel, 312–320. Walnut Creek, Calif.: Altamira.

D'Andrea, A. C., M. Klee, and J. Casey (2001): Archaeobotanical evidence for pearl millet *(Pennisetum glaucum)* in sub-Saharan West Africa. *Antiquity* 75:341–348.

David, N. (1980): History of crops and peoples in north Cameroon to A.D. 1900. In *West African Culture Dynamics,* edited by B. K. Swartz and R. A. Dumet, 139–182. The Hague: Mouton.

Davies, O. (n.d.): Field notes. Mimeographed manuscript, Legon, Ghana.

———. (1957a): The climatic and cultural sequence in the late Pleistocene of the Gold Coast. In *Third Annual Panafrican Congress on Prehistory (1955),* edited by J. D. Clark and S. Cole, 1–5. London: Chatto and Windus.

———. (1957b): The old Stone Age between the Volta and the Niger. *Bulletin de l'IFAN,* ser. B, 19(3–4):592–616.

———. (1957c): The Sangoan culture in Africa. *South African Journal of Science* 50:273–277.

———. (1959): Distribution of old Stone Age material in Guinea. *Bulletin de l'IFAN*, ser, B, 21(1–2):102–108.

———. (1961a): *Archaeology in Ghana*. Edinburgh: Thomas Nelson.

———. (1961b): Geological and archaeological evidence for the late Quaternary climatic sequence in West Africa. *Ghana Journal of Science* 1:69–73.

———. (1964): *The Quaternary in the coastlands of Guinea*. Glasgow: Jackson.

———. (1967): *West Africa before the Europeans*. London: Methuen.

———. (1968): Mesoneolithic excavations at Legon and New Todzi (Ghana). *Bulletin de l'IFAN*, ser. B, 30(3):1147–1194.

———. (1971): *The archaeology of the flooded Volta Basin*. University of Ghana, Occasional Papers in Archaeology No. 1.

———. (1973): Excavations at Ntereso, Gonja, Northern Ghana, final report. Mimeographed manuscript.

———. (1976): The "Sangoan" industries. *Annals of the Natal Museum* 22(3):885–911.

DeBarros, P. (1990): Changing paradigms, goals, and methods in the archaeology of Francophone West Africa. In *A history of African archaeology*, edited by P. Robertshaw, 155–172. London: James Currey.

———. (1992): Preliminary report on excavations at Agarade rockshelter, Togo, West Africa. Paper presented at the 11th Biennial Society of Africanist Archaeologists Conference, March 26–29, 1992, Los Angeles, California.

Delafosse, M. (1900): Sur des traces probables e civilisation égyptienne et d'hommes de race blanche à la Côte d'Ivoire. *L'Anthropologie* 11:677–683.

———. (1922): *Les Noirs de l'Afrique*. Paris: Payot.

Delcroix, R., and R. Vaufrey (1939): Le toumbien de Guinée Française. *L'Anthropologie* 49:265–312.

deMenocal, P. B, J. Oritz, T. Guilderson, J. Adkins, M. Sarnthein, L. Baker, and M. Yarusinsky (2000): Abrupt onset and termination of the African humid period: Rapid climate responses to gradual insolation forcing. *Quaternary Science Reviews* 19:347–361.

Dombrowski, J. C. (1977): Preliminary note on excavations at a small midden near Tema, Ghana. *Nyame Akuma* 10:31–34.

Dupont, L. M., and H. Hooghiemstra (1989): The Saharan-Sahelian boundary during the Brunhes chron. *Acta Botanica* 38:405–415.

Dupont, L. M., F. Marret, and K. Winn (1998): Land-sea correlation by means of terrestrial and marine palynomorphs from the equatorial East Atlantic: Phasing of SE trade winds and the oceanic productivity. *Palaeogeography, Palaeoclimatology, Palaeoecology* 142:51–84.

Dupont, L. M., S. Jahns, F. Marret, and S. Ning (2000): Vegetation change in equatorial West Africa: Time slices for the last 150 ka. *Palaeogeography, Palaeoclimatology, Palaeoecology* 155:95–122.

Eyo, E. (1972): Rop rock shelter excavations, 1964. *West African Journal of Archaeology* 2:13–16.

Fagg, B. (1945): Preliminary report on a microlithic industry at Rop rock shelter, northern Nigeria. *Proceedings of the Prehistoric Society*, n.s., 10:68–69.

———. (1956): An outline of the Stone Age of the Plateau Minesfield. In *Proceedings of the III International West African Conference, Ibadan, Nigeria, 1949*, 203–222. Lagos: Nigerian Museum.

———. (1972): Rop rock shelter excavations, 1944. *West African Journal of Archaeology* 2:1–12.

Flenley, J. R. (1979): *Equatorial rainforest: A geological history*. London: Butterworths.

Folster, H. (1968): Review of *The Quaternary in the coastlands of Guinea*, by Oliver Davies. *West African Archaeological Newsletter* 9:45–47.

Fritsch, P. (1969): Note préliminaire sur la morphologie du piédmont nord de l'Adamaoua dans la région de Kontcha (Cameroun). *Annales de la Faculté des Sciences de l'Université Fédérale du Cameroun* 3:101–111.

Gabel, C. (1976): Microlithic occurrences in the Republic of Liberia. *West African Journal of Archaeology* 6:21–35.

Gregory, S. (1962): The raised beaches of the peninsular area of Sierra Leone. *Transactions of the Institute of British Geographers* 31.

Gritzner, J. A. (1988): *The West African Sahel*. University of Chicago Geography Research Paper No. 226. Chicago: University of Chicago Press.

Hall, J. B., and M. D. Swaine (1981): *Distribution and ecology of vascular plants in a tropical rainforest: Forest vegetation in Ghana*. The Hague: Junk.

Hall, J. B., M. D. Swaine, and M. R. Talbot (1978): An early Holocene leaf flora from Lake Bosumtwi, Ghana. *Palaeogeography, Palaeoclimatology, Palaeoecology* 24: 247–261.

Hamy, E. T. (1904): L'age de pierre à la Côte d'Ivoire. *Bulletin du musée d'histoire naturelle de Paris* 10:534–536.

Hartle, D. D. (1980): Archaeology east of the Niger: A review of cultural-historical developments. In *West African culture dynamics*, edited by B. K. Swartz and R. E. Dumett, 195–203. Mouton: The Hague.

Hawkins, A., J. Casey, D. Godfrey-Smith, and A. C. D'Andrea (1996): A Middle Stone Age component at the Birimi site, northern region, Ghana. *Nyame Akuma* 46:34–36.

Hébrard, L., P. Elouard, and H. Faure (1971): Quaternaire du littoral Mauritanien entre Nouakchott et Nouadhibou (Port Etienne). *Quaternaria* 15:297–304.

Hervieu, J. (1970a): Le Quaternaire du Nord-Cameroun: Schéma d'evolution géomorphique et relations avec la pédogenèse. *Cahiers ORSTROM*, série pédologie, 8(3):295–317.

———. (1970b): Influences des changements de climat Quaternaires dur le relief et les sols du Nord-Cameroun. *Bulletin de l'ASEQUA* 25:97–105.

Holmes, J. A., F. A. Street-Perrott, M. J. Allen, P. A. Fothergill, D. D. Harkness, D. Kroon, and R. A. Perrott (1997): Holocene palaeolimnology of Kajemarum Oasis, northern Nigeria: An isotopic study of ostracodes, bulk carbonate, and organic carbon. *Journal of the Geological Society* (London) 154:311–319.

Hooghiemstra, H. (1989): Variations of the NW African trade wind regime during the last 140,000 years: Changes in pollen flux evidenced by marine sediment records. In *Paleoclimatology and Paleometeorology*, edited by M. Leinen and M. Sarnthein, 733–770. Dordrecht: Kluwer.

Houessou, A., and J. Lang (1978): Contribution à l'étude du 'Continental Terminal' dans le Bénin méridional. *Science and Géology Bulletin* 31:137–139.

Huang, Y., L. Dupont, M. Sarnthein, J. M. Hayes, and G. Eglington (2000): Mapping of C_4 plant input from northwest Africa into northeast Atlantic sediments. *Geochimica et Cosmochimica Acta* 64(20):3305–3513.

Huysecom, E., and O. Sow (2001): Peuplement humain et paléoenvironnement en Afrique de l'Ouest: Un programme de recherche international au Mali. Paper presented at the 11ème Congrès de l'Association Panafricaine De Phéhistoire et Disciplines Assimilées, February 2001, Bamako, Mali.

Isaac, G. L. (1982): The earliest traces. In *The Cambridge history of Africa*, vol. 1, edited by J. D. Clark, 157–247. London: Cambridge University Press.

Jahns, S., M. Huls, and M. Sarnthein (1998): Vegetation and climate history of Africa based on a marine pollen record off the coast of Liberia (Site GIK 16776) covering the last 400,000 years. *Review of Paleobotany and Palynology* 102:277–288.

Kense, F. J. (1990): Archaeology in Anglophone West Africa. In *A history of African archaeology*, edited by P. Robertshaw, 135–154. London: James Currey.

Kitson, A. E. (1913): Southern Nigeria: Some considerations of its structure, people, and history. *Geographical Journal* 41:16–38.

———. (1916): Southern Nigeria: Some considerations of its structure, people, and history. *Geographical Journal* 48:369–392.

Kleindienst, M. R. (1962): Components of the East African Acheulian assemblage: An analytic approach. In *Actes de IV Congrés Panafricaine de Préhistoire et de l'étude du Quaternaire*, edited by G. Mortelmans and J. Nenquin, section 111:81–112. Tervuren: Musée Royal de l'Afrique Centrale.

Laforgue, P. (1925): État actuel de nos connaissances sur la pr"ehistoire en Afrique Occidentale Francaise. *Bulletin du comité d'Études historiques et scientifiquis de l'A.O.F.* 8:105–171.

———. (1936): Note bibliographique sur la préhistoire de L'Ouest africain. *Bulletin du comité d'Études historiques et scientifiquis de l'A.O.F.* 20:113–130.

Lane, D. A. (1962): The forest vegetation. In *Agriculture and land use in Ghana*, edited by J. B. Wills, 160–169. London: Oxford University Press.

Lauer, W., and P. Frankenberg (1980): Modelling of climate and plant cover in the Sahara for 5500 and 18000 B.P. *The Palaeoecology of Africa* 12:307–314.

Lavachery, P., E. Cornelissen, J. Moeyersons, and P. de Maret, (1996): 30 000 ans d'occupation, 6 mois de fouilles: Shum Laka, un site exceptionnel en Afrique centrale. *Anthropologie et Préhistoire* 107:197–211.

Lawson, G. W. (1966): *Plant life in West Africa*. Accra: Ghana Universities Press.

———. (1986): Vegetation and environment in West Africa. In *Plant ecology in West Africa*, edited by G. W. Lawson, 1–11. London: J. Wiley and Sons.

Lézine, A.-M. (1991): West African paleoclimates during the last climatic cycle inferred from an Atlantic deep-sea pollen record. *Quaternary Research* 35:456–463.

Lioubine, V. P., and F. Y. Guede (2000): *The Paleolithic of Republic of Côte d'Ivoire (West Africa)*. St. Petersburg: Russian Academy of Sciences.

MacDonald, K. C. (1997): Korounkorokalé revisited: The Pays Mande and the West African microlithic technocomplex. *African Archaeological Review* 14(3):161–200.

MacDonald, K. C., and P. Allsworth-Jones (1994): A reconsideration of the West African macrolithic conundrum: New factory sites and an associated settlement in the Valée du Serpant, Mali. *African Archaeological Review* 12:73–104.

Maley, J. (1981): *Etudes Palynologiques Dans Le Bassin Du Tchad Et Paléoclimatologie De l'Afrique Nord-Tropicale De 30,000 Ans À L'époque Actuelle.* Paris: ORSTOM.

———. (1987): Fragmentation de la fôret dense humide africaine et extension des biotopes montagnards au Quaternaire récent: Nouvelles données polliniques et chronologiques: implications paléoclimatiques et biogéographiques. *Paleoecology of Africa* 18:307–334.

———. (1989): Late quaternary climatic changes in the African rainforest: Forest refugia and the major role of sea surface temperature variations. In *Palaeoclimatology and palaeometerology: Modern and past patterns of global atmospheric transport,* edited by M. Leinen and M. Sarnthein, 585–616. Dordrecht: Kluwer Academic Publications.

Marliac, A. (1974) Prospection archéologique au Cameroun Septentrional. *West African Journal of Archaeology* 4:83–97.

———. (1975): Analyse morphologique des industries du Mayo Tsanga. *Travaux et Documents de l'ORSTROM* 43:75–104.

———. (1978): L'industrie de la haute terrasse du Mayo Louti: Note préliminaire sur le site de Mokorvong au Cameroun septentrional. *Cahiers ORSTROM, ser. sciences humaines,* 15(4):367–377.

Marliac, A., and M. Gavaud (1975): Premiers éléments d'une séquence paléolithique au Cameroun septentrional. *Bulletin de l'ASEQUA* 46:53–66.

Maron, P. (1969): Stratigraphical aspects of the Niger delta. *Journal of Mining and Geology* 4(1–2).

Martin, L. (1972): Variation du niveau de la mer et du climat en Côte d'Ivoire depuis 25,000 ans. *Cahiers ORSTOM, ser. geologie,* 4(2):73–103.

Martin, L., and P. Assemien (1970): Études sedimentologiques et palynologiques des sondages de Bogue (Basse vallee du Sénégal) et leur interpretation morphoclimatique. *Review of Geomorphological Dynamics* 19(3):98–113.

Mauny, R. (1947): Une route préhistorique à travers le Sahara occidental. *Bulletin de l'Institut Français d'Afrique Noire* 9:341–357.

———. (1949): État actuel de nos connaissances sur la préhistoire de la Colonie du Niger. *Bulletin de l'Institut Français d'Afrique Noire* 11:141–158.

———. (1950): État actuel de nos connaissances sur la préhistoire du Dahomey et du Togo. *Études Dahoméens* 4:5–11.

———. (1952): Essai sur l'histoire des métaux en Afrique occidentale. *Bulletin de l'Institut Français d'Afrique Noire* 14:545–595.

———. (1957): État actuel de nos connaissances sur la préhistoire et l'archéologie de la Haute Volta. *Notes Africaines* 73:16–25.

McBrearty, S. (1987): Une évaluation du Sangoen: Son age, son environnement et son rapport avec l'origine de l'Homo sapiens. *L'Anthropologie* 91:127–140.

———. (1988): The Sangoan-Lupemban and Middle Stone Age sequence at the Muguruk site, western Kenya. *World Archaeology* 19(3):388–420.

McBrearty, S., and A. Brooks (2000): The revolution that wasn't: A new interpretation of the origin of modern human behavior. *Journal of Human Evolution* 39: 453–563.

McIntosh, S. K. (1994): Changing perspectives on Africa's past: Archaeological research since 1988. *Journal of Archaeological Research* 2(2):165–198.

McIntosh, S. K., and R. J. McIntosh (1983): Current directions in West African prehistory. *Annual Review of Anthropology* 12:73–98.

McMaster, R. L., T. P. La Chance, A. Ashraf, and J. De Boer (1970a): Continental shelf geomorphic features of Portuguese Guinea, Guinea, and Sierra Leone. *Marine Geology* 9.

McMaster, R. L., J. De Boer, and A. Ashraf (1970b): Magnetic and seismic reflections studies on the West African continental shelf. *Bulletin of the American Association of Petroleum Geologists* 54:158–167.

Mercader, J., and A. S. Brooks (2001): Across forests and savannas: Later Stone Age assemblages from Ituri and Semliki, Democratic Republic of Congo. *Journal of Anthropological Research* 57:197–217.

Mercader, J., and R. Martí (1999): Middle Stone Age sites in the tropical forests of Equatorial Guinea. *Nyame Akuma* 51:14–24.

Migeod, F.W.H. (1916): Discovery of presumed palaeolith. *Man* 36:56–58.

———. (1919): Discovery of palaeoliths and pierced stones. *Man* 5–6:13.

———. (1926): *A new history of Sierra Leone.* London: Keegan Paul.

Morton, J. K. (1986): Montane vegetation. In *Plant ecology in West Africa,* edited by G. W. Lawson, 247–271. New York: John Wiley and Sons.

Moss, R. P., and W. B. Morgan (1970): Soils, plants, and farmers in West Africa. In *Human ecology in the tropics,* edited by J. P. Garlick and R.W.J. Keay, 1–31. Oxford: Pergamon Press.

Moeyersons, J. (1997): Geomorphological processes and their palaeoenvironmental significance at the Shum Laka cave (Bamenda, western Cameroon). *Palaeogeography, Palaeoclimatology, Palaeoecology* 133:103–116.

Moeyersons, J., E. Cornelissen, P. Lavachery, and H. Doutrelepont (1996): L'abri sous-roche de Shum Laka (Cameroun Occidental) données climatologiques et occupation humaine depuis 30,000 ans. *Geo-Eco-Trop* 20(1–4):39–60.

Nygaard, S., and M. Talbot (1976a): Coastal Ghana: Archaeology and geology. *Nyame Akuma* 8:36–37.

———. (1976b): Interim report on Asokrochona, Ghana. *West African Journal of Archaeology* 6:13–19.

———. (1977): First dates from the coastal sites near Kpone, Ghana. *Nyame Akuma* 11:39–30.

———. (1984) Stone Age archaeology and environment on the southern Accra plains, Ghana. *Norwegian Archaeological Review* 17(1):19–38.

Pasty, J.-F. (1999): *Contribution à l'Étude de l'Aterian du Nord Mauritanian.* Oxford: British Archaeological Reports International, 758.

Petit-Maire, N. (1988): Climatic change and man in the Sahara. In *Prehistoric cultures and environments in the late Quaternary of Africa,* edited by J. Bower and D. Lubell, 19–42. Cambridge, BAR International Series, 405.

———. (1991): Recent Quaternary climatic change and man in the Sahara. *Journal of African Earth Sciences* 12(1–2):125–132.

Pokras, E. M., and A. C. Mix (1987): Earth's precession cycle and Quaternary climatic change in tropical Africa. *Nature* 326:486–87.

Posnansky, M. (1982): African archaeology comes of age. *World Archaeology* 13: 345–358.

Robert, A., and S. Soriano (2000): Preliminary results on the settlement of the Dogon Country during the Early and Middle Stone Age. Paper presented at the Society of Africanist Archaeologists Conference, Cambridge.

Rognon, P. (1980): Pluvial and arid phases in the Sahara: The role of non-climatic factors. *Palaeoecology of Africa* 12:45–62.

Rosenfeld, A. (1972): The microlithic industries of Rop rock shelter. *West African Journal of Archaeology* 2:17–28.

Rossignol-Strick, M. (1983): African monsoons, an immediate climate response to orbital insolation. *Nature* 304:46–49.

Rossignol-Strick, M., and D. Duzer (1979): West African vegetation and climate since 22,500 B.P. from deep sea cores. *Pollen et Spores* 21:105–134.

Rossignol-Strick, M., M. Paterne, F. C. Bassinot, K.-C. Emeis, and G. J. DeLange (1998): An unusual mid-Pleistocene monsoon period over Africa and Asia. *Nature* 392:269–272.

Sarnthein, M. (1978): Sand deserts during glacial maximum and climatic optimum. *Nature* 272:43–46.

Servant, M., and S. Servant-Vildary (1980): L'environment quaternaire du bassin du Tchad. In *The Sahara and the Nile,* edited by M.A.J. Williams and H. Faure, 133–162. Rotterdam: Balkema.

Shaw, C. T. (1944): Report on investigations carried out in the cave known as "Bosumpra" at Abetifi, Kwahu, Gold Coast Colony. *Proceedings of the Prehistoric Society* 10:1–67.

———. (1977): Hunters, gatherers, and first farmers in West Africa. In *Hunters, gatherers, and first farmers beyond Europe,* edited by J.V.S. Megaw, 69–125. London: Leicester University Press.

———. (1978): *Nigeria: Its archaeology and early history.* London: Thames and Hudson.

———. (1981): The prehistory of West Africa. In *General history of Africa,* vol. 1, edited by J. Ki-Zerbo, 611–630. London: Heinemann/UNESCO.

———. (1984): Archaeological evidence and effects of food producing in Nigeria. In *From hunters to farmers,* edited by J. D. Clark and S. A. Brandt, 152–157. Berkeley: University of California Press.

Shaw, C. T., and S.G.H. Daniels (1984): Excavations at Iwo Eleru. *West African Journal of Archaeology* 14.

Sheppard, P. J., and M. R. Kleindienst (1996): Technological change in the earlier and Middle Stone Age of Kalambo Falls (Zambia). *African Archaeological Review* 13(3):171–196.

Smith, A. B. (1975): Radiocarbon dates from Bosumpra Cave, Abetifi, Ghana. *Proceedings of the Prehistoric Society* 41:179–182.

Soper, R. (1965): The Stone Age in northern Nigeria. *Journal of the Historical Society of Nigeria* 3(2):175–194.

Sowunmi, M. A. (1981): Late Quaternary environmental changes in Nigeria. *Pollen et Spores* 13(1):125–148.

———. (1986): Change of vegetation with time. In *Plant ecology in West Africa*, edited by G. W. Lawson, 273–307. London: J. Wiley and Sons.

Stahl, A. B. (1985): Reinvestigation of Kintampo 6 rockshelter, Ghana: Implications for the nature of culture change. *African Archaeological Review* 3:117–150.

———. (1994): Innovation, diffusion, and culture contact: The Holocene archaeology of Ghana. *Journal of World Prehistory* 8(1):1994.

Steentoft, M. (1988): *Flowering plants in West Africa*. London: Cambridge University Press.

Street, F., and A. T. Grove (1976): Environmental and climatic implications of late Quaternary lake-level fluctuations in Africa. *Nature* 261:385–390.

Sutton, J.E.G. (1981): Middle Stone Age. *Archaeology in Ghana* 2:1–4.

Swartz, B. K. (1972): An analysis and evaluation of the Yapei pebble tool industry, Ghana. *International Journal of African Historical Studies* 2:265–270.

———. (1974): A stratified succession of Stone Age assemblages at Hohoe, Volta region, Ghana. *West African Journal of Archaeology* 4:57–81.

———. (1980): The status of Guinea Coast paleoarchaeological knowledge as seen from Legon. In *West African Culture Dynamics*, edited by B. K. Swartz and R. E. Dumett, 37–40. The Hague: Mouton.

Szuomowski, G. (1957) Fouilles au Nord du Macina et dans la région de Segou. *Bulletin du Comité d'Études historiques et scientifiques d l'A.O.F.* 19:224–258.

Talbot, M. R. (1980): Environmental responses to climatic change in the West African Sahel over the past 20,000 years. In *The Sahara and the Nile*, edited by M.A.J. Williams and H. Faure, 37–62. Rotterdam: Balkema.

———. (1981): Changes in tropical wind intensity and rainfall: Evidence from southeast Ghana. *Quaternary Research* 16:201–220.

———. (1983): Lake Bosumtwi, Ghana. *Nyame Akuma* 23:11–12.

Talbot, M. R., and J. B. Hall (1981): Further late Quaternary leaf fossils from Lake Bosumtwi, Ghana. *Palaeoecology of Africa* 13:83–92.

Talbot, M. R., D. A. Livingstone, P. A. Palmer, J. Maley, J. M. Melack, G. Delibrias, and S. Gulliksen (1984): Preliminary results from sediment cores from Lake Bosumtwi, Ghana. *Palaeoecology of Africa* 16:173–192.

Tillet, T. (1983): *Le Paléolithique du Bassin Tchadien Septentrional (Niger-Tchad)*. Paris: CNRS.

———. (1985): The Palaeolithic and its environment in the northern part of the Chad Basin. *African Archaeological Review* 3:163–177.

Todd (1903): Notes on stone circles in Gambia. *Man* 3:164–165.

Van Noten, F. (1977): Excavations at Matupi Cave. *Antiquity* 51:35–40.

Vaufrey, R. (1947): Le néolithique para-Toumbien: Une civilisation agricole primitive du Soudan. *La Revue Scientifique,* 205–232.

———. (1953): L'age de la pierre en Afrique. *Journal de la société des Africanistes* 23:103–138.

Wai-Ogosu, B. (1971): Quaternary climatic changes in West Africa—A review. *Ghana Journal of Social Science* 1(2):94–116.

———. (1973): Was there a Sangoan industry in West Africa? *West African Journal of Archaeology* 3:9191–196.

Wild, R. P. (1927): Stone artifacts of the Gold Coast and Ashanti. *Gold Coast Review* 3(1–2).

———. (1931): A stone implement of Paleolithic type from Gold Coast Colony. *Gold Coast Review* 5(2).

———. (1934): Ashanti: Baked clay heads from graves. *Man* 34:1–4.

———. (1935a): An ancient pot from Tarkwa, Gold Coast. *Man* 35:136–7.

———. (1935b): Bone dagger and sheath from Obuasi, Ashanti. *Man* 35:10–11.

———. (1935c): The inhabitants of the Gold Coast and Ashanti before the Akan invasion. *Gold Coast Teachers' Journal* 6–7:1–19.

Willett, F. (1962): The microlithic industry from Old Oyo, western Nigeria. In *Actes du V Congrés Panafricaine de Prèhistoire et de l'Etude du Quaternaire Section 11,* edited by G. Mortelmans and J. Nenquin, 261–271. Tervuren: Musée Royal de l'Afrique Centrale.

York, R. N. (1978): Excavations at Dutsen Kongba, Plateau State, Nigeria. *West African Journal of Archaeology* 8:139–163.

Zhao, M., G. Eglinton, S. K. Haslett, R. W. Jordan, M. Sarnthein, and Z. Zhang (2000): Marine and terrestrial biomarker records for the last 35,000 years at ODP site 658C off NW Africa. *Organic Geochemistry* 31:919–930.

CHAPTER 2

The Middle Stone Age Occupation of Atlantic Central Africa

New Evidence from Equatorial Guinea and Cameroon

Julio Mercader and Raquel Martí

The Middle Stone Age of tropical Africa is sandwiched between the Acheulian and the Later Stone Age. Outside the tropical forest, elsewhere in equatorial Africa, the Early to Middle Stone Age boundary and the behavioral shift toward modernity have been placed circa 250,000 B.P. (McBrearty and Brooks, 2000), while the earliest Later Stone Age of tropical Africa may date from 40,000 B.P. to 30,000 B.P. (Van Noten, 1977; Lavachery et al., 1996; Ambrose, 1998). In Central Africa, the Middle Stone Age is conventionally divided in two horizons: the Sangoan and the Lupemban. These techno-complexes are characterized by scarcity or absence of Mode 2 (bifacial hand axes and cleavers) and Mode 5 (microlithism), which typify the Early and Later Stone Age, respectively. In the tropical forest, these two horizons represent the earliest widespread archaeological manifestations. However, it should be noted that industries of Acheulian affiliation have been found in the evergreen rain forests of the High Sangha, Central African Republic (Bayle des Hermens, 1975).

The Sangoan complex has been subject of intense debate (McBrearty, 1988). It was originally defined as a heavy-duty industry characterized by picks, core-axes, large core-scrapers, some hand axes, pebble tools, and spheroids (Clark, 1964). The Sangoan complex appears across the African tropics in both the semi-open terrains of the Côte d'Ivoire (Paradis, 1980; Chernokian, 1983; Chernokian and Paradis, 1982), Ghana (Davies, 1954; Nygaard and Talbot, 1976; Casey, chapter 1 in this book), Nigeria (Soper, 1965), Rwanda (Nenquin, 1967), Angola (Clark, 1963), Kenya (McBrearty, 1988), Tanzania (Mehlman, 1987), Zimbabwe (Cooke, 1963), and Zambia (Clark, 1969) and in regions currently covered by tropical forest, such as Gabon (Clist, 1995; Bayle des Hermens et al., 1987), Cameroon (Omi, 1977), and the People's Republic of Congo (Lanfranchi, 1990). The chronological position and cultural meaning of the Sangoan are unclear, but it precedes the Middle Stone Age assemblages of the Lupemban (McBrearty, 1988). There seems to be some indication, however, that the Sangoan complex was associated with "archaic" hominid morphologies and behavioral repertoires (McBrearty and Brooks, 2000: 485–486) and could represent part of the technological variability of the Early Stone Age in Central Africa.

Lupemban assemblages of the Middle Stone Age are characterized by carefully crafted lanceolate stone points reduced by hard percussion and sometimes by pressure flaking. Lupemban sites appear both outside the forest, in countries such as Rwanda-Burundi (Nenquin, 1967; Van Noten et al., 1972), Malawi (Clark and Haynes, 1970), Kenya (McBrearty, 1988), Angola (Clark, 1963), Zambia (Clark, 1969, 1974; Barham, 2000; Clark and Brown, 2001), and the Democratic Republic of Congo (Alimen, 1957; Van Moorsel, 1968, 1970; Cahen and Moeyersons, 1969–71; Cahen, 1976; Van Noten, 1982; Cabot, cited in de Maret, 1990), and in the lowland forests of Gabon (Pommeret, 1965, 1966) and Equatorial Guinea (this study). The Lupemban techno-complex seems to coexist with some Sangoan ones and the earliest manifestations of the Later Stone Age (Lavachery et al., 1996; Van Noten, 1977).

Many archaeological finds related to both the Sangoan and the Lupemban come from present-day nonforest environments or from insecurely dated sites with dubious stratigraphic validity. However, their extensive representation and widespread geographical dispersion across tropical Africa indicate beyond doubt that humans were everywhere in or near the African forest during the Middle Stone Age. In this context, this chapter reports the excavation of two Middle Stone Age (MSA) assemblages from the forested core of the African continent. These assemblages come from the sites of Mosumu, Equatorial Guinea, dated to a minimum age of 30.000 B.P., and Njuinye, Cameroon, first inhabited considerably earlier than 35,000 B.P. The aim of this chapter is to show that there was a pre–Later Stone Age settlement of Central African regions currently covered by humid evergreen forest. Unfortunately, direct paleobotanical evidence in the form of phytoliths, starch grains, and pollen from these Middle Stone Age sites could not be retrieved in sufficient amount. However, direct paleoenvironmental data from neighboring lake sites allow us to suggest that tropical forest settlement by Middle Stone Age humans was possible.

Setting

The open-air site of Mosumu is located in continental Equatorial Guinea (fig. 2.1), within Monte Alen National Park. Mosumu is near the Uoro River, which flows through the Uoro Tectonic Rift (Martinez-Torres and Riaza, 1996). This basin, formed after northeast-southwest rifting, is 90 km long and 14 km wide and is filled with Miocene clays and Quaternary fluvial deposits. Topography within this rift is flat or undulating, with gentle slopes and

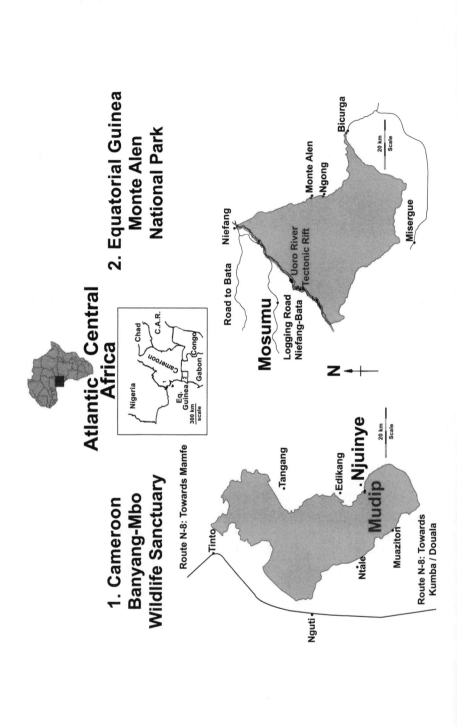

Atlantic Central Africa

1. Cameroon Banyang-Mbo Wildlife Sanctuary

2. Equatorial Guinea Monte Alen National Park

Nigeria

Chad

C.A.R.

Cameroon

Eq. Guinea

Gabon

Congo

300 km scale

Route N-8: Towards Mamfe

Tinto

Tangang

Nguti

Edikang

Njuinye

Ntale

Mudip

Muaziton

Route N-8: Towards Kumba / Douala

20 km Scale

Road to Bata

Niefang

Mosumu

Logging Road Niefang-Bata

Uoro River Tectonic Rift

Monte Alen

Ngong

Bicurga

Misergue

20 km Scale

N

altitudes of 300 to 550 m. Although little rain falls in January-February and July-August, an annual rainfall of 2000 mm to 3500 mm (Bultot and Griffiths, 1972: 286; CUREF, Bata, unpublished climate report) supports Guineo-Congolian dense tropical forest (White, 1983; Van Reeth and Ngomo, 1997; Lejoly, 1998) characterized by different proportions of trees of the Burseraceae, Euphorbiaceae, and Caesalpiniaceae. The most common species of lowland trees in the Mosumu area are *Aukomea Klaineana* (Burseraceae), *Lophira Alata* (Ochnaceae), and *Pycnanthus angolensis* (Myristicaceae). The most common herbaceous families in this area include the Zingiberaceae (*Costus lucasianus*), Marantaceae (*Megaphrynium macrostachyum*) and Poaceae (*Olyra latifolia*) (D. Ngomo, Ecofac, personal communication).

The open-air site of Njuinye is located in southwest Cameroon (fig. 2.1), within Banyang-Mbo Wildlife Sanctuary, near the Mfi River. Banyang-Mbo shows considerable geological diversity (Dumort, 1968), with Cretaceous outcrops of gres and conglomerates, Tertiary vulcanism, and plutonic outcrops of granite and gneiss. The wildlife sanctuary comprises a large number of altitudinal contexts ranging from 150 m above sea level in the north to 1750 m in the south, near the foothills of Mount Cameroon. In spite of a relatively dry period in December-January, the region enclosed between Mount Cameroon, Kribi, and Douala has average precipitation in the order of 3000 mm to 4000 mm (Dumort, 1968; Bultot and Griffiths, 1972: 296, 301). This wet climatic regime allows Banyang-Mbo Wildlife Sanctuary to support Guineo-Congolian dense tropical forest (White, 1983). The lowlands comprise species such as *Lophira Alata* (Ochnaceae), *Pycnanthus angolensis* (Myristicaceae), and *Uapaca staudtii* (Euphorbiaceae). Herbaceous families in this sector include the Zingiberaceae, Marantaceae, and Poaceae. However, it is impossible to quantify these species' relative importance in the overall ecosystem, as detailed floristic inventories are not available.

Mosumu, Continental Equatorial Guinea

A surface-road survey east of the Uoro Rift, from Niefang to Misergue through Monte Alen, Ngong, and Bicurga, reported 118 stone artifacts from 9 open-air sites along 130 km (Mercader and Martí, 1999a; Martí, 2001). These stone assemblages are nondiagnostic and consist of quartz industries made up of cores, flakes, and debitage from simple reduction with little platform preparation, very rarely from radial centripetal percussion (n = 2). These finds east of the Uoro River could belong to the Middle or Later Stone Age. A road-surface survey inside the Uoro Rift yielded a promising stone assemblage made up of 210 stone artifacts from 34 archaeological sites along

Fig. 2.1. Site location in the central African rain forest

40 km (Mercader and Martí, 1999a; Martí, 2001). Inside the rift, quartzite is the most common raw material. Reduction modes include simple debitage, bifacial reduction, and radial technology. Diagnostic tools include 12 bifacial lanceolate fragments and 2 whole lanceolate points. Although the whole specimens measure 6.0–7.2 cm, the estimated maximum length for fragments is 10 cm. Bifacial "scrapers," 4–8 cm in maximum length, were also found. These surface finds inside the Uoro Rift have a Middle Stone Age affiliation and suggest an extensive pre–last glacial maximum settlement of continental Equatorial Guinea.

The site of Mosumu is one of these 34 open-air sites with Middle Stone Age affiliation (Martí, 2001; Mercader et al., 2002). Mosumu is located in the village of Mosumu II (1°N 43' 37" 10°E 04' 39"), right outside the northwestern corner of Monte Alen National Park, at 425 m above sea level. Surface lithic debitage at Mosumu scatters over more than 1000 m². Archaeological excavation consisted of 22 m² in 10 pits, their dimensions ranging from 0.5 m² to 6 m².

A detailed geoarchaeological, taphonomic, and chronometric study of the site has been presented by Mercader et al. (2002), and a thorough techno-typological analysis is available from Martí (2001). Mosumu yielded 6455 artifacts distributed in two layers: one from the Middle Stone Age (layer 1) and another from the Later Stone Age (layer 2, cap). The total depth of archaeological deposits at the site of Mosumu is 2 m, and two stratigraphic units can be distinguished (fig. 2.2). The lower level, or layer 1 (1.5–2.0 m), is made up of ferralitic gravel, sand, silt, and clay. Gravel size follows a vertical gradient, so that smaller fractions, 1–2 cm, are represented in the lower part of layer 1, while larger fractions, 2–5 cm, rarely up to 15 cm, exist in the upper part of layer 1. The upper level, or layer 2 (0.0–1.5 m), is composed of sand, silt, clay, and, eventually, small gravel.

Archaeological remains appear throughout the entire sequence, but reach their highest horizontal and vertical concentrations in layer 1 (average number of artifacts per m² in layer 1: ~251; layer 2: ~39) through 50 ± 20 cm of sedimentary matrix. Layer 1 contained 86.5% of the total assemblage; layer 2 contained 13.5%. Preferred axis alignment of the artifacts or cardinal orientation could not be detected. Unstable positions of repose of the artifacts in the matrix, such as those represented by vertical dips, are infrequent (6.3%). Strongly abraded artifacts are completely absent. Microdebitage, 1–10 mm (28%) and 11–20 mm (46%) in maximum dimension, forms the bulk (~75%) of our sample. All elements of the technological chain are represented in level 1 and include cores (2.06%), debris (76.30%), flakes (18.01%), and tools (0.60%). The existence of 25 conjoining refitted breaks indicates an average horizontal mobility of 25.45 cm, while vertical

Stratigraphic Section at Njuinye, Cameroon

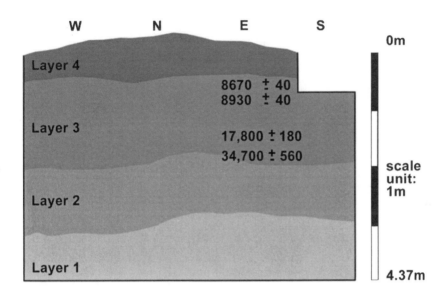

W   N   E   S

Layer 4

Layer 3

Layer 2

Layer 1

8670 ± 40
8930 ± 40

17,800 ± 180
34,700 ± 560

0m

scale
unit:
1m

4.37m

Stratigraphic Section at Mosumu, Equatorial Guinea

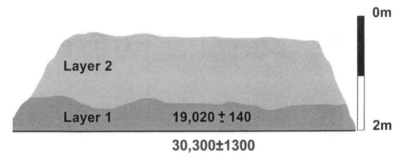

Layer 2

Layer 1   19,020 ± 140

30,300±1300

0m

2m

Fig. 2.2. Stratigraphic sections and radiocarbon dates from Njuinye, Cameroon, and Mosumu, Equatorial Guinea

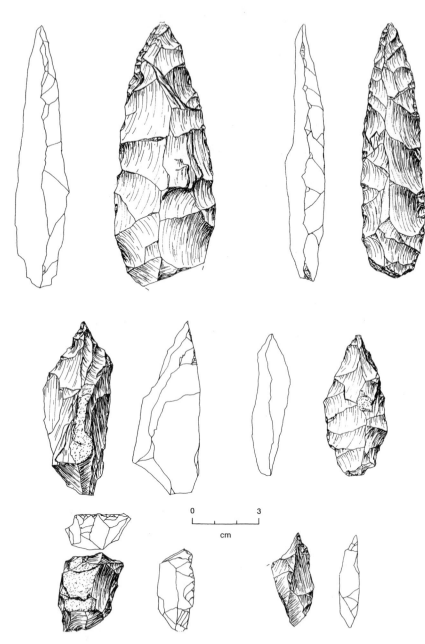

Fig. 2.3. Selected Lupemban tools from Mosumu, Equatorial Guinea, layer 1. Drawings by Dennis Knepper.

displacements average 3.88 cm. Thus, it is safe to assume that Mosumu is mostly a stone-knapping assemblage or workshop with an acceptable archaeological integrity.

For the purposes of this chapter, we focus on the Middle Stone Age level, or layer 1. Raw materials are quartzite (72.17%) and quartz (27.70%). Flint (0.13%) was used very infrequently. The assemblage is dominated by microdebris derived from the knapping of pebbles brought to the site from nearby riverbeds within the Uoro Tectonic Rift. Formal tools make up a low proportion of the total industry. Among these tools (n = 37), bifacial stone reduction (n = 22, 64.7%) forms a high proportion of the total tool kit. Formal tools include lanceolates, "core-axes" (*sensu* Clark, 1963), large "bifacial scrapers" (*sensu* McBrearty, 1988), perforators, steep-edge core-scrapers, side-scrapers, and small chopping tools (fig. 2.3). Three knapping sequences have been distinguished. The most abundant one is simple quartzite/quartz debitage (cores, flakes, cortical flakes, preparation by-products, debitage: 96.3%), in which cores show little or no platform preparation. In importance, the second knapping sequence is bifacial shaping (n = 227; lanceolates, bifacials, bifacial thinning flakes, and undetermined thinning flakes: 3.5%), which in Mosumu is mostly directed to making lanceolate points and other types of bifacials by applying steep, hard percussion. In fact, finished pressure flaking has been documented in two cases. Large daggers (25 cm to over 30 cm maximum dimension) do not appear in Mosumu. Instead, 5 out of 6 whole bifacial points measure 7.5 cm to 11.2 cm. One of them measures 5.4 cm, and the rest of them (10 specimens) are fragments whose reconstructed maximum length would be larger than 10 cm. In spite of the large number and careful manufacture of these lanceolate points, rarely do they present any evidence of basal thinning for hafting, nor impact scars suggestive of use. Unifacial radial centripetal percussion (n = 7, radial cores and flakes: 1.3%) appears in cores, in which one side preserves its cortex while the other side provides pseudo-Levallois removals (Wendorf and Schild, 1974: plates 26, 28, 30, 36, 37; Bayle des Hermens, 1981: 343; Cahen, 1976: fig. 3; Emphoux, 1970: plates 1 and 2; Clark, 1954: 206). Neither heavy-duty picks nor microliths are present.

Njuinye, Southwest Cameroon

The context of fieldwork, methods, and a preliminary report are available elsewhere (Mercader and Martí, 1999b). In this region, only two sites yielded Stone Age materials. These two sites are Mudip (5°N 15' 213" 9°E 42' 559", 364 m above sea level) and Njuinye (5°N 15' 213" 9°E 43' 38", 366 m above sea level). (See fig. 2.1.) Mudip is located by the Mfi River on a hilltop currently settled by villagers. The site of Mudip provided 627 artifacts over

NJUINYE

MUDIP

600 m². The bedrock crops out over most of the site. It is possible that Mudip's surface industries represent a shallow occupation horizon currently dismantled by erosion and recent human occupation. The surface assemblage from Mudip comprises 10 raw materials (quartz: 91%), including exotic materials such as obsidian, chert, flint, quartzite, and quartz crystal. This industry is heavily dominated by debris (95%). Tools (5%) have a microlithic component, with perforators (n = 19), core-scrapers (n = 7), and side-scrapers (n = 4). Cores (n = 10) indicate simple debitage and peripheral flake extraction in which artisans reduce pebbles in a circumferential manner, with the area used as a handle remaining untouched; these exhibit almost perpendicular blows to obtain usually cortical flakes with unprepared platforms. Other cores show radial centripetal percussion (n = 1).

The site of Njuinye is found in a lowland intra-montane plateau that lies 366 m above sea level and is surrounded by highland massifs up to 1483 m. This site rendered a surface assemblage of 561 artifacts scattered across 11,900 m². Archaeological materials are visible along the road cuts that traverse the village of Njuinye. The weathering of these cuts yielded prehistoric stone industries from the two uppermost layers of Njuinye. The surface assemblage from Njuinye comprises 8 raw materials (quartz: 96%), including exotic materials such as obsidian, chert, flint, quartzite, and quartz crystal. This surface industry is also dominated by shatters (97%). Tools (3%) have a microlithic composition almost identical to that of Mudip, with perforators (n = 9), core-scrapers (n = 6), and side-scrapers (n = 3). Cores (n = 7), as is the case in Mudip, show a majority of simple debitage and peripheral flake extraction. Radial centripetal percussion is very rare (n = 1).

Archaeological excavation of Njuinye revealed the existence of deep stratification in four levels (fig. 2.2). We retrieved 258 artifactual remains from a 2 m² excavation pit (see a selected sample in fig. 2.4; the whole assemblage appears in table 2.1). The total depth of archaeological deposits at the site of Njuinye is unknown, but artifactual evidence was still present in the deepest excavation unit at a depth of 4.37 m. The two lower levels, comprising layer 1 (2.40–4.37 m) and layer 2 (1.90–2.40 m), are clayey deposits with abundant angular clasts, rare gravel, and sand. Sediments are highly compacted. Layer 3 (0.80–1.90 m) is made up of clay, silt, and fine sand. Layer 4 (0.00–0.80 m) has a porous nature with abundant organic matter. Both lowermost layers are of unknown chronology and unclear

Fig. 2.4. Selected middle and later Stone Age stone tools from Njuinye, Cameroon: *a*, spit 22; *b*, spit 24; *c*, spit 19; *d*, spit 8; *e*, spit 16; *f*, spit 38; *g*, spit 40; *h*, spit 3; *i–o*, surface; *p*, spit 39 (see table 2.1 for stratigraphic affiliation and depths at Njuinye); and from Mudip, Cameroon: *q–v*, surface. Drawings by José Martínez and Nuria Castañedo.

Table 2.1. Lithic Assemblage at the Site of Njuinye

Level/technology	Chronological period	Spit #	Depth (cm)	Debris	Flakes	Cores	By-prod.	Scrapers	Drills	Choppers	Subtotal
LSA	LATE HOLOCENE	1	16	28	3	–	1	–	–	–	32
		2	26	8	6	–	3	–	–	–	17
		3	36	6	–	–	3	3	5	–	17
		4	40	8	2	–	4	–	–	–	14
		5	47	5	–	–	4	–	1	–	10
		6	55	–	–	–	3	–	–	–	3
		7	59	–	1	–	2	–	1	–	4
		8	65	–	1	–	–	–	–	–	1
		9	71	1	–	–	1	–	–	–	2
		10	78	3	–	–	2	–	–	–	5
		11	84	4	1	–	1	–	–	–	6
Total				**63**	**14**	**–**	**24**	**3**	**7**	**–**	**111**
LSA	HOLOCENE	12	88	3	–	–	–	–	–	–	3
		13	94	1	1	–	–	–	1	–	3
		14	99	3	1	–	–	–	–	–	4
		15	105	1	1	–	–	–	–	–	2
		16	111	2	1	–	–	–	–	–	3
		17	117	2	3	1	1	–	–	–	7
		18	122	7	4	2	1	1	–	–	15
		19	127	6	2	–	4	–	–	–	12
		20	132	4	–	1	3	–	–	–	8
Total				**29**	**13**	**4**	**9**	**1**	**1**	**–**	**57**
LSA	PLEIST./HOL.	21	138	1	3	–	2	–	–	–	6
		22	144	3	5	1	5	1	–	–	15
		23	151	5	–	1	1	1	–	–	8
Total				**9**	**8**	**2**	**8**	**2**	**–**	**–**	**29**
LSA	LGM	24	152	6	4	3	3	–	–	–	16
		25	162	5	2	–	–	–	–	–	7
		26	168	3	1	1	1	–	–	–	6
		27	174	1	–	1	1	–	–	–	3
		28	182	6	–	1	1	–	–	–	8
		29	188	1	–	–	1	–	–	–	2
Total				**22**	**7**	**6**	**7**	**–**	**–**	**–**	**42**

Table 2.1. *Continued*

				Lithics							
Level/technology	Chronological period	Spit #	Depth (cm)	Debris	Flakes	Cores	By-prod.	Scrapers	Drills	Choppers	Subtotal
EARLY LSA/MSA?	PRE-LGM/NJILIAN	30	196	3	–	–	1	1	–	–	5
		31	203	4	–	–	–	–	–	–	4
		32	207	1	–	–	–	–	–	–	1
Total				8	–	–	1	1	–	–	10
MSA/ESA?	PRE-NJILIAN	38	300	–	–	–	1	–	–	–	1
		39	318	–	–	1	–	–	–	1	2
		40	331	–	–	–	1	–	–	–	1
		41	350	–	1	–	–	1	–	–	2
		45	437	–	–	–	–	1	–	–	1
Total				–	1	1	2	2	–	1	7

cultural affiliation, possibly Middle or Early Stone Age. All debitage derived from quartz reduction. Artifactual remains include debris, preparation by-products, cores, flakes, side-scrapers, and what could be interpreted as a large size pick or "chopper." No microlithic technologies were found in these basal layers.

Layer 3 has a bottom chronology of 34,700 ± 560 B.P., through the last glacial maximum (17,800 ± 180 B.P.), onto the Pleistocene to Holocene transition, with early Holocene (8,930 ± 40 B.P.; 8,670 ± 40 B.P.) atop of the layer. Industrial remains increased, compared to those represented in layers 1 and 2. Industries from layer 3 include quartz debitage and tool production of microlithic affiliation very similar to surface occurrences from this site and from neighboring Mudip. For the first time in the sequence, exotic raw materials such as obsidian, flint, chert, and quartzite appear (table 2.2), although exotic occurrences become common only in the upper subunit of layer 3, that is, in layers deposited from the onset of the Holocene on. Layer 4 comprises lithics and ceramic remains of the late Holocene (Mercader and Martí, 1999b).

Table 2.2. Vertical Distribution of Exotic Raw Materials at the Site of Njuinye

Spit #	# exotic specimens	Level #	Total per level	% exotic raw materials	Total # exotic Holocene	Total # exotic Pleistocene
1	6	4				
2	1	4				
3	1	4				
5	2	4				
6	2	4				
8	1	4				
9	1	4				
10	2	4				
11	6	4	22	19.8 (n = 111)		
12	3	3				
13	2	3				
14	2	3				
16	1	3				
17	2	3				
18	2	3				
19	3	3	*Pleistocene to Holocene boundary*		37	
24	1	3				
25	2	3				
28	1	3				
29	1	3	20	15.5 (n = 129)		
38	1	1	1	?		6

Discussion

Classification and Chronology of the Mosumu Industry

In spite of the large number of Sangoan and Lupemban sites reported in savannas and forest-savanna mosaics of equatorial Africa (fig. 2.5), very little is known regarding the chronostratigraphic position and cultural identity of these two complexes (McBrearty, 1987, 1988) inside the tropical lowland forest. Traditionally, the Sangoan has been defined as a large-size industry with a heavy-duty tool component, even though light-duty equipment in classical Sangoan sites outnumbers heavy-duty equipment (McBrearty, 1988: 389). McBrearty and Brooks (2000) indicate that this complex may predate 250,000 B.P. The appearance of Middle Stone Age techno-complexes characterized by large bifacial lanceolates like those of the Lupemban, among others, is considered part of the behavioral package that comes with anatomically modern humans and may date from 100,000 B.P. to over 300,000 B.P. (Barham and Smart, 1996; Barham, 2000; McBrearty and Brooks, 2000; Clark and Brown, 2001; Henshilwood et al., 2001). A late persistence into

Fig. 2.5. Map of Lupemban sites in tropical Africa. Lupemban sites in lowland rain forest environments: *1*, Ogooue River (Pommeret, 1965, 1966), Gabon, *2*, Mosumu (this study), Equatorial Guinea. Lupemban sites in open terrain or forest-woodland mosaics: *3*, Nzako, Lobaye River (Bayle des Hermens, 1975), Central African Republic; *4*, Lodjo (Van Noten, 1982), Democratic Republic of Congo; *5*, Muguruk (McBrearty, 1988), Kenya; *6*, Masango (Cahen and Van Noten, 1982), Burundi; *7*, Kalambo Falls (Clark, 1969, 1974), Zambia; *8*, Kamoa (Cahen and Moeyersons, 1969–71), Democratic Republic of Congo; *9*, Lunda (Clark, 1963), Angola; *10*, Lupemba (Alimen, 1957), Democratic Republic of Congo; *11*, Bandundu area (Van Moorsel, 1970), Democratic Republic of Congo; *12*, Kinshasa Plain (Van Moorsel, 1968; Cabot, cited in de Maret, 1990), Democratic Republic of Congo; and *13*, Pointe de la Gombe (Cahen, 1976, 1978), Democratic Republic of Congo.

the final Pleistocene could, however, be possible, which would imply some degree of overlap between the Lupemban and the earliest LSA techno-complexes (Van Noten, 1977; Lavachery et al., 1996; Mercader and Brooks, 2001). Conventionally, Lupemban tools were perceived as wood-working industries adapted to life in woodlands and tropical forests, even though most Lupemban sites are found in open environments (McBrearty, 1987), with two exceptions: Gabon (Pommeret, 1965, 1966) and Equatorial Guinea (this study).

There are very few excavated samples with which to compare our data. Probably, the best baseline data comes from the excavations by McBrearty in western Kenya (McBrearty, 1988). The site of Muguruk yielded two MSA stratified industries: the Ojolla industry and the overlaying Pundo-Makwar industry. They both belong to the MSA, but only the underlaying Ojolla industry yields large lanceolates. In spite of this major difference between upper and lower levels, McBrearty shows that radial flake technology remains virtually unchanged over the entire sequence (McBrearty, 1988: 414), while single-platform cores for direct simple percussion are rare. In contrast, the stone industry from Equatorial Guinea is characterized by an overwhelming majority of single-platform cores for simple percussion, a minor representation of bifacial reduction in the total assemblage, and a very slim component of radial centripetal percussion. There can be little doubt that Mosumu's inhabitants had an MSA bifacial approach to stone manufacture of lanceolate points and "bifacials." However, lanceolate points from Equatorial Guinea differ from those reported in the classical Lupemban sites mentioned above in one important respect. Namely, typical Lupemban lanceolates measure approximately 25–40 cm in maximum dimension, while Mosumu's measure approximately 7–11 cm. Thus, in size, Mosumu's points are similar to those reported in old excavations by Pommeret (1965, 1966) in the lowland forest of Gabon (7–18 cm in maximum length), not to the large daggers from more open terrain. Mosumu, on the other hand, does not have Tshitolian microliths, bladelets, geometrics, 4–7 cm foliated, tanged spear points with serrated edges, or tranchets (Bayle des Hermes, 1981), all of them typical LSA markers of this part of Central Africa. Yet, the industry from Mosumu shares with most LSA industries of the Congo Basin (Mercader, 1997; Mercader and Brooks, 2001) and periphery (Lavachery et al., 1996; Mercader and Brooks, 2001; Bayle des Hermens, 1981) the very large percentage of single-platform cores, small-size tools, and, in some cases, the late survival of ancestral MSA radial centripetal cores with cortex on the base (Bayle des Hermens, 1981; Emphoux, 1970).

The Lupemban industries from Zambia have been dated to the middle Pleistocene (Barham, 2000) through the late Pleistocene, at 95,000 B.P. and later (Clark and Brown, 2001). In western Kenya, with radial technologies,

McBrearty (1991) proposed a minimum age of 65,000 B.P. for the Lupemban. The "Upper Lupemban" of Lunda and Kalambo Falls (Clark, 1963; Clark and Bakker, 1964; Clark, 1982) comprises no heavy-duty equipment, but has ellipsoidal "core-axes," a great morphological lanceolate diversity, and rare radial centripetal technology. These assemblages are very similar to the Mosumu industry and were bracketed between 38,000 ± 2500 B.P. and 14,503 ± 560 B.P., although these dates could be minimum-age estimates instead of actual dates (McBrearty, 1988) because of potential contamination with recent carbon (Gowlett and Hedges, 1986). The earliest Tshitolian industries of the lower Congo's watershed, for their part, are dated to the close of the Pleistocene and onset of the Holocene, between 15,000 B.P. and 10,000 B.P. (Cahen, 1976; Pincon, 1991; Lanfranchi, 1987), but no Tshitolian industries have been found in Mosumu.

The slim representation of MSA radial cores with cortex on the base (Wendorf and Schild, 1974), the close parallel between the neighboring "Upper Lupemban" industries from Angola and those of Mosumu, and various geoarchaeological, taphonomic, and radiometric considerations (Mercader et al., 2002) lead us to propose a minimum *ante quem* age of greater than 30,000 B.P. for the MSA industry from Mosumu. This industry shares typical Lupemban features, such as the bifacial approach to tool production, with radial core industries, small flakes, a high representation of cores without platform preparation (Barut, 1994), and an incipient diminution in tool size.

The Archaeological Sequence at the Site of Njuinye: Classification and Chronology

The archaeological site of Njuinye in Cameroon yielded a limited number of industrial remains. Nonetheless, Njuinye has provided the longest archaeological sequence of human occupation of a region presently covered by dense lowland forest in Central Africa. Layers 1 and 2 yielded Stone Age industries in which most of the technological attributes that typify the LSA are not represented. Basal layers at the site of Njuinye provided quartz stone industries with large flakes, side-scrapers, and heavy-duty equipment, such as one large pick or chopping tool. Very tentatively, these features could indicate a Middle Stone Age or perhaps Early Stone Age component in these industries. Layers 1 and 2 may comprise the MSA to early LSA transition. Unfortunately, we possess very little evidence to support this claim. Only one cultural feature speaks of potentially relevant changes that could mark this transition; namely, the import of exotic raw materials. Exotic stone is not represented in layers 1 and 2, but appears for the first time at the bottom of layer 3 around 35,000 B.P. and then gradually increases during the Pleistocene to Holocene transition and early Holocene. This feature seems to indicate wider social networks and landscape exploitation by LSA peoples, compared with previous groups. The archaeological evidence from layer 3 is of great interest. Chronologically, it covers a great part of the late Pleistocene, including the middling mild phase of the "Njilian" (approximately 40,000 B.P. to 30,000 B.P.), the periodically cooler Leopoldvillian (approximately 30,000 B.P. to 10,000 B.P.), and the very wet and humid early Holocene. The materials recovered from excavation of layer 3, as well as those from the exposed road cuts, indicate a Later Stone Age affiliation of these industries as suggested by parallels with other LSA industries from equatorial Africa. These LSA industries are characterized by a microlithic nature and quartz predilection (Lavachery et al., 1996; Mercader, 1997), even in contexts in which better knapping-quality materials are available (cf. Lavachery et al., 1996; Moeyersons, 1997). Njuinye shares with other LSA sites the single-platform core technology, peripheral flake extraction, bipolar percussion, and rare radial centripetal percussion. It also shares a low representation of formal tools (Cahen, 1976; Van Noten, 1977; de Maret, 1982; Mercader and Brooks, 2001; Bayle des Hermens and Lanfranchi, 1978; Locko, 1988; Pincon, 1991; Mercader, 1997). Therefore, we propose a late Pleistocene through Holocene affiliation for this LSA level.

In short, the total depth of this site is unknown, but radiometric ages point to initial occupation episodes considerably older than 34,700 ± 560 B.P. (This age was obtained on a charcoal sample retrieved 2.2 m above the deepest layer bearing artifactual evidence.) We suggest a tentative Early Stone Age and/or Middle Stone Age classification for layers 1 and 2, and LSA affinity for layer 3. Given the reduced extent of our excavations, we cannot determine whether the small number of industrial remains excavated corresponds to a very slim and highly discontinuous settlement or if these reduced numbers are simply a sampling artifact. Yet, this region supported a continued human occupation by both pre-LSA and LSA humans throughout the late Pleistocene and the entire Holocene (see table 2.3).

Palaeoenvironment

Rainfall and seasonality determine the variability in botanical formations within the African forest vegetation. The amount of rainfall varies greatly from one region to another. Although most rain forests are supported by climatic regimes with annual precipitations from 1500 mm to over 2500 mm and dry seasons of less than 60 days, there are variations to this pattern. Accordingly, the African forest encompasses at least two major types of forest (White, 1983). First, there is the seasonal forest, which grows in places with annual rainfall below 1500 mm and a prolonged dry season of at least three to four months, clearly above the 60-dry-day limit. This forest type is common in parts of the central Congo Basin and especially abundant throughout the southern Congo Basin. The seasonal forest is characterized by semi-deciduous or fully deciduous taxa and open canopies. Second, there is the non-seasonal or humid rain forest, which grows in regions with a short dry season, less than two months. This forest type abounds in the eastern and northern Congo Basin and Atlantic Central Africa. Although detailed botanical inventories are not available in all regions, the data from the northeast Congo Basin (unpublished data from Cefrecof; Hart and Hart, personal communication, 1995) suggest that the humid rain forests yield 86%–94% of evergreen trees and that evergreen species may form either mono-dominant or mixed forest stocks (Hart, 1985; Hart et al., 1989).

Increasing paleoenvironmental and archaeological evidence indicates that current distribution, structure, and composition of the African forest have changed dramatically during the Quaternary as a result of climatic change and perhaps human interference. There is clear evidence through-out the wet tropics to propose a 4–5°C decrease in temperature (Bonne-fille et al., 1990; Jolly et al., 1998) and a 20–30% reduction in precipitation (Haberle and Maslin, 1999) during glacial times.

For decades, the "refugia model" has supported the idea that during the Pleistocene droughts fragmented the tropical forest into a few isolated refugia or havens in which tropical forests persisted during the Ice Age (as in Haffer, 1969; or Hamilton, 1972). Thus, areas now covered by tropical forest were colonized by variably deforested savannas during most of the Pleistocene.

However, a growing body of paleoenvironmental evidence worldwide in-dicates that larger tropical areas than just refugia pockets remained forested during cooler and probably dryer glacial phases (Colinvaux et al., 1996; Piperno, 1997; Haberle and Maslim, 1999: fig. 7; Jahns et al., 1998; Newsome and Flenley, 1988; Van der Kaars and Dam, 1995; Hope and Tulip, 1994; Van der Hammen and Absy, 1994; Hooghiemstra and Van der Hammen, 1998: 153 for bibliographic summary; Jolly et al., 1998). Recent research suggests instead that paleoecological history was markedly different depending on the ecological and climatic conditions at different geographic locations; some places observed partial deforestation, others did not (Hooghiemstra and Van der Hammen, 1998: 154).

Forests under different ecological and climatic conditions underwent dif-ferential changes in structure and botanical composition during Pleistocene perturbations. In general, the areas currently covered by evergreen forest and wetter climates yielded a combination of lowland taxa, montane ele-ments that extended to lower elevations as a result of cooling, and deciduous species that adapted to higher seasonality. This environment yielded eco-logical associations not seen today (Colinvaux et al., 1996; Maley and Brenac, 1998). On the other hand, from areas currently occupied by seasonal and

Table 2.3. Central African Sites of the Middle and Later Stone Age

Site	Country	Location	Radiocarbon dates, B.P.	Lab. #	Environment	source
Middle Stone Age						
P. de la Gombe	D. R. Congo	Open	>43,800 to 14,430 ± 20	GrN 7277 / GrN 8054	Savanna	Cahen, 1976
Mosumu	E. Guinea	Open	30,300 ± 1300	AA 36793	Forest	This study
Kinshasa Plain	D. R. Congo	Open	>30,000 to 15,080 ± 480	Lv 44/47 / Lv 166	Savanna	Van Moorsel, 1968
Njuinye	Cameroon	Open	34,700 ± 560	OS-21497	Forest	This study
Mosumu	E. Guinea	Open	>19,020 ± 140	Beta-127080	Forest	This study
Later Stone Age						
Matupi Cave	D. R. Congo	Cave	>40,700 to 12,050 ± 250	GrN 7246 / GrN 7243	Mosaic	Van Noten, 1977, 1978
Shum Laka	Cameroon	Shelter	31,700 ± 750 to 12,800 ± 110	OxA 4945 / OxA 5200	Mosaic	Lavachery et al.,1996
Makubasi SE	D. R. Congo	Shelter	18,800 ± 100	Os-21250	Forest	Mercader et al., 2000
Njuinye	Cameroon	Open	17,800 ± 180	Os-22245	Forest	This study
Loukoko II	P. R. Congo	Open	13,600 + 400	Bondy 540	Savanna	Pinçon, 1991
Matangai T. NW	D. R. Congo	Shelter	10,530 ± 50 to 715 ± 45	UtC Nr 5075 / Os 21251	Forest	Mercader et al., 2000
Lopé 2	Gabon	Open	10,320 ± 110 / 6760 ± 120	Gif A95561 / Gif 9864	Savanna	Oslisly in press
Isak Baite SW	D. R. Congo	Shelter	10,015 ± 55	AA 33225	Forest	Mercader et al., 2000
Shum Laka	Cameroon	Shelter	9880 ± 100 to 6070 ± 340	OxA 5635 / Hv 8963	Mosaic	Van Moorsel, 1968
Kinshasa Plain	D. R. Congo	Open	9730 ± 200 to 3870 ± 90	Lv 164 / Lv 288	Savanna	Van Moorsel, 1968
Mbi crater	Cameroon	Shelter	9050 ± 100	OxA 1139	Mosaic	Assombang, 1992
Njuinye	Cameroon	Open	8930 ± 40 / 8670 ± 40	b-8494 / b-128493	Forest	This study
P. de la Gombe	D. R. Congo	Open	8095 ± 50 to 3525 ± 35	GrN 7220 / GrN 7279	Savanna	Cahen, 1976

Site	Country	Type	Date	Lab number	Environment	Reference
Case Barnier II	P. R. Congo	Open	7690 ± 70	Gif 8157	Savanna	Pinçon, 1991
Sablières	Gabon	Open	7500 ± 90 to 4400 ± 70	Gif 6175 / Beta 14829	Forest	Peyrot et al.,1990
Ntadi Yomba	P. R. Congo	Shelter	7090 ± 140	Gif 4392	Savanna	Bayle des H. & Lanfranchi, 1978
Iguela	Gabon	Open	6300 ± 60 / 3680 ± 60	b-74285 / b-74286	Forest	Clist, 1995b
Nzogobeyok	Gabon	Open	6190 ± 80	b-25547	Coastal	Clist, 1989
Wataka G. E.	D. R. Congo	Shelter	6025 ± 70	AA 33226	Forest	Mercader et al., 2000
Kamoa	D. R. Congo	Open	6025 ± 70	GrN 6287	Savanna	Cahen, 1982
Obobogo	Cameroon	Open	6020 ± 505	Hv 10581	Forest	Maret, 1982
Okala 1	Gabon	Open	5580 ± 60	Gif 8614	Coastal	Clist, 1995a
Pahouin cave	Gabon	Cave	5570 ± 70 / 4000 ± 70	No Date	No data	Clist, 1995a
Abeke	Cameroon	Shelter	5565 ± 120	Hv 10586	Mosaic	Maret et al.,1987
Ndendé	Gabon	Open	5530 ± 230 / 5420 ± 120	b-29772 / b-22081	Forest	Locko, 1991; 1989
Ikengué 1	Gabon	Open	5160 ± 100 to 3250 ± 70	b-18734 / b-17518	Forest	Locko, 1988
Owendo	Gabon	Open	5040 ± 130	Gif 4157	Coastal	Cahen, 1978
Bitorri cave	P. R. Congo	Cave	4060 ± 200 / 3939 ± 200	Gif 459 / Gif 460	Savanna	Emphoux, 1970
Mandilou 2	Gabon	Open	3890 ± 140	b- 20068	Forest	Digombe et al.,1987
Ntsimou	P. R. Congo	Open	3700 ± 90	Gif 6504	Savanna	Pinçon, 1991
Lengbe	D. R. Congo	Shelter	2970 ± 70 / 840 ± 70	b-127079 / b-127078	Forest	Mercader et al., 2000
Galintsoa	Gabon	Open	2680 ± 50	Gif 8150	Savanna	Pinçon, 1991
Makubasi NW o.	D. R. Congo	Shelter	1080 ± 41 to 971 ± 33	UtC Nr 5076 / UtC Nr 5077	Forest	Mercader et al., 2000

dry tropical forests the dryer climates produced mixtures of savanna and thorny scrub (Van der Hammen and Absy, 1994), even a replacement of seasonal forests by open vegetation.

African climates and ecosystems similar to the present-day were established rather late during the Holocene, probably only within the last two to three millennia (e.g., Jolly et al., 1998; Maley and Brenac, 1998). Before 3000 B.P. until the onset of the Holocene, circa 10,000 B.P., the climate over equatorial Africa was warmer and wetter than at present (Elenga et al., 1994; Jolly et al., 1998; Maley and Brenac, 1998.). Thus, dense tropical forests probably extended beyond their current limits to areas nowadays covered by savannas.

From the last glacial maximum until 10,000 B.P., cooling and, perhaps, drying was common all over tropical Africa (Talbot and Johannessen, 1992). Notice, however, that wet and cool environments have been documented at sites with pollen records between 30,000 B.P. and 20,000 B.P. (Elenga et al., 1994; Maley and Brenac, 1998). All of them suggest that during the later part of a climatic phase known as the "Leopoldvillian" West and Central Africa, even the Western Rift (Beuning et al., 1997), may have undergone dry excursions (West Africa: Talbot and Johannessen, 1992) but also very humid oscillations (e.g., southwestern Cameroon: Maley and Brenac, 1998; People's Republic of Congo: Elenga et al., 1994).

Further back in time, the ten-thousand-year period between 40,000 B.P. and 30,000 B.P. could correspond to a brief warm and humid interstadial called "Njilian," in which climatic conditions were somewhat similar to those of today (Moeyersons, 1997) and in which forested conditions probably reigned. Pollen studies from Kouilou (Caratini and Giresse, 1979) suggest that mangroves and forest taxa abounded in coastal Gabon between 42,000 B.P. and 32,700 B.P., while Poaceae grasses are largely absent. West African savanna locations such as those of the Niger delta (Sowumni, 1984) present a parallel scenario in which mangroves and forest taxa expanded while grasses become scarce in coastal Nigeria before 35,000 B.P. Wood anatomical identification of charcoal samples older than 30,000 B.P. (Dechamps et al., 1988) suggests that the Bateke region in the People's Republic of Congo, currently covered by savannas, was forested by tree species from the Caesalpiniaceae, a family which currently requires approximately 2000 mm of rainfall annually.

Beyond 40,000 B.P., the lengthiest proxy records are the pollen data from marine cores retrieved off Atlantic Central Africa (Bengo and Maley, 1991; Dupont et al., 2000) and West Africa (Jahns et al., 1998; Dupont et al., 2000), covering the last 130,000 years and 400,000 years, respectively. These records show a continuous presence of rain forest during glacial and interglacial periods. Nevertheless, montane species such as those from the Podocarpus were pervasively present in the lowlands during cooler and drier phases (see Maley and Brenac, 1998, for similar scenarios during the late Pleistocene).

In sum, glacial periods of the Pleistocene may have seen substantial changes in the extent and nature of the African tropical forest. However, direct conclusive evidence for a simultaneous elimination of almost the entire African tropical forest is simply nonexistent. In spite of the large geographical extent of the area under study, as large as the United States east of the Mississippi River, paleobotanical evidence that could indicate a major forest retreat comes from three sites alone: Lake Bosumtwi (Ghana: Maley, 1991; Talbot and Johannessen, 1992), Bokuma-Isoku/Imbonga, in the central Congo Basin (Preuss, 1990), and Ngamakala (Elenga et al., 1994). Lake Bosumtwi is located in a seasonal forest near the savanna border (Maley, 1991). Bokuma-Isoku/Imbonga is located in the seasonal zone with semi-deciduous forest from the central Congo Basin, where Marantaceae and flood-adapted species are very common. Ngamakala pond falls within a seasonal forest that is near the savanna border, with mean annual rainfall as low as 1300 mm and a dry season of three to four months (Elenga et al., 1994: 346). Paleoecological interpretations of major forest loss at these three sites are hardly surprising, given their location well outside the evergreen forest, and indeed match current predictions of ecological perturbation during glacial times of the Pleistocene (Van der Hammen and Absy, 1994). These predictions include replacement of seasonal forests by mixtures of forest, grasslands, and xerophytic vegetation under 4–5°C cooler conditions and about 25% decrease in rainfall. Thus, the replacement pattern observed in the dryer seasonal forests is by no means representative of the transformation patterns that took place within the wetter evergreen forest. Indeed, the available data from two areas, within the evergreen forests of southwestern Cameroon (Maley and Brenac, 1998) and Ituri, Democratic Republic of Congo (Mercader et al., 2000), confirm the persistence of forest during the last glacial maximum.

Conclusion

Although grassland encroachment occurred in some seasonal forests during glacial periods (Maley, 1991; Talbot and Johannessen, 1992; Preuss, 1990; Elenga et al., 1994), only more data from sites located throughout the diverse climatic and vegetational environments of the African forest will solve the question of what type of tropical forests there were (or were not) in central and West Africa during the cooler glacial periods of the Pleistocene. It is important to realize that paleoecological data within the evergreen lowlands are limited to two sites separated by 2000 km of unbroken forest. These sites are Barombi-Mbo in the northwest corner of the Central African rain forest (Maley and Brenac, 1998), and the Ituri region in the northeast corner of the Congo Basin (Mercader et al., 2000).

At the site of Mosumu, Equatorial Guinea, and at Njuinye, Cameroon, the Middle Stone Age occupation started well before 35,000 B.P. to 30,000 B.P., to continue throughout the late Pleistocene and Holocene. Thus, human occupation of both regions evolved through variable climatic conditions that, at least at certain times, included wet conditions conducive to rain forest growth. Pollen data from Lake Barombi-Mbo, southwestern Cameroon (Maley and Brenac, 1998), lends support to the belief that Middle Stone Age and Later Stone Age groups from Equatorial Guinea and Cameroon subsisted in a tropical forest context. This lake is located only 50 km southeast of Njuinye and 350 km northwest of Mosumu. In chronometric terms, the pollen record relevant to our study is that from pollen zone 1, circa 28,000 B.P. to 20,000 B.P. This record is characterized by the presence of tropical forest families such as the Caesalpiniaceae and montane elements that indicate the existence of a wet but cooler forested environment between approximately 30,000 B.P. to 20,000 B.P. Additional pollen and wood records, dated from 42,000 B.P. to over 30,000 B.P., from neighboring countries suggest that during the Njilian Central African climates were warm and humid and, thus, supported tropical forests even in areas presently occupied by savannas (Caratini and Giresse, 1979; Dechamps et al., 1988).

Not only does the idea of tropical forest settlement by Middle Stone Age and Later Stone Age groups deserve further testing, but archaeological and sequential data from Mosumu and Njuinye indicate that both sites could in part represent the Middle to Later Stone Age transition, to include the African wet tropics in the cradle of the earliest Later Stone Age (Ambrose, 1998: 389; Mercader and Brooks, 2001; Lavachery et al., 1996; Oslisly, et al., 2001) of the African continent.

References

Alimen, H. (1957): *The prehistory of Africa*. London: Hutchinson.

Ambrose, S. (1998): Chronology of the Later Stone Age and food production in East Africa. *Journal of Archaeological Science* 25:377–392.

Barham, L. S., and P. Smart (1996): Early date for the Middle Stone Age of central Zambia. *Journal of Human Evolution* 30:287–290.

Barut, S. (1994): Middle and Later Stone Age lithic technology and land use in East African savannas. *African Archaeological Review* 12:43–72.

Bayle des Hermens, R. (1975): *Recherches préhistoriques en Republique Centrafricaine*. Paris: Labethno.

————. (1981): Note typologique sur le Tshitolien du Basin du Congo. In *Prehistoire africaine: Melangues offerts au doyen Balout*, 341–348. Paris: ADPF.

Bayle des Hermens, R., and R. Lanfranchi (1978): L'abri Tshitolien de Ntadi Yomba (Republique Populaire du Congo). *L'Anthropologie* 82(4):539–564.

Bayle des Hermens, R., R. Oslisly, and B. Peyrot (1987): Prèmières séries de Pierres Taillées du Paléolithique Inférieur découvertes au Gabon, Afrique Central. *L'Anthropologie* 91(2):693–698.

Bengo, M., and J. Maley (1991): Analyses des flux pollinique sur le marge du golfe de Guinee depuis 135,000 ans. *Comptes rendus de l'Academie des Sciences du Paris*, ser. 2, 113:843–849.

Beuning, K., M. Talbot, and K. Kelts (1997): A revised 30,000-year paleoclimatic and paleohydrologic history of Lake Albert, East Africa. *Palaeogeography, Palaeoclimatology, Palaeoecology* 136:259–279.

Bonnefille, R., J. Roeland, and J. Guiot (1990): Temperature and rainfall estimates for the past 40,000 years in equatorial Africa. *Nature*, 347–349.

Bultot, F., and J. Griffiths (1972): The equatorial wet zone. In *Climates of Africa*, edited by J. Griffiths, vol. 10, *World Survey of Climatology*. Amsterdam: Elsevier.

Cahen, D. (1976): Nouvelles fouilles a la Pointe de la Gombe (ex-pointe de Galina), Kinshasa, Zaire. *L'Anthropologie* 80:573–602.

Cahen, D. (1982): The Stone Age in the south and west. In *The archaeology of Central Africa*, edited by F. Van Noten, 41–56. Verlagsanstalt, Gras, Austria: Akademische Druck-u.

Cahen, D., and J. Moeyersons (1969–71): Le site prehistorique de la Kamoa, Katanga, Zaire. *Palaeoecology of Africa* 6:237–238.

Caratini, C., and P. Giresse (1979): Contribution palynologique a la connaisance des environnements continentaux et marins du Congo a la fin du Quaternaire. *C. R. Academy of Science* (Paris) 288:379–382.

Chernokian, R. (1983): Ivory Coast prehistory: Recent developments. *African Archaeological Review* 1:127–142.

Chernokian, R., and G. Paradis (1982): Une industrie Paleolithique decouverte dans la "Terre de Barre" d'une Terrasse proche d'Anyama (Region d'Abidjan). *Nyame Akuma* 21:18–26.

Clark, D. (1954): Upper Sangoan industries from northern Nyasaland and the Luangwa Valley: A case of environmental differentiation? *South African Journal of Science* (March):201–208.

————. (1963): *Prehistoric cultures of northeast Angola and their significance in tropical Africa*. Museo do Dundo Publicacoes culturais, 62. Lisbon: Companhia de Diamantes de Angola.

————. (1964): *The Sangoan culture of Equatoria: The implication of its stone equipment*. Miscellanea en Homenaje al Abate Henri Breuil. Diputación Provincial de Barcelona, Instituto de Prehistoria y Arqueología, Monografías, 9:309–325.

————. (1969): *The Kalambo Falls prehistoric site*. Vol. 1, *The geology, palaeoecology, and detailed stratigraphy of the excavations*. Cambridge: Cambridge University Press.

————. (1974): *The Kalambo Falls prehistoric site*. Vol. 2, *The later Pleistocene cultures*. Cambridge: Cambridge University Press.

————. (1982): The cultures of the middle Palaeolithic and Middle Stone Age. In *The Cambridge history of Africa;* vol. 1, *From the earliest times to 500 B.C.*, edited by D. Clark, 248–341. Cambridge: Cambridge University Press.

Clark, D., and C. Haynes (1970): An elephant butchery site at Mwanganda's village, Korongo, Malawi, and its relevance for Paleolithic archaeology. *World Archaeology* 1:390–411.

Clark, D., and K. Brown (2001): The twin rivers, Kopje, Zambia: Stratigraphy, fauna, and artefact assemblages from the 1954 and 1956 excavations. *Journal of Archaeological Science* 28:305–330.

Clark, D., and V. Z. Bakker (1964): Prehistoric culture and Pleistocene vegetation at the Kalambo Falls, northern Rhodesia. *Nature* 4923(March):971–975.

Clist, B. (1995): *Gabon: 100,000 ans d'Histoire.* Collection Découvertes du Gabon. Libreville: Centre Culturel Français Saint Exupery.

Colinvaux, P., P. Oliveira, E. Moreno, C. Miller, and M. Bush (1996): A long pollen record from lowland Amazonia: Forest and cooling in glacial times. *Science* 274:85–88.

Cooke, C. (1963): Report on excavations at Pomongwe and Tshangula Caves. *South African Archaeological Bulletin* 18:73–151.

Davies, O. (1954): The Sangoan culture in Africa. *South African Journal of Science* 50(10):273–277.

Dechamps, R., R. Lanfranchi, L.C.A. Schwartz, and D. Schwartz (1988): Reconstitution d'environnements quaternaires par l'etude de macrorestes vegetaux (pays Bateke, R.P. du Congo). *Palaeogeography, Palaeoclimatology, Palaeoecology* 66:33–44.

de Maret, P. (1982): Belgian archaeological project in Cameroon (July–August 1981 field work). *Nyame Akuma* 20:11–12.

de Maret, P. (1990): Phases and facies in the archaeology of Central Africa. In *A history of African archaeology,* edited by P. Robertshaw, 109–134. London: J. Currey.

Dumort, J. (1968): *Carte geologique de reconnaissance a l'echelle du 1:500,000.* Yaounde: Direction des Mines et de la Geologie du Cameroun.

Dupont, L., S. Jahns, F. Marret, and S. Ning (2000): Vegetation change in equatorial West Africa: Time slices for the 150ka. *Palaeogeography, Palaeoclimatology, Palaeoecology* 155:95–122.

Elenga, H., D. Schwartz, and A. Vincens (1994): Pollen evidence of Late Quaternary vegetation and inferred climate changes in Congo. *Palaeogeography, Palaeoclimatology, Palaeoecology* 109:345–356.

Emphoux, J. (1970): La grotte de Bitorri au Congo-Brazaville. Cah. *ORSTOM,* Science Humanities ser., 7(1):3–20.

Gowlett, J., and R. Hedges (1986): Lessons of context and contamination in dating the upper Palaeolithic. In *Archaeological results from accelerator dating,* edited by J. Gowlett and R. Hedges, 63–72. Oxford: Oxford University Press.

Haberle, S., and M. Maslin (1999): Late Quaternary vegetation and climate change in the Amazon Basin based on a 50,000-year pollen record from the Amazon fan, ODP site 932. *Quaternary Research* 51:27–38.

Haffer, J. (1969): Speciation in Amazonian forest birds. *Science* 165 (11 July, 3889): 131–136.

Hamilton, A. (1972): The significance of patterns of distribution shown by forest plants and animals in tropical Africa for the reconstruction of upper Pleistocene palaeoenvironments. *Palaeoecology of Africa* 9:63–97.

Hart, T. (1985): The ecology of a single-species dominant forest and mixed forest in Zaire. Ph.D. diss., Michigan State University, East Lansing.

Hart, T., and J. Hart (1986): The ecological basis of hunter-gatherer subsistence in African rain forests: The Mbuti of eastern Zaire. *Human Ecology* 14(1):29–55.

Hart, T., J. Hart, and P. Murphy (1989): Monodominant and species-rich forests of the humid tropics: Causes for their co-ocurrence. *American Naturalist* 133: 613–633.

Henshilwood, C., J. Sealy, R. Yates, K. Cruz-Uribe, P. Goldberg, F. Grine, R. Klein, C. Poggenpoel, V. Niekerk, and I. Watts (2001): Blombos Cave, Southern Cape, South Africa: Preliminary report on the 1992–1999 excavations of the Middle Stone Age levels. *Journal of Archaeological Science* 28:421–448.

Hooghiemstra, H., and T. Van der Hammen (1998) Neogene and Quaternary developments of the Neotropical rain forest: The forest refugia hypothesis, and a literature overview. *Earth-Science Reviews* 44:147–183.

Hope, G., and J. Tulip (1994): A long vegetation history from lowland Irian Jaya, Indonesia. *Palaeogeography, Palaeoclimatology, Palaeoecology* 109:385–398.

Jahns, S., M. Huls, and M. Sarnthein (1998): Vegetation and climate history of west equatorial Africa based on a marine pollen record off Liberia (site GIK 16776) covering the last 400,000 years. *Review of Palaeobotany and Palynology* 102:277–288.

Jolly, D., S. Harrison, B. Damnati, and B. Bonnefille (1998): Simulated climate and biomes of Africa during the late Quaternary: Comparison with pollen and lake status data. *Quaternary Science Reviews* 17:629–658.

Lanfranchi, R. (1987): Recherches prehistoriques en Republique Populaire du Congo, 1984–86. *Nsi* 1:6–8.

——. (1990): Les industries préhistoriques en R. P. du Congo et leur contexte paléogéographique. In *Paysages quaternaires de l'Afrique centrale Atlantique,* edited by R. Lanfranchi and D. Schwartz, 439–446. Collection Didactiques. Paris: ORSTOM.

Lavachery, P., E. Cornelissen, J. Moeyersons, and P. de Maret (1996): 30,000 ans d'occupation, 6 mois de fouilles: Shum Laka, un site exceptionnel en Afrique centrale. *Anthropologie et Prehistoire* 107:197–211.

Lejoly, J. (1998): Contribution a la typologie des forest du Rio Muni (Guinee Equatoriale). C.U.R.E.F.

Lioubine, V., and F. Guede (2000): *The Paleolithic of Republic Côte d'Ivoire (West Africa).* In Russian with an English summary. St. Petersburg: Russian Academy of Sciences.

Locko, M. (1988): La recherche archéologique à l'Université Omar Bongo: Bilan scientifique. *Muntu* 8:26–44.

Maley, J. (1991): The African rain forest vegetation and palaeoenvironments during late Quaternary. *Climatic Change* 19:79–98.

Maley, J., and P. Brenac (1998): Vegetation dynamics, palaeoenvironments, and climatic changes in the forests of western Cameroon during the last 28,000 years B.P. *Review of Palaeobotany and Palynology* 99:157–187.

Martí, R. (2001): Guinea Ecuatorial: El asentamiento humano del Pleistoceno final en el cinturón forestal centroafricano. Ph.D. diss., Universidad Nacional de Educación a Distancia, Madrid.

Martinez-Torres, L., and A. Riaza (1996): *Mapa geologico de Guinea Ecuatorial continental.* Alava: Asociacion Africanista Manuel Iradier.

McBrearty, S. (1987): Une evaluation du Sangoen: Son age, son environement et son rapport avec l'origine de l'Homo sapiens. *L'Anthropologie* 91:127–140.

———. (1988): The Sangoan-Lupemban and Middle Stone Age sequence at the Muguruk site, western Kenya. *World Archaeology* 19:379–420.

———. (1991): Recent research in western Kenya and its implications for the status of the Sangoan industry. In *Cultural beginnings: Approaches to understanding early hominid life-ways in the African savanna,* edited by D. Clark, 159–176. Bonn: Romisch-Germanisches zentralmuseum.

McBrearty, S., and S. Brooks (2000): The revolution that wasn't: A new interpretation of the origin of modern human behavior. *Journal of Human Evolution* 39(5): 453–563.

Mehlman, M. (1987): Provenience, age, and associations of archaic *Homo sapiens* crania from Lake Eyasi, Tanzania. *Journal of Archaeological Science* 14:133–162.

Mercader, J. (1997): Bajo el techo forestal: La evolución del poblamiento en el bosque ecuatorial del Ituri, Zaire. Ph.D. diss., Universidad Complutense, Madrid.

Mercader, J., and A. Brooks (2001): Across forests and savannas: A comparison of Later Stone Age assemblages from Ituri and Semliki, northeast Democratic Republic of Congo. *Journal of Anthropological Research* 57(2):197–217.

Mercader, J., and R. Martí (1999a): Middle Stone Age site in the tropical forests of Equatorial Guinea. *Nyame Akuma* 51(1):14–24.

———. (1999b): Archaeology in the tropical forest of Banyang-Mbo, SW Cameroon. *Nyame Akuma* 51(2):17–24.

Mercader, J., F. Runge, L. Vrydaghs, H. Doutrelepont, E. Corneille, and J. Juan-Tresseras (2000): Phytoliths from archaeological sites in the tropical forest of Ituri, Democratic Republic of Congo. *Quaternary Research* 54:102–112.

Mercader, J., R. Martí, J. Martínez, and A. Brooks (2002): The nature of 'stone-lines' in the African Quaternary record: Archaeological resolution at the rainforest site of Mosumu, Equatorial Guinea. *Quaternary International* 89:71–96.

Moeyersons, J. (1997): Geomorphological processes and their palaeoenvironmental significance at the Shum Laka cave (Bamenda, western Cameroon). *Palaeogeography, Palaeoclimatology, Palaeoecology* 133:103–116.

Nenquin, J. (1967): *Contribution to the study of the prehistoric cultures of Rwanda and Burundi.* Annales du Musée Royal de l'Afrique Centrale, 59. Tervuren: Musée Royal de l'Afrique Centrale.

Newsome, J., and R. Flenley (1988): Late Quaternary vegetational history of the central highlands of Sumatra; II, Palaeopalynology and vegetational history. *Journal of Biogeography* 15:555–578.

Nygaard, S., and M. Talbot (1976): Interim report on excavation at Asokrochona, Ghana. *West African Journal of Archaeology* 6:13–19.

Omi, G. (19977): Prehistoric sites and implements in Cameroon: An annex to the interim report of the Tropical African Geomorphology Research Project, 1975/76. Nagoya, Japan.

Oslisly, R., H. Doutrelepont, M. Fontugne, H. Forestier, P. Giresse, C. Hatte, and L. White (2001): Le site de Maboué 5 dans la réserve de la Lopé au Gabon: Premier résultats pluridisciplinaires d'une stratigraphie vielle de plus de 40,000 ans. Paper presented at the Congress of the Panafrican Association for Prehistory and Related Studies, Bamako, Mali.

Paradis, G. (1980): Découverte d'une industrie paléolithique d'âge Sangoen dans les sables argileux "néogènes" (ou "terre de barre") de la basse Côte-d'Ivoire. *Compte Rendus Academy of Science* (Paris), ser. D, 290:1393–1395.

Pincon, B. (1991): Archeologie des plateux et collines Teke (Republique Populaire du Congo): De nouvelle donnees. *Nsi* 8–9:29–32.

Piperno, D. (1997): Phytoliths and microscopic charcoal from leg 155: A vegetational and fire history of the Amazon basin during the last 75 k.y. *Proceedings of the ocean-drilling program, scientific results* (edited by R. Flood, D. Piper, A. Klaus, and L. Peterson) 155:411–418.

Pommeret, Y. (1965): *Civilisations prehistoriques au Gabon, Tome 1; Vallee du Moyen Ogooue: Notes preliminaires a propos du gisement neolithique et lupembien de Ndjole.* Memoires de la Societe prehistorique et protohistorique Gabonaise, 2. Libreville: Centre Culturel Francaise Saint Exupery.

———. (1966): *Civilisations prehistoriques au Gabon, Tome 2; Vallee du Moyen Ogooue: Presentation de l'industrie lithiques de traditions sangoenne, lupembienne et neolithique.* Memoires de la Societe prehistorique et protohistorique Gabonaise, 2. Libreville: Centre Culturel Francaise Saint Exupery.

Preuss, J. (1990): L'evolution des paysages du bassin interieur du Zaire pendant les quarante dernier millenaires. In *Paysages quaternaires de l'Afrique centrale Atlantique,* edited by R. Lanfranchi and D. Schwartz, 260–270. Paris: Orstom.

Soper, R. (1965): The Stone Age in northern Nigeria. *Journal of History and Sociology* (Nigeria) 3(2):175–194.

Sowumni, A. (1984): Nigerian vegetational history from the late Quaternary to the present day. *Palaeoecology of Africa* 16:217–234.

Talbot, M., and T. Johannessen (1992): A high-resolution palaeoclimatic record for the last 27,500 years in tropical West Africa from the carbon and nitrogen isotopic composition of lacustrine organic matter. *Earth and Planetary Science Letters* 110:23–37.

Van der Hammen, T., and M. Absy (1994): Amazonia during the last glacial. *Palaeogeography, Palaeoclimatology, Palaeoecology* 109:247–261.

Van der Kaars, W., and M. Dam (1995): A 135,000-year record of vegetational and climatic change from the Bandung area, West Java, Indonesia. *Palaeogeography, Palaeoclimatology, Palaeoecology* 117:55–72.

Van Moorsel, H. (1968): *Atlas de prehistoire de la plaine de Kinshasa.* Kinshasa: Publications Universitaires, Universite Lovanium.

———. (1970): Recherches prehistoriques au pays de l'entre-fleuves Lukenie-Kasai. *Etudes d'Histoire Africaine* 1:7–36.

Van Noten, F. (1977): Excavations at Matupi Cave. *Antiquity* 51:35–40.

———. (1982): *The archaeology of Central Africa.* Graz: Akademische Druck unds Verlagsanstalt.

Van Noten, F., P. Haesaerts, and D. Cahen (1972): Un habitat lupembien a Masango, Burundi, rapport preliminaire. *Africa Tervuren* 18:78–85.

Van Reeth, L., and D. Ngomo (1997): Biodiversite vegetale des ligneaux sur le transect de Mosumu dans le Parc de Monte Alen (Guinee Equatoriale). E.C.O.F.A.C.

Wendorf, F., and R. Schild (1974): *A Middle Stone Age sequence from the central Rift Valley, Ethiopia.* Warsaw: Polska Akademia Nauk, Instytut Historii Kultury Materialnej.

White, F., ed. (1983): *The vegetation of Africa: A descriptive memoir to acompany the Unesco/Aetfat/Unso vegetation map of Africa.* Paris: Unesco.

CHAPTER 3

Foragers of the Congo
The Early Settlement of the Ituri Forest

Julio Mercader

Was the prehistoric settlement of tropical forests by hunter-gatherers possible without symbiosis with farmers? Some scholars argue that present-day forest dwellers do not live independently of farming, nor are wild carbohydrates bountiful enough to have optimally supported ancient foraging groups (Hart and Hart, 1986; Headland, 1987; Bailey and Peacock, 1988; Bailey et al., 1989; Bailey and Headland, 1991; Gamble, 1993; Headland, 1997). Dense tropical forests of the Holocene are portrayed as unfriendly environments unable to support prehistoric foragers before the late Holocene advent of farming (Headland, 1987; Bailey et al., 1989; Bailey and Headland, 1991; Headland, 1997). Prehistoric hunter-gatherers thus lived in tropical forests for the last few millennia, only after farmers colonized rain forests and enhanced the low productivity of the forest environment by farming and subsequent alteration of closed-canopy forests. This farming modification of the forest, in turn, made hunting and gathering feasible.

This scenario has very important implications. The question being investigated is not whether humans had the biological capacity to occupy the tropical forest at very early dates, but whether humans knew how to successfully exploit the food resources and whether their technologies would allow them to use tropical forest environments. Archaeological and sequential evidence for pre-farming occupation of tropical forests would provide a definite answer. Evidence for pre-farming occupation from wet tropical regions of the African continent has been regarded as inconclusive due to (1) lack of precise provenance, (2) dating or stratigraphic uncertainty, and, especially, (3) lack of paleoenvironmental evidence for tropical forests. The majority of Paleolithic sites in the African wet tropics are located near current forest fringes, in forest-savanna mosaics, or in variably forested highlands, not in evergreen lowlands. Recently, however, the discovery of Later Stone Age (this chapter) and Middle Stone Age sites (Mercader and Martí, 1999a,b; Mercader and Martí, chapter 2 in this book) in the African forest have raised several important issues, including (1) whether ancient *Homo sapiens* and predecessors had the ability to settle tropical forests, (2) whether early humans invaded rain forest regions during their Pleistocene global territorial expansion, and (3) whether tropical lowlands were a backwater for human development. This chapter considers the implications for these issues of archaeological data from ten Stone Age sites in the tropical forests of Ituri, Democratic Republic of Congo.

Previous Archaeological Research

Investigations in the Ituri district started in the mid-1970s (Van Noten, 1977) in Mount Hoyo, a transitional montane domain at 1450 m above sea level, located between the tropical forest border, 40 km to the west, and the savanna, 10 km to the east. Although Matupi Cave falls within the administrative limits of the Ituri region, faunal and pollen analyses (Van Noten, 1977; Van Neer, 1989; Brook et al., 1990) indicate that this site was never located within the lowland forest, either at present or in the past, but rather was situated in a transitional habitat dominated by woodlands and savannas. This type of mosaic environment was settled by Later Stone Age foragers since at least the last glacial maximum and by early LSA groups (Kamuanga, 1985) or, perhaps, late Middle Stone Age peoples as early as 40,000 B.P. to 50,000 B.P. (McBrearty and Brooks, 2000).

The first archaeological investigations in the lowland forest per se were undertaken by J. Fisher (1988) and G. Laden (1992) under the auspices of the Ituri Project (Bailey and DeVore, 1989). Both Fisher and Laden focused on ethnoarchaeological work, but complemented it with test excavations at several open-air and rock shelter sites (Fisher and Laden, personal communication).

From 1993 to 1995, multidisciplinary archaeological work was pursued by J. Mercader (1997) under the supervision of R. Bailey and the auspices of the Ituri Project. Archaeological surveys in the lowland forest of Ituri identified over 50 archaeological sites in just 8.5 km². Over a period of 12 months, ten rock shelter sites were studied, with up to 2 m² excavated per shelter. Mercader retrieved archaeological evidence extending back over the last 20,000 years and suggested that prehistoric foragers may have inhabited tropical forest environments since at least the late Pleistocene. Early LSA and Middle Stone Age occupations are known in the bordering transitional zones to highlands and savannas (Van Noten, 1977; Van Noten, 1982: 30, fig. 15). However, occupations from the initial phases of the Later Stone Age, Middle Stone Age, or Early Stone Age, as well as the number and diversity of the hunter-gatherer populations who lived in this part of Central Africa (Mercader, et al., 2001), remain unknown.

Geology, Flora, and Cultural Context

The equatorial forest of Ituri is located between the African Western Rift and the central Congo Basin (fig. 3.1). Central African lowland forests develop on two distinct geological substrates. Extending 1,200,000 km^2, the intracratonic depression of the Congo (Goodwin, 1991) interdigitates with granitic and karstic domains around its periphery (Cahen et al., 1984). The High Congo-Zaire Granitoid Massif (Lavreau, 1982) is partly covered by the Ituri rain forest. This plutonic domain is surrounded by the Nile formation to the north, the Lindian limestone belt to the southwest, the Bomu complex to the west, and the Cenozoic Rift, 100 km to the east of the present forest-savanna border (Goodwin, 1991). The Ituri forest spreads over a granitic landscape whose altitudinal, topographical, and geomorphological characteristics have been shaped by rift neotectonics (fig. 3.1). Altitudes range from 1000 m in the east to 600 m in the west. Sites studied fall within 700 to 800 m. Two topographic zones can be distinguished: the hilly "uplands" of the north and east and the flat terrains of the center, south, and west. *Grabens* and *Inselbergs* are common in that section of the forest nearest to the rift (the northeast), and so are tors, boulders, and rock shelters (figs. 3.2, 3.3; table 3.1) that attracted human occupation up to the present. A selected sample of the type of outcrops in which rock shelters form and plans of available sheltered spaces are presented in figure 3.3.

In spite of a substantial rainfall decrease during January and February, an almost evenly distributed precipitation, up to 1900 mm annually (Bultot, 1971), supports a Guineo-Congolian tropical forest (White, 1983). A smaller heterogeneous annual rainfall sustains a transitional belt where mosaics and savannas are common, north and east of the forest zone, toward the Kilo-Moto mining region and through the inter-lacustrine rim of the Ituri district, respectively. The phytogeographical setting of the Ituri region is varied. It comprises two basic vegetation formations: monodominant and mixed (Hart, 1985; Hart and Hart, 1986; Hart et al., 1989). Monodominant forests, common in the south and west, are characteristic of the Ituri region. They represent mature evergreen stands (94.2%, Hart and Hart, unpublished data) in which one species *(Gilbertiodendron dewevrei)* accounts for more than 90% of the canopy. In this type of forest, tree species diversity is low (10 per hectare, Hart and Hart, 1986: 34–35), canopies tend to interlock, undergrowth rarely develops, and overall productivity of potential foodstuffs for present humans is low. Mixed forests are scattered through the region. They represent variably mature stands of a combination of evergreen (85.9%, Hart and Hart, unpublished data) and semi-evergreen species dominated by the Caesalpiniaceae family. Never does one species represent more than

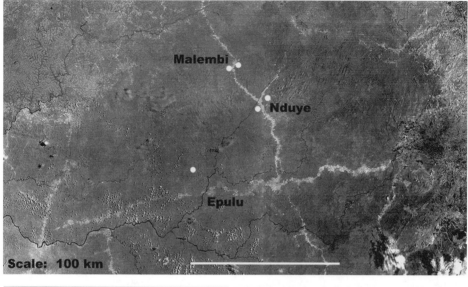

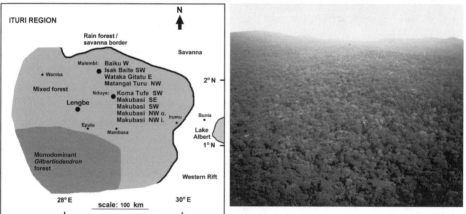

Fig. 3.1. The Ituri region. *Upper box:* aerial image showing the rain forest–savanna boundary, impact of neotectonics on local geomorphology, major watersheds, and location of research areas. *Lower right:* aerial view of the rain forest canopy near Mambasa. *Lower left:* diagram of the Ituri region indicating the forest-savanna border, forest domains, and location of archaeological sites studied in the text.

40% of the canopy. Tree species diversity is higher (53 per hectare, Hart and Hart, 1986: 34–35) than in monodominant forests, canopies are open, there is abundant undergrowth, and there is greater potential for human exploitation. The most common tree species are *Gilbertiodendron dewevrei* (Caesalpiniaceae), *Cynometra alexandrii* (Caesalpiniaceae), *Julbernardia seretii* (Caesalpiniaceae), *Klainedoxa gabonensis* (Irvingiaceae), *Cleistanthus michel-*

Fig. 3.2. Rock shelter views: *1,* Makubasi Southeast, Nduye; *2,* Koma Tufe Southwest, Nduye; *3,* Makubasi Southwest, Nduye; and *4,* Makubasi Northeast, Nduye.

sonii (Euphorbiaceae), *Erythrophleum suaveolens* (Caesalpiniaceae), *Fagara macrophylla* (Rutaceae), and *Canarium schweinfurthii* (Burseraceae). The most common grasses include the Marantaceae and the Poaceae (T. Hart and E. Corneille, personal communication).

The present populations living in the Ituri district are diverse (Vansina, 1990: 167–177), as four linguistic groups overlap: Ubanguian, Central-Sudanic, Bantu, and Nilotic. Central Sudanic– and Bantu-speaking populations have settled the forested lowlands, but Nilotic speakers are not found in the forest. A very small Ubanguian-speaking population is currently distributed through the western Ituri. Present population density in the inner forest is very low, with less than 0.25 people per 1 km^2 (Peterson, 1991: 6; Doumenge, 1990; Bailey, 1991: 13), but dense settlement with more than 100 people per 1 km^2 occurs in the northeast ecotones. Forest populations are classified as hunter-gatherers (Efe, Mbuti, Tswa, Sua) or farmers (Lese, Bira, Baali, Mbo, Mdaaka, Budu, Nande). Currently, interaction between

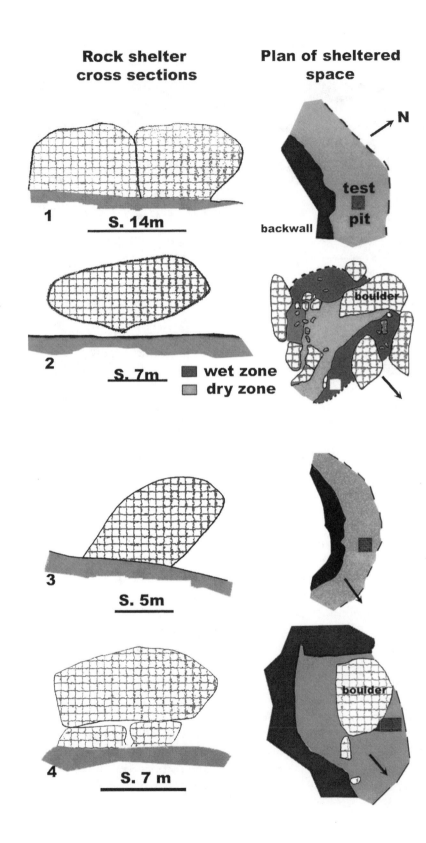

Rock shelter cross sections

Plan of sheltered space

N

test pit

backwall

1 S. 14m

boulder

2 S. 7m ■ wet zone ☐ dry zone

3 S. 5m

boulder

4 S. 7 m

Table 3.1. Features of Rock Shelter Sites in the Ituri Rain Forest

Site	Distance to water, m	Area, m²	Orienta- tion of mouth	Artifact density/m² (surface)	Pit size, m	Depth, m	Soil profile classification
WGE	50	67	NW/NE	1.04	1 by 1	1.68	Humic cambisol
BW	105	60	SE	1.15	1 by 1	1.43	Humic cambisol
IBSW	130	125	NE/SW	24.2	1 by 1	1.71	Humic acrisol
MTNW	95	47	NE	0.5	1 by 1	2.15	Humic nitisol
KTSW	250	30	SW	0.4	2 by 1	Unknown	Unknown
MSE	250	50	NW	26.4	2 by 1	1.04	Humic acrisol
MSW	250	36	NW	5.08	3 by 0.5	0.28	Distric leptosol
MNW I.	250	13	SW	4.4	1 by 0.5	0.9	Humic cambisol
MNW o.	250	15	NE	3.8	1 by 0.5	0.9	Distric cambisol
L.	290	28	NE	2	2 by 0.5	0.74	Ferric acrisol

foragers and farmers is extensive and includes labor exchange and economic, social, ritual, and marriage relationships (Hart and Hart, 1986; Bailey and Peacock, 1988; Bailey, 1991; Bailey and DeVore, 1989; Wilkie, 1987; Grinker, 1994; Rösler, 1997).

Field Methods

The lack of data from a vast region of 60,000 km² favored a regional field strategy consisting of test digging at ten sites. Vertical excavation of small trenches 1–2 m² was used to penetrate to the basal strata of rock shelter sites, establish the extent of archaeological deposits, and reveal stratigraphic sequences. This excavation approach, although of limited aerial value, proved adequate to recover specific sequential and environmental information from a region for which no background archaeological data was available prior to our work. The total excavated area amounts to 11.5 m². Test pits were divided into 16 squares per 1 m², 25 by 25 cm each, so that any positioning data lost during the excavation had a maximum relocation error of 12.5 cm. Deposits were dug by arbitrary spits thinner than 3 cm. All pieces recovered were plotted in three dimensions, together with orientation, position of repose of the artifact, and dip, to obtain site formation data regarding sedimentary environment, the potential effects of water action in the configuration of the assemblage, and possible disturbances (Mercader

Fig. 3.3. Cross sections and plan of inhabitable space in the rock shelters of *1*, Wataka Gitatu East, Malembi; *2*, Isak Baite Southwest, Malembi; *3*, Matangai Turu Northwest, Malembi; and *4*, Makubasi Southeast, Nduye.

et al., 2002). Lithics of the Ituri rock shelters were made on quartz, which generated abundant microdebitage, 2–5 mm. The generally clayey matrix made recovery of microdebitage difficult. Therefore, all deposits were wet-sieved through a 1.810 mm mesh, and the sediment from a selected 25-by-25-cm square was sieved through a 0.513 mm mesh for closer inspection. In addition, one 25-by-25-cm square per artificial layer was weighed in order to control the amount of sediment being excavated. For every 25 cm of depth, pH, water content, and color were assessed; pH was determined on-site with an electronic pH meter (Crison 507.3350) on a paste derived from a 1:1 mix of sediment and distilled water. Water content was estimated on-site by taking 100 g of sediment, letting it air dry for 48 hours, then weighing the dried sample to establish percent water as weight loss after drying. Color was measured by comparing dry and wet samples with the Munsell soil color chart. Samples for phytolith analysis were taken at depth intervals of 10 cm.

Forager Settlement Prior to Agriculture

Chronostratigraphic Sequences

Radiometric dating (table 3.2) shows that human occupation occurred before 18,800 ± 100 B.P., prior to the last glacial maximum. Approximately 8000 years later, by the beginnings of the Holocene, human groups continued to occupy the Ituri region. Human settlement continued through the mid and late Holocene to date. Radiometric ages and repeated stratigraphic sequences detected throughout the region indicate long-term settlement prior to farming:

1. All sites present a capping unit with an average thickness of 23.3 cm. Material remains comprise unstandardized quartz shatter (lithics may reach 85% of the total artifact inventory); macrobotanical remains from the African Canarium and to a lesser extent from the African oil palm; faunal remains from *Atherurus africanus*-porcupine, *Cephalophus sp*-antelope, primate, and small bovids; a single human skeleton (Mercader et al., 2001); molluscs *(Achatina* and *Limicolaria); late Iron Age ceramics; archaeometallurgical debris and iron tools (Mercader et al., 2000a); and high concentrations of organic matter. Seven AMS (Accelerator Mass Spectrometry) radiocarbon dates indicate an age for these thin capping units between 1080 ± 41 B.P. and 715 ± 45 B.P.

2. Below this late Holocene cap there is a variably deeper basal unit that may be as much as 175 cm thick (average thickness 82.5 cm). Material remains include no evidence of farming technologies, but abundant stone industries, less charcoal, fewer phytoliths, and fewer *Canarium*

Table 3.2. Radiocarbon Dates from the Ituri Region

C14	B.P.	Lab. #	Type of sample (max. length)	Site	Level
AMS	18,800 ± 100	Os-21250	1 mm charcoal	MSE	Basal
AMS	10,530 ± 50	UtCnr5075	<1 mm charcoal	MTNW	Basal
AMS	10,015 ± 55	AA33225	Endocarpus Canarium schweinfurthii	IBSW	Basal
AMS	6025 ± 70	AA 33226	1 mm charcoal	WGE	Basal
AMS	2970 ± 70	Beta 127079	Endocarpus Canarium schweinfurthii	L	Basal
AMS	1080 ± 41	UtC Nr 5076	2 mm charcoal	MNW o.	Cap
AMS	1010 ± 60	Os-22251	2mm charcoal	MNW o.	Cap
AMS	971 ± 33	UtC Nr 5077	2 mm charcoal	MNW o.	Cap
AMS	850 ± 70	Beta 127078	Endocarpus Elaeis guineensis	L	Cap
AMS	825 ± 30	Os-21235	Endocarpus Canarium schweinfurthii	MTNW	Cap
AMS	813 ± 35	UtC Nr 5074	Several 1–5 mm fragments of human bone	MTNW	Cap
AMS	715 ± 45	Os-21251	2 mm charcoal	MTNW	Cap

endocarps. No faunal remains appear in these preceramic levels. Five AMS radiocarbon dates indicate an age between 2970 ± 70 B.P. and 18,800 ± 100 B.P.

Archaeological research summarized above provides four independent but complementary lines of evidence indicating that foragers settled the northeast Congo Basin many millennia before farming appeared in the Ituri region around 1000 B.P. These lines of evidence are as follows:

1. Stone industries are found beneath late Holocene assemblages that include materials conventionally associated with farming (e.g., ceramics, metal).
2. The matrix in which preceramic assemblages occur is on average fourfold thicker than the capping layers.
3. Stone industries make up all or most of the technological resources used prior to, during, and after farming appearance.
4. Five radiocarbon age determinations much predate the oldest documented presence of farming in the region (Mercader et al., 2000b). Indeed, the oldest of these dates go back to the early Holocene and late Pleistocene, a time in which no farming activity has been reported anywhere in equatorial Africa.

Material Culture from Basal Units

The total lithic assemblage excavated yielded over 10,000 artifacts. Selected artifacts are shown in figure 3.4. A detailed analysis of these industries is provided by Mercader and Brooks (2001). Briefly, lithic tools were made of local quartz from pebbles, reef quartz, and, eventually, crystal. Quartzite and chert were flaked infrequently. Several knapping sequences were documented, including simple debitage and peripheral flake extraction, bipolar and centripetal percussion, bladelet/geometric techniques, small chopping-tool reduction, and flaw propagation. Retouch is rare. Formal tools include core-scrapers, side-scrapers, points, backed tool, and perforators. Lithic assemblages from Ituri show strong similarities with Later Stone Age (LSA) assemblages from sites located through the inter-lacustrine area, such as Matupi Cave (Van Noten, 1977; Kamuanga, 1985) and Ishango 11 (Brooks and Smith, 1987; Mercader and Brooks, 2001). The techno-typological characteristics of the stone industries from Ituri thus indicate LSA affiliation. An extensive occupation of the Ituri region by LSA groups is indicated by the pervasive appearance of LSA artifacts in all caves surveyed (n > 50) and the dense LSA artifactual concentrations detected during preceramic occupation (spit thickness of 2–3 cm; in Lengbe there were more than 400 specimens per 1 m^2; in Matangai Turu Northwest, 360; in Isak Baite Southwest, 140).

Materials from Capping Units

POTTERY (FIGURE 3.5)

Ceramics retrieved from the uppermost layers have been studied by Mercader et al. (2000b). The type of pottery found in the Ituri rock shelters belongs to the Late Iron Age. In spite of a well-documented presence of earlier ceramic traditions in the central basin and in forest-bordering locations of Atlantic Central Africa (Imbonga/Batalimo in the Central Cuvette: Eggert, 1993; Ngovo in Bas-Zaire, Democratic Republic of Congo: de Maret, 1986; Tschissanga/Madingo in People's Republic of Congo: Denbow, 1990; and Shum Laka in Cameroon: de Maret et al., 1987), Early Iron Age ceramics were never encountered in the rock shelter sites of the Ituri forest. Mercader et al. (2000b) proposed that these Late Iron Age ceramics could in fact represent the onset of ceramic technology in the Ituri rain forest; some of these ceramics were produced, used, and discarded by primarily foraging groups who interacted with and received ceramics and ceramic technology from neighboring ceramic-using peoples. Ceramics were made of local clays, without adding temper, by coiling. Firing techniques reached limited temperatures in open bonfires around 600°C. These Late Iron Age assemblages have roulette and non-roulette decoration and present shape and stylistic

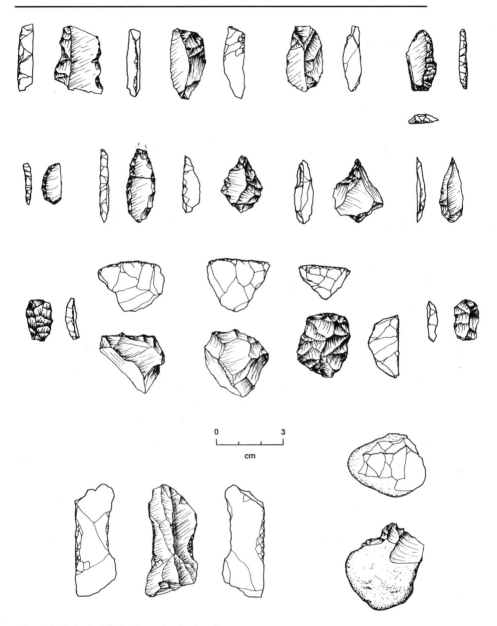

Fig. 3.4. Selected lithics from the Ituri region

similarities (e.g., roulette decoration schemes, "boudine"; see fig. 3.5) with those from the late Holocene layers of Matupi Cave (unpublished material deposited at the Musee Royal de l'Afrique Centrale, Tervuren, Belgium) and especially with those from sites in Kibiro (Connah, 1997) and Chobi, in western Uganda (Soper, 1971).

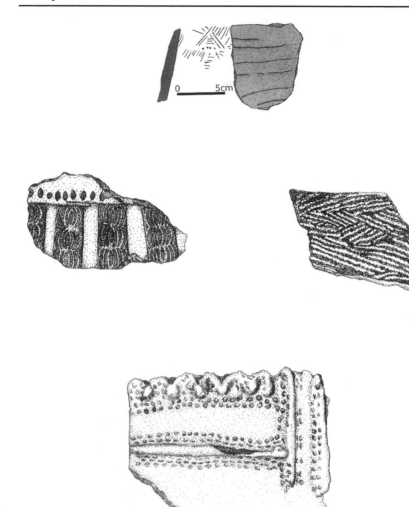

Fig. 3.5. Selected decorated pottery sherds from the Ituri region. The *middle* and *lower* pieces were previously reproduced in the *Journal of Archaeological Science*. Courtesy of Academic Press.

IRON (FIGURE 3.6)

The archaeometallurgical evidence excavated in Ituri was studied by Mercader et al. (2000a). Types of debris recovered include slag pieces, three finished artifacts (fig. 3.6), abundant vitrified clay, tuyere fragments, and ore specimens. The comparison of iron artifacts with those of neighboring sites leads to the inter-lacustrine area of Chobi (Soper, 1971: fig. 14) and Kibiro

Fig. 3.6. Iron artifacts from the Ituri region. This figure was previously published in *Azania*. Courtesy of the British Institute in East Africa.

(Connah, 1997: fig. 12), in which penannular rings very similar to those found in Ituri (Mercader et al., 2000a) were discovered.

HUMAN REMAINS

Little is known about the number and diversity of hunter-gatherer populations in the Central African lowland tropical forest due to a paucity of human remains. Mercader et al. (2001) have reported a single skeleton from the rock shelter site of Matangai Turu Northwest. The skeleton dates from approximately 810 B.P. and is associated with Later Stone Age lithics, animal bone and shell remains from wild taxa, fruit endocarps from forest trees, phytoliths from tropical forest plants, Late Iron Age ceramics, and a single iron artifact. Phytolith analysis indicates the habitat was dense tropical forest, and there is no evidence of domesticated food. The skull of the Matangai Turu Northwest individual is small and gracile, with an ovoid shape and a dolichocranial cephalic index. The cranium has a narrow face and a vertical forehead. The teeth are large, have no evidence of caries, and have moderate attrition on their occlusal surfaces and light accumulations of dental calculus. Indications of dental pathologies include slight periodontal disease in the alveolar region, moderate amount of dental calculus, and enamel hypoplasias. The postcranial remains consist of part of the left humerus, ulna, and radius. The sex is indeterminate. Comparisons between the Matangai Turu Northwest specimen and other African populations do not conclusively establish the morphological or ethnic affinities of the find. Based on material culture, faunal remains, macrobotanical data, and phytolith analysis, this individual seems to have lived primarily by hunting and gathering in the lowland forest.

USE OF TROPICAL OILS AND FAUNAL RESOURCES

Nuts and fruits of several species of the African forest are ethnographically attested as used for extraction of vegetable fat; these include *Canarium schweinfurthii, Elaeis guineensis, Baillonella toxisperma, Coula edulis, Dacryoides edulis, Poga oleosa, Panda oleosa,* and, to a degree, *Irvingia gabonensis* and *Ricinodendron heudelotii. Canarium schweinfurthii* is a flexible species able to thrive in close canopy conditions, like some other oil trees cited above, and in semi-open spaces and humanly and naturally disturbed areas under the canopy: gaps, trails, camps, and garbage mounds (Laden, 1992). Archaeological evidence indicates that prehistoric foragers exploited tropical forest oils at early dates. At Matangai Turu Northwest, Isak Baite Southwest, and Lengbe, the remains from *Canarium* endocarps are distributed throughout the entire preceramic sediment packages (table 3.3). A long tradition of tropical forest arboriculture by foragers is indicated by two direct radiometric ages on *Canarium* endocarps: one of them dated to the early Holocene, the other one to the late Holocene (table 3.2). Although *Canarium* has been repeatedly found at rock shelter sites located in transitional zones, outside the lowland tropical rain forest per se (Smith, 1975; Stahl, 1985; Lavachery et al., 1996: table 2), the Ituri assemblages are associated with lithics only, not with pottery (cf. Smith, 1975; Stahl, 1985; Lavachery et al., 1996). The earliest direct dates for foraging use of this species in the Ituri forest is 10,015 ± 55 B.P.

Elaeis guineensis is an indigenous species with a long history of natural residence near forest environments and forest galleries (e.g., Sowunmi, 1991; Elenga et al., 1994; Maley and Brenac, 1998). However, openings in the canopy are needed if *Elaeis* is to survive in a closed forest. *Elaeis guineensis* is not abundant in the rock shelter sites of the Ituri forest, neither as macro-

Table 3.3. Occurrence of *Canarium* Endocarps at the Sites of Isak Baite Southwest, Matangai Turu Northwest, and Lengbe

Site	Spit no.	Total assemblage, gr	Maximum depth, m
IBSW	18,23,25	4.2	0.6
MTNW	30,40,50,55	2.7	1.41
Lengbe	7,8,9,10,11,14,15,17	3.62	0.74

botanical remains nor as phytoliths (Mercader et al., 2000c). It appears sporadically, within the late Holocene capping unit. This may suggest low reliance on the African oil palm by the foragers of the late Holocene who inhabited these rock shelters. In fact, current cave-dwelling foragers seem to rely on *Elaeis* to a limited extent, especially if settlement occurs far away from farming villages. Today, during the temporary occupation of cave sites, Efe "pygmy" archers bring in a number of oily species for consumption, including *Canarium, Elaeis, Poga,* and *Panda.* This consumption often translates into a 10-cm thick refuse layer of unburned, broken kernels. Mercader excavated one of these "ethnoarchaeological" macrobotanical assemblages (n = 707, 553.7 gr) in an attempt to look at the archaeological signature of Efe consumption of tropical oils during the occupation of the satellite rock shelter post of Koma Tufe Southwest, Nduye. The excavation of a 2-by-1-m test pit indicates that up to 80% of the oily species eaten and discarded at this site derived from *Canarium,* while 20% derived from *Elaeis.* The oldest *Elaeis* remain excavated in the Ituri forest has been directly dated to 840 ± 70 B.P., but older oil palm remains have been found throughout the African wet tropics. Direct dates for *Elaeis* endocarps are available at two sites: Kintampo, Ghana, West Africa (Stahl, 1985: table 3, 127) and Lope 12 (Oslisly and Peyrot, 1988), Gabon, Atlantic Central Africa. The exploitation of the African oil palm in Ghana's forest-savanna transitional zones dates to 6100 ± 250 B.P. through 3495 ± 100 B.P. (Stahl, 1985), while Gabon's seasonal forest at Lope 12 provides direct chronological evidence at 2280 ± 80 B.P. (Oslisly and Peyrot, 1988).

Mammal bones and teeth and snail shells were retrieved from seven sites (total assemblage, n = 67). All bones excavated were found in the uppermost stratigraphic level dated to the late Holocene and belong to small- to medium-size wild game of forest taxa (table 3.4), mostly bovids (42%), primates (37%), rodents (10%), and carnivores (5%). Animal bones were highly fragmented (80% fall within the 1–5 cm fraction) and many of them were affected by fire. Bones do not occur below a depth of 30 cm, although average depth at which they occur is 15.1 cm. Bone remains appear in a variety of geological, textural, and soil types and consist of teeth and phalanges. Other tropical forest faunal resources exploited during late Holocene times included large molluscs (Achatina and Limicolaria) dated to 715 ± 45 B.P.

Paleoenvironment

Few paleoenvironmental data are available from the lowland evergreen forests of the Congo's watershed. Reconstructions of Quaternary environments are based on data from lakes located in mosaic areas of the southern Congo

Table 3.4. Faunal Remains from Baiku West, Matangai Turu Northwest, Koma Tufe Southwest, Makubasi Southeast, Makubasi Northwest Outer Cave, Makubasi Northwest Inner Cave, and Lengbe

Site	Identification
BW	Primate
MTNW	*Atherurus africanus (porcupine)* *Cephalophus sp.* Bovidae
KTSW	*Cephalophus monticola* *Atherurus africanus (porcupine)* Primate
MSE	*Cephalophus callipygus* Primate
MNW o.	*Cephalophus sp.*
MNW I.	Non-identifiable
L.	*Cephalophus monticola*

Basin (Elenga et al., 1994), in the far northwestern corner of the Central African forest at its transition to the West African domain (Maley and Brenac, 1998), and in West Africa (Talbot and Johannessen, 1992). Other complementary contexts include pollen data from the neighboring Western Rift (Sowunmi, 1991; Bergonzini et al., 1997; Marchant et al., 1997), cave pollen from savanna-forest mosaic environments in their transition to both Western Rift savannas and the Afro-montane belt (Brook et al., 1990), and paleoenvironmental interpretations of pedogenesis in "stone-lines" (Runge, 1995). Therefore, reconstructions of Pleistocene–early Holocene human settlement and distributions of tropical rain and seasonal and dry forests in the Central African lowlands currently covered by evergreen and semi-evergreen rain forests remain conjectural. Moreover, the fact that ancient forests were not like modern analogues, not even like late Holocene counterparts that existed only a few thousand years ago (Hart et al., 1996), suggests that prehistoric finds within areas currently covered by dense forest cannot be assumed to originate from tropical forest settlements unless paleoenvironmental indicators are recovered in association with artifacts to indicate the type of tropical forest (or savanna) in which humans lived. Thus, archaeological inquiry into the ability of humans to occupy the African tropical forests is necessarily interrelated with a long climatic and vegetational history.

The phytolith data retrieved from the Ituri region record late Quaternary and Holocene vegetation changes at three archaeological sites. A summary of the results presented by Mercader et al. (2000c) is presented below.[1] The oldest direct date available in a site in which phytolith analysis was carried out is 10,530 ± 50 B.P., obtained on charcoal from a depth of 1.25 m from the site of Matangai Turu Northwest. Because there is 0.90 m of cultural deposits (with chipped stone representative of the Later Stone Age) beneath this layer, and because major changes in the Matangai Turu Northwest paleoenvironmental assemblages occur just below this culturally dated stratum, we propose that the basal layers of this site are no younger than the late glacial maximum. The ancient phytoliths from the site of Matangai Turu Northwest (Malembi) include phytoliths from grasses (e.g., Poaceae, category G2-1) which are most common in strata deposited before 10,530 B.P. In contrast, phytoliths from rain forest trees and herbs (category D1) increase greatly during the Holocene, reaching their maximum values in the last 2000 B.P. Some types of phytoliths derived from trees (category B) are most abundant in the Holocene but are present throughout the sequence. Category C, another arboreal indicator, is common during both the late Pleistocene and the Holocene. At Makubasi Northwest (Nduye), arboreal phytoliths characterize this sequence, which may be entirely Holocene. The sequence lacks major variations in the phytolith assemblage. The morphotypes include arboreal (B, C, and D) and herbaceous plants (e.g., D1) typical of rain forest environments. Lastly, at the site of Lengbe (Epulu), we observed a pervasive dominance of phytoliths typical of the rain forest environment. Changes in forest composition during the Holocene may be indicated by phytolith C1, from unknown trees, which is present at the base of the site, disappears at about the middle of the sequence, and then reappears toward the end of the sequence. Also, phytoliths C3, D, and F peak during the last millennium.

Phytoliths found in association with LSA industries of northeastern Congo generally occurred in high quantities in all of the sites examined (Mercader et al., 2000c) and appear to derive from a diverse set of plants and plant structures (e.g., tree and herbaceous seeds, tree and herbaceous leaves, barks, wood). As detailed elsewhere (Mercader et al., 2000c), in cave sediments of equatorial regions most phytoliths represent local and *in situ* deposition, as dense vegetation prevents long-distance transport by wind. Human activity (e.g., foodstuffs, thatching material, sleeping mattresses, medicinal and ritual plants, leaves brought in with branches collected for firewood) and colluvial influx (plant material growing and decomposing around the rock shelter) brought in through the sides of the caves by rain wash likely account for the great majority of phytoliths present at these sites.

Although all sites suggest diachronic changes in plant use and, perhaps, in local forest composition, only Matangai Turu Northwest, the only site with a proposed late Pleistocene through early Holocene human occupation, shows abrupt changes in tree-grass ratios. Sample 4, dated to 10,530 B.P., is characterized by significant increases in the percentages of arboreal phytoliths and dramatic decreases in grass phytoliths relative to the percentages of phytoliths from trees and grasses prior to that date. Thus, the oldest deposits from Matangai Turu Northwest with late Pleistocene age contain abundant phytoliths of grasses, but also enough arboreal forms to show that the landscape was forested. The late-glacial forests may have had a more open canopy than today's forests. Younger phytolith assemblages show that the northeast Congo Basin was densely forested throughout the Holocene. Archaeological materials among the phytoliths show that people lived in this region during the Pleistocene and the Holocene.

Rock Shelter Function and Socioeconomic Affiliation

Late Pleistocene and early to mid-Holocene preceramic peoples used these rock shelters as stone knapping places. During the late Holocene rock shelters continued to be used as knapping spots, but people brought in a wider variety of items, including ceramics, metal, several types of fruits, firewood and diverse plant remains, molluscs, and forest game; they even used rock shelters as burial sites. Cave settlements probably were satellite posts that complemented longer-term habitation sites in the open. Jack Fisher (personal communication) has retrieved sequences and materials from open-air sites that are similar to those found in these rock shelters.

Even though archaeologists cannot distinguish whether the refuse discarded by prehistoric occupants of the rock shelters was generated during brief occupations by one single large group over time or by a number of smaller groups during multiple occupations, the variations in stone discard rates through time have been used elsewhere in the tropics to indicate intensity of site use and stone reduction (e.g., Morwood and Hobbs, 1995; cf. Hiscock, 1990: 43). The rock shelters seem to have been more actively occupied earlier on in the sequence. With regard to the peoples represented in the basal layers, the only evidence for their hunting-gathering affiliation derives from the presence of *Canarium* endocarps. The type of archaeological record excavated in these basal units hints at sporadic and highly discontinuous settlements by small and mobile groups. Nonetheless, dense occupation occurred at some sites, as suggested by high artifact densities (see above) and the projection of human settlement to adjacent areas outside the cave. At Lengbe preceramic settlements included occupation surfaces located 34 m away from the present cave mouth. The peoples represented in the capping units primarily lived as foraging groups, not as farmers. Faunal resources were wild game resources. All of them from tropical forest animals. Plantains and other typical indicators of farming economies are minimal or absent in both microbotanical and macrobotanical assemblages from Ituri.

Some features of present-day Pygmy people can be traced back in the archaeological record for several millennia. Foragers continue to use these caves today, while farmers live in rock shelters very sporadically and often in the company of foragers. Other features, however, may have a recent origin. Thus, the archaeological correlates of some of the current hunting-gathering practices observed today among Pygmy groups can be traced back only eight centuries. These correlates include (1) hunting with bow and iron-tipped arrow (one iron arrow point was found in the late Holocene levels of Matangai Turu Northwest); (2) prey species (primate and bovid bone remains make up the majority of faunal taxa in archaeological assemblages); and (3) gathering of big molluscs (*Achatina* and *Limicolaria* were discovered at Matangai Turu Northwest but could not be detected in layers older than 825 ± 30 B.P. at any site in which there were no other remains except stone, charcoal, *Canarium* endocarps, and phytoliths).

Conclusion

The Ituri rock shelters broaden the scope of archaeological research in Central African forests because excavations of sites occupied by prehistoric hunter-gatherers had been previously restricted to bordering savannas, transitional zones, or montane areas rather than evergreen lowland domains. Abundant cultural debris in the form of quartz stone tools, charcoal, and phytoliths was found in association and indicates that humans settled and exploited both the variably open forest environments of the late Pleistocene and the closed forest environments of the early Holocene period, when dense forest formations went beyond their current geographical limits (Elenga et al., 1994; Jolly et al., 1998; Maley and Brenac, 1998). Regional stratigraphic evidence, radiometric ages, and direct dates on *Canarium* endocarps circa 10,000 B.P. are of utmost relevance, for they show (1) use of tropical oils during the very early Holocene; (2) exploitation of tropical forest species; and (3) presence of humans in a tropical forest environment at a time of maximum forest closeness and expansion. Archaeological and paleoenvironmental data indicate that forager groups of the late Pleistocene inhabited a different kind of tropical forest, with open canopies and increased grass cover. This type of environment was part of a wider regional landscape that, in the northeast Congo Basin, included not only open tropical forests (Mercader et al., 2000c), but also mosaics of grasslands and woodlands (Brook et al., 1990; Sowunmi, 1991; Runge, 1995). Lack of stratigraphic hiatus between the late Pleistocene cultural layers and the early Holocene ones indicates that after the onset of the Holocene humans stayed where they were and adapted to dense tropical forest environments. This shows that early foragers did not shun the encroaching tropical forest. While sociocultural anthropologists promote the hypothesis that humans could not survive in lowland tropical rain forests without some dependence on agriculture (Headland, 1997), the archaeological evidence retrieved from Ituri disproves this hypothesis.

Archaeological and paleoenvironmental data from the northeast Congo Basin have important implications for the study of the early colonization of tropical forest regions worldwide; namely, (1) there is a direct association of early stone technologies with microbotanical and macrobotanical indicators of lowland tropical forest environments of late Pleistocene and early Holocene chronology; (2) ancient humans may have been able to settle all environments since early on in the process of global colonization; (3) extensive population deserts did not exist throughout the tropical belt; (4) the global colonization of the Pleistocene did not exclude tropical forests; and (5) ancient hunter-gatherers were capable of successful procurement in tropical forests.

Acknowledgments

The Spanish Ministry of Education and Culture provided funds for research. The cultural and military authorities of the Democratic Republic of Congo kindly issued authorizations. Robert Bailey made this work possible. Fieldwork could have never been undertaken without the peoples of Malembi, Nduye, and Epulu and the efforts of Raquel Martí, Olga Carreras, Ignacio González, and Julián Guerrero, who often lived under rugged conditions. The missionaries in Mambasa and Nduye gave us their warmth and logistical support, so did Terese Hart, John Hart, Bryan Curran, Karl Ruff, Rossie Ruff, and Cefrecof.

I am greatly indebted to Alison Brooks, who delivered many suggestions, showed confidence in my work, and helped with resources. Stanley Ambrose provided many helpful suggestions for revision and improvement of this chapter. The Department of Anthropology at George Washington University provided the logistical support and encouragement to accomplish this research. Jack Fisher and Greg Laden shared unpublished data. Dennis Knepper did the illustrations of stone; Jody Clark did the pottery.

Multidisciplinary work was undertaken in collaboration with 21 colleagues from 13 different institutions in four countries (Spain, Belgium, Germany, and the Democratic Republic of Congo). These include Raquel Martí (taphonomy); Ignacio González and Julián Guerrero (geology); Manuel García-Heras (ceramic archaeometry); María Dolores Garralda (physical anthropology); Freya Runge, Luc Vrydaghs, Hughes Doutrelepont, and Jordi Juan (phytolith analysis); Almudera Sánchez and Pilar García (edaphology); Salvador Rovira and Pablo Gómez (archaeometallurgy); Bernardo Pérez (faunal taxonomic identifications); Hans Beeckman and Hughes Doutrelepont (charcoal anatomy identification); Emile Roche (test pollen analysis); Lucia Lopez-Polín, Alicia Pardo, and Esther Jiménez (restoration); Ewanjo Corneille (botanical inventories and sampling); and C. Cornet (diatoms).

Note

1. This summary was previously reproduced in *Quaternary Research* 54 (2000) and is here reproduced with permission from Academic Press.

References

Bailey, R. (1991): *The behavioral ecology of Efe Pygmy men in the Ituri forest, Zaire.* Vol. 86. Ann Arbor: University of Michigan.

Bailey, R., and I. DeVore (1989): Research on the Efe and Lese populations of the Ituri forest, Zaire. *American Journal of Physical Anthropology* 78:459–471.

Bailey, R., and N. Peacock (1988): Efe Pygmies of northeast Zaire: Subsistence strategies in the Ituri forest. In *Copying with uncertainty in food supply,* edited by I. de Garine and G. Harrison, 88–117.

Bailey, R., and T. Headland (1991): The tropical rain forest: Is it a productive environment for human foragers? *Human Ecology* 19(2):261–285.

Bailey, R., G. Head, M. Jenike, B. Owen, R. Rechtman, and E. Zechenter (1989): Hunting and gathering in tropical rainforest: Is it possible? *American Anthropologist* 91:59–82.

Bergonzini, L., F. Chalie, and F. Gasse (1997): Paleoevaporation and paleoprecipitation in the Tanganyika Basin at 18,000 years B.P. inferred from hydrologic and vegetation proxies. *Quaternary Research* 47:295–305.

Brook, G., D. Burney, and J. Cowart (1990): Paleoenvironmental data for Ituri, Zaire, from sediments in Matupi Cave, Mount Hoyo. In *Evolution of environments and Hominidae in the African Western Rift valley,* edited by N. Boaz, 49–70. Martinsville: Virginia Museum of Natural History.

Brooks, A., and C. Smith (1987): Ishango revisited: New age determinations and cultural interpretations. *African Archaeological Review* 5:65–78.

Bultot, F. (1971): *Atlas climatique du Bassin Congolase.* Brussels: INEAC.

Cahen, L., N. Snelling, J. Delhal, and J. Vail (1984): *The geochronology and evolution of Africa.* Oxford: Clarendon.

Connah, G. (1997): The cultural and chronological context of Kibiro, Uganda. *African Archaeological Review* 14:25–67.

de Maret, P. (1986): The Ngovo group: An industry with polished stone tools and pottery in Lower Zaire. *African Archaeological Review* 4:103–133.

de Maret, P., W. Van Neer, and B. Clist (1987): Résultats des premiéres fouilles dans les abris de Shum Laka et d'Abeke au Nord-Ouest du Cameroun. *L'Anthropologie* 91:559–584.

Denbow, J. (1990): Congo to Kalahari. *African Archaeological Review* 8:139–176.

Doumenge, C. (1990): La conservation des ecosystemes forestiers du Zaire: UICN.

Eggert, M. (1993): Central Africa and the archaeology of equatorial rain forest: Reflections of some major topics. In *The Archaeology of Africa: Foods, metals, and towns,* edited by T. Shaw, 289–329. O.W.A., 20. London: Routledge.

Elenga, H., D. Schwartz, and A. Vincens (1994): Pollen evidence of late Quaternary vegetation and inferred climate changes in Congo. *Palaeogeography, Palaeoclimatology, Palaeoecology* 109:345–356.

Fisher, J. (1988): Shadows in the forest: Ethnoarchaeology among the Efe Pygmies. Ph.D. diss., University of California, Berkeley.

Gamble, C. (1993): *Timewalkers: The prehistory of global colonization.* Stroud: Alan Sutton.

Goodwin, A. (1991): *Precambrian geology: The dynamic evolution of the continental crust.* London: Academic Press.

Grinker, R. (1994): Houses in the rainforest. Berkeley: University of California Press.

Hart, T. (1985): *The ecology of a single-species dominant forest and mixed forest in Zaire.* Ph.D. diss. Ann Arbor: University Microfilms International.

Hart, T., and J. Hart (1986): The ecological basis of hunter-gatherer subsistence in African rain forests: The Mbuti of eastern Zaire. *Human Ecology* 14(1):29–55.

Hart, T., J. Hart, and P. Murphy (1989): Monodominant and species-rich forests of the humid tropics: Causes for their co-ocurrence. *American Naturalist* 133:613–633.

Hart, T., J. Hart, M. Dechamps, M. Fournier, and M. Ataholo (1996): Changes in forest composition over the last 4000 years in the Ituri Basin, Zaire. In *The biodiversity of African plants,* edited by L. Van der Maesen, 545–563. Dordrecht: Kluwer Academic Pub.

Headland, T. (1987): The wild yam question: How well could independent hunter-gatherers live in a tropical rain forest ecosystem? *Human Ecology* 15(4):463–491.

Headland, T. (1997): Revisionism in ecological anthropology. *Current Anthropology* 38(4):605–630.

Hiscock, P. (1990): A study in scarlet: Taphonomy and inorganic artefacts. In *Problem solving in taphonomy: Archaeological and paleontological studies from Europe, Africa, and Oceania,* edited by S. Solomon, 34–46. Tempus, 2. St. Lucia: University of Queensland.

Jolly, D., S. Harrison, B. Damnati, and B. Bonnefille (1998): Simulated climate and biomes of Africa during the late Quaternary: Comparison with pollen and lake status data. *Quaternary Science Reviews* 17:629–658.

Kamuanga, M. (1985): Prehistoire du Zaïre oriental: Essai de synthése des âges de la pierre taillée. Ph.D. diss., Université Catholique de Louvain.

Laden, G. (1992): *Ethnoarchaeology and land-use ecology of the Efe (Pygmies) of the Ituri rain forest, Zaire: A behavioral ecological study of land-use patterns and foraging behavior.* Harvard: Department of Anthropology.

Lavachery, P., E. Cornelissen, J. Moeyersons, and P. de Maret (1996): 30,000 ans d'occupation, 6 mois de fouilles: Shum Laka, un site exceptionnel en Afrique centrale. *Anthropologie et Prehistoire* 107:197–211.

Lavreau, J. (1982): *Etude geologique du Haut-Zaire. Genesse et evolution d'un segment lithospherique Archeen.* Tervuren: MRAC.

Maley, J., and P. Brenac (1998): Vegetation dynamics, palaeoenvironments, and climatic changes in the forests of western Cameroon during the last 28,000 years B.P. *Review of Palaeobotany and Palynology* 99:157–187.

Marchant, R., D. Taylor, and A. Hamilton (1997): Late Pleistocene and Holocene history at Mubwindi Swamp, southwest Uganda. *Quaternary Research* 47:316–328.

McBrearty, S., and A. Brooks (2000): The revolution that wasn't: A new interpretation of the origin of modern human behavior. *Journal of Human Evolution* 39(5): 453–563.

Mercader, J. (1997): Bajo el techo forestal: La evolucion del poblamiento en el bosque ecuatorial del Ituri, Zaire. Ph.D. diss., Universidad Complutense, Madrid, Spain.

Mercader, J., and A. Brooks (2001): Across forest and savannas: Later Stone Age assemblages from Ituri and Semliki, Northeast Democratic Republic of Congo. *Journal of Anthropological Research* 57(2):197–217.

Mercader, J., and R. Martí (1999a): Middle Stone Age site in the tropical forests of Equatorial Guinea. *Nyame Akuma* 51:14–24.

———. (1999b): Archaeology in the tropical forest of Banyang-Mbo, SW Cameroon. *Nyame Akuma* 52:17–24.

Mercader, J., S. Rovira, and P. Gómez (2000a): Shared technologies: Forager-farmer interaction and ancient iron metallurgy in the Ituri rain forest, Democratic Republic of Congo. *Azania* 35:107–122.

Mercader, J., M. García-Heras, and I. González (2000b): Ceramic tradition in the African rain forest: Characterization analysis of ancient and modern pottery from Ituri, D. R. Congo. *Journal of Archaeological Science* 27:163–182.

Mercader, J., F. Runge, L. Vrydaghs, H. Doutrelepont, E. Corneille, and J. Juan (2000c): Phytoliths from archaeological sites in the tropical forest of Ituri, Democratic Republic of Congo. *Quaternary Research* 54:102–112.

Mercader, J., M. D. Garralda, O. Pearson, R. Bailey (2001): 800-year-old human remains from the Ituri tropical forest, Democratic Republic of Congo: The rock shelter site of Matangai Turu NW. *American Journal of Physical Anthropology* 115: 24–37.

Mercader, J., R. Martí, I. González, A. Sánchez, P. García (2002): Archaeological site formation in tropical forests: Insights from the Ituri rock shelters, Congo. *Journal of Archaeological Science.*

Morwood, M., and D. Hobbs (1995): Themes in the prehistory of tropical Australia. Transitions: Pleistocene to Holocene in Australia and Papua New Guinea, edited by J. Allen and F. O'Connell. *Antiquity* 69:747–768.

Oslisly, R., and B. Peyrot (1988): Synthèse des données archéologiques des sites de la Moyenne Vallée de l'Ogooue (Provience du Moyen Ogooué et Ogooué Ivindo) Gabon. *Nsi* 3:63–68.

Peterson, R. (1991): To search for life. Ph.D. diss., University of Michigan.

Rösler, M. (1997): Shifting cultivation in the Ituri forest (Haut-Zaire): Colonial intervention, present situation, economic and ecological aspects. In *Les peuples des forets tropicales,* edited by D. Joiris and D. Laveleye, 44–61. Brussels: Guyot.

Runge, J. (1995): New results on late Quaternary landscape and vegetation dynamics in eastern Zaire (Central Africa). *Zeitschrift für Geomorphologie* 99:65–74.

Smith, A. (1975): Radiocarbon dates from Bosumpra Cave, Abetifi, Ghana. *Proceedings of the Prehistoric Society* 41:179–182.

Soper, R. (1971): Iron Age archaeological sites in the Chobi sector of Murchison Falls National Park, Uganda. *Azania* 6:59–88.

Sowunmi, M. (1991): Late Quaternary environments in equatorial Africa: Palynological evidence. *Palaeoecology of Africa* 22:213–238.

Stahl, A. (1985): Reinvestigation of Kintampo 6 rock shelter, Ghana: Implications for the nature of culture change. *African Archaeological Review* 3:117–150.

Talbot, M., and T. Johannessen (1992): A high-resolution palaeoclimatic record for the last 27,500 years in tropical West Africa from the carbon and nitrogen isotopic composition of lacustrine organic matter. *Earth and Planetary Science Letters* 110:23–37.

Van Neer, W. (1989): *Contribution to the archeozoology of Central Africa.* Tervuren: MRAC.

Van Noten, F. (1977): Excavations at Matupi Cave. *Antiquity,* 51:35–40.

———. (1982): *The archaeology of Central Africa.* Graz: Akademische Druck unds Verlagsanstalt.Vansina, J. (1990): *Paths in the rainforest.* London: James Curvey.

White, F., ed. (1983): *Vegetation of Africa: A descriptive memoir to accompany the Unesco/ Aetfat/Unso vegetation map of Africa.* Paris: Unesco.

Wilkie, D. (1987): Impact of swidden agriculture and subsistence hunting on diversity and abundance of exploited fauna in the Ituri forest of northeastern Zaire. Ph.D. diss., University of Wisconsin.

PART II

AUSTRALASIAN SETTLERS

Hunter-Gatherer Occupation of the Malay Peninsula from the Ice Age to the Iron Age

F. David Bulbeck

Bailey et al. (1989: 71) exempted the Malay Peninsula from their controversial generalization that tropical rain forests are too poor in edible resources to fully sustain hunter-gatherers. Soon afterwards, Endicott and Bellwood (1991) combined ethnographic data on Malaya's rain forest foragers, the Semang, with evidence of a long sequence of hunter-gatherer sites to aver the feasibility of independent foraging within this environment. Their conclusions included the cautionary note: "Given the numerous virtues of cultivated foods and their widespread availability in the modern world, it would be surprising if there were still pure 'foragers' in the world's tropical rain forests, but their absence is not in itself evidence that independent foraging is impossible there" (181).

This chapter extends Endicott and Bellwood's recognition that the Semang have adapted dynamically to late Holocene change. As argued here, their ethnographically recorded foraging strategies were attuned to acquiring useful chattels and tasty foodstuffs through trade, while reducing exposure to slave raiding. The associated organizational features of the Semang probably evolved quite recently and may be irrelevant to Malaya's ancient past. Rather, the theoretical framework of cultural ecology, including due consideration of sea-level fluctuations and related environmental change, is recommended for interpreting the hunter-gatherer record in the long term. The resulting perspective suggests that, prior to the late Holocene, hunter-gatherer occupation was focused on the coast and subcoastal lowlands. Transhumant parties who learned to specialize in the jungle's portfolio of resources probably effected the colonization of the interior rain forest, including the rolling hinterland inhabited by the Semang.

The Semang are not the only Orang Asli Aborigines to share Malaya with the Malays. Early ethnographers in the Malay Peninsula recognized three indigenous, non-Islamic cultures and explained them in terms of successive waves of migration. Hunter-gatherers were assumed to be the most primitive inhabitants, engendering the myth of a pristine Negrito occupation. The "Sakai" or Senoi were seen as more advanced, based on their swidden economy and certain physical differences from the Negritos (lankier build, lighter complexion, and wavy rather than woolly hair). Then came the so-called Jakun or proto-Malays, who approached the Malays in various ways, as indicated by their current epithet, the Aboriginal Malays. Today, few scholars still endorse a model of increasingly evolved immigrants who repeatedly displaced older inhabitants, but the tripartite system remains intact (e.g., Endicott, 1983; Bellwood, 1993; Fix, 1999). Classificatory schemes based on human ecology (Rambo, 1984) and societal tradition (Benjamin, 1985) largely correspond to the original somatic classification (table 4.1). And all three approaches agree on the territory ascribed to the Semang, Senoi, and Aboriginal Malays across an area larger than half of West Malaysia (fig. 4.1).

    While these mutually agreed areas might indicate core territories for the three Orang Asli divisions, the boundaries of the cores could still have been pushed outwards through recent expansion (see Fix, 1999: 192–194). Most Orang Asli speak languages within the Aslian branch of Austro-Asiatic, even if dialects of Malay, an Austronesian language, are widespread among Aboriginal Malays (Benjamin, 1985; Bellwood, 1993: 38). (Elsewhere, Austro-Asiatic languages occur in the Nicobar Islands to the north of Sumatra, northeast India, and throughout Mainland Southeast Asia.) Within the Aslian branch, North Aslian is strongly related to the Semang; Central Aslian languages are spoken mainly by the Senoi; and Southern Aslian is associated with Aboriginal Malays (table 4.1). Had these three separate traditions crystallized before the late Holocene, we would hardly expect such neat linguistic correlations. Although Bellwood (1997: 265) assumes the ancestors of the Semang had switched to speaking Austro-Asiatic languages in the late Holocene, multiple independent switches to a branch of Aslian spoken almost exclusively by the Semang would appear unlikely. Indeed, linguistic evidence for a late Holocene expansion of the Semang looks stronger than it is for the Senoi or Aboriginal Malays (table 4.1).

    The Semang core comprises a block of remote lowlands typically between 100 m and 1000 m above sea level (fig. 4.2). Most of our records on the Semang in terms of ethnography (e.g., Schebesta, 1928; Endicott, 1984), anthropometrics, and skeletal anatomy (Bulbeck, 1996: fig. 3) come from this area. But its archaeological record is paltry: merely four excavated rock shelters (Matthews, 1961: 29–31; Azman, 1998) and a further two noted only after their deposit had been taken as fertilizer (Matthews, 1961: 32–33). The focus of the Holocene hunter-gatherer record lies elsewhere, on the limestone massifs which contain the vast majority of Malaya's rock shelters. Accordingly, the Semang tradition could well reflect the late Holocene expansion of a successful foraging adaptation across a remote lowland region with a scanty archaeological record. Ethnographic analogies drawn from the Semang may thus be of oblique relevance, at best, to interpreting the

Table 4.1. Semang, Senoi, and Aboriginal Malay Classifications

	Hair form[a]	Human ecology[b]	Societal tradition[c]
Semang	Mainly woolly	Foragers	Egalitarian/ patrilineal
	Kensiu (N)	Kensiu (N)	Kensiu (N)
	Kentaq (N)	Kentaq (N)	Kentaq (N)
	Jehai (N)	Jehai (N)	Jehai (N)
	Menriq (N)	Menriq (N)	Menriq (N)
	All Batek (N)	Batek Deq/Batek Iga' (N)	All Batek[d] (N)
	Che Wong (N)		Che Wong[d] (N)
	Lanoh (C)	Lanoh (C)	
	Temoq (S)	Temoq (S)	
		Inland Semaq Beri (S)	
Senoi	Mainly wavy	Swidden farmers	Egalitarian/cognatic
	Temiar (C)	Temiar (C)	Temiar (C)
	All Semai (C)	All Semai (C)	Highland Semai (C)
		Jah Hut (C)	Jah Hut[d] (C)
		Che Wong (N)	Lanoh (C)
		Batek Nong (N)	
		Batek Happen (N)	
		Semelai (S)	
Aboriginal Malay	Mainly straight	Horticulturalists	Ranked tribe/ peasantry
	Semelai (S)		Semelai (S)
	Mah Meri (S)	Mah Meri (S)	Mah Meri (S)
	Temuan (A)	Temuan (A)	Temuan (A)
	Tribes to the south	Semaq Beri (S)	Semaq Beri[d] (S)
		Orang Hulu (A)	Orang Hulu (A)
		Orang Kanaq (A)	Orang Kanaq (A)
		Duano (A)	Temoq (S)
		Orang Selatar (A)	Lowland Semai (S)

SOURCE: From Benjamin (1985). Ethnolinguistic distributions are taken from Benjamin (1985, 1996) for Aslian groups and from Bellwood (1993) for Austronesian groups.

NOTE: (N) = North Aslian; (C) = Central Aslian; (S) = South Aslian; (A) = Austronesian.

[a] Schebesta and Lebzelter (1928). Not all groups are classified; Temuan are called Kenaboi.

[b] Rambo (1984). Ethnolinguistic groups deduced from Rambo's distribution map.

[c] Benjamin (1985).

[d] Mixed traditions are assigned to the Semang if North Aslian in linguistic affiliation, to the Senoi if Central Aslian, and to Aboriginal Malay if South Aslian.

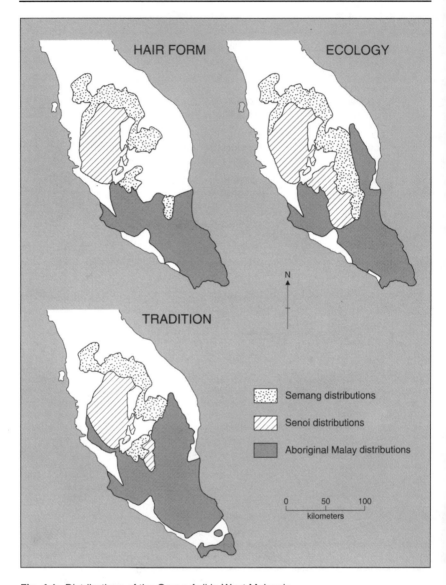

Fig. 4.1. Distributions of the Orang Asli in West Malaysia

Malay Peninsula's hunter-gatherer sites. For instance, the lack of reported use of stone to knap stone among the Semang (Rambo, 1979) hardly implies any deficiency of stone knapping in the past.

Pebble tools are prominent in Holocene prehistoric hunter-gatherer assemblages in West Malaysia. As river pebbles provide the most accessible source of flaking stone (albeit of indifferent quality) in the heavily forested

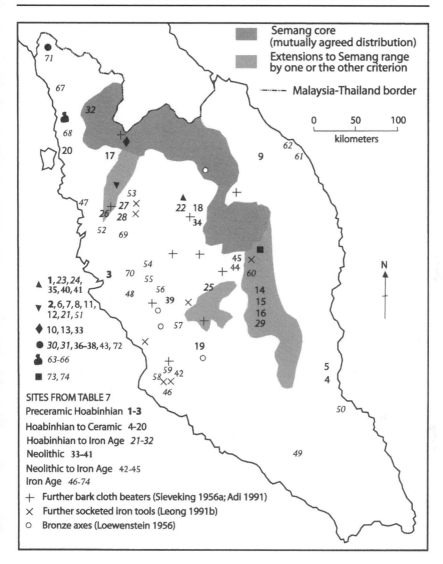

Fig. 4.2. Holocene archaeological sites in West Malaysia compared to Semang distribution

terrain of Malaya, the focus on pebble tools represents a straightforward adaptation. The assemblages are called Hoabinhian (after Hoa Binh, in North Vietnam) by most researchers (e.g., Bellwood, 1993, 1997; Adi, 2000). However, Zuraina and her students (e.g., Zuraina, 1994; Zolkurnian, 1998) prefer the term *epi-Paleolithic,* to counter any idea of a sweeping migration from the north. Certainly, there are major problems in applying the Hoabinhian label automatically to Holocene flaked stone assemblages in Mainland Southeast Asia (Anderson, 1990: 72–74; Shoocongdej, 2000). Indeed, geographical variation pertains to the pebble tools in Malaya, as in the apparent absence of unifacially flaked "sumatraliths" along the eastern peninsula, where oval bifaces prevail (Adi, 2000: 157). It lies beyond the scope of this chapter to speculate on whether Malaya's Holocene lithics might have a cultural-historical relationship with local or external late Pleistocene assemblages. In the absence of a comprehensive review and terminological clarification of Southeast Asian flaked stone assemblages, I shall persevere with the familiarity of the Hoabinhian label in the Malay Peninsula case. As a last introductory point, for reasons to be explained, all radiocarbon dates referred to in this chapter are uncalibrated.

Late Pleistocene and Holocene Environmental Changes

The Malay Peninsula borders the vast Sunda Shelf, which has emerged as a low-lying continent, then fragmented into islands and shallow sea, with each major change in late Quaternary sea levels. Dramatic changes have ensued in marine currents, atmospheric humidity, the strength of the monsoons, and even snowfall and glacial cover. In general, mangroves and rain forest expanded with higher sea levels, and more open vegetation spread as the sea dropped, but with tremendous regional variability and no detectable change in certain refuges of lowland rain forest (Hope, in press). The slim evidence currently available from the Malay Peninsula, especially the 10,000-year-old pollen core from Nong Thale Song Hong in south Thailand, suggests marked changes in the peninsula's vegetation, related to sea-level fluctuations.

The late Pleistocene coastlines prior to 14,000 B.P. may be inferred from the generalized sea-level curve in Chappell et al. (1996). Between approximately 120,000 B.P. and 75,000 B.P., the coast would have vacillated between its present position and the 50-m sea-depth contour which lies tens of kilometers offshore. Between approximately 75,000 B.P. and 30,000 B.P., the coast would have fluctuated between today's 50 m and 90 m sea-depth contours. Vast lowland drainage systems would have emerged in the Strait of Melaka and, especially, between the Gulf of Thailand and the South China Sea during low sea-level intervals (fig. 4.3). After 30,000 B.P., the sea rapidly receded and would have breached today's 100-m sea-depth contour at around 25,000 B.P. The ocean now lay approximately 100 km to the west and 600 km to the east of the peninsula's present-day shores. During the last glacial maximum, at approximately 18,000 B.P., the sea dropped to 120 m below its present level before recovering sharply to the 100-m sea-depth contour by approximately 14,000 B.P.

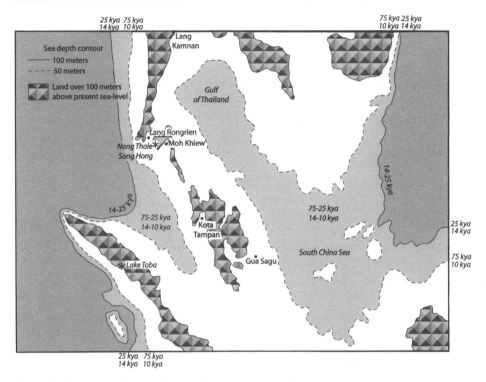

Fig. 4.3. Sites clearly dated to the late Pleistocene in the Malay Peninsula and approximate coastlines modeled on sea-depth contours

The sea would have sped toward present-day land at the end of the Pleistocene but would still have lain tens of kilometers offshore at 10,000 B.P. (fig. 4.3). It continued to advance during the early Holocene, passing its present level at about 7000 B.P. and peaking at 4 m above the present at about 6000 B.P. The land connection to Sumatra would have been sundered at some stage between 9000 and 7000 B.P., which is relevant to understanding the continuous distribution of the Hoabinhian across the Melaka Strait into northern Sumatra. After 7000 B.P. the sea level dipped and rose several times, but always within 2 m of its present height (Tjia, 1987, 1991; Sinsakul, 1994).

Shifting sea levels may be largely responsible for the vegetation changes that can be inferred from Maloney's (1999) Nong Thale Song Hong core. The core should show evidence of significantly drier conditions at about 10,000 B.P. because depressed sea levels should be associated with reduced rainfall and, especially, the loss of the mid-year wet season along what is now the peninsula's east coast (Anderson, 1990: 26). This expectation is confirmed by the rising ratio of pteridophyte spores to pollen between 10,000 B.P. and the middle Holocene (Maloney, 1999: fig. 4); pteridophyte spores are attributed to fluvial transport, so their low incidence in the early

Table 4.2. Model of Paleoenvironmental Change in the Malay Peninsula

Period B.P.	Sea level	Vegetation	Predominant fauna
120,000–75,000	0–50 m below present	Mainly rain forest	Arboreal/rain forest
75,000–25,000	50–100 m below present	Rain forest/mosaic flux	Fluctuating
25,000–14,000	>100 m below present	Mainly savanna mosaic	Open forest/woodland
14,000–10,000	50–100 m below present	Spreading rain forest	Increasingly rain forest
6000	4 m above present	Rain forest (intense?)	Arboreal/rain forest
5000–0	+2 to −2 m cf. present	Rain forest + clearance	Arboreal/rain forest (except disturbed areas)

Holocene suggests drier conditions. A peak in the incidence of spores from dry-land ferns at around 10,000 B.P. supports this interpretation. Finally, the dry-forest taxon *Lagerstroemia* exhibits a consistent presence only in the levels of the core dated to about 10,000 B.P. (Maloney, 1999: 132, fig. 6). Hence we may have some confidence in linking the rise and fall of the sea to wetter and drier conditions, respectively (see also Shoocongdej, 2000: 25).

From his study of sedimentary sequences in the south of West Malaysia, De Dapper identified an interlude of savanna between two phases of dense forest. Summarizing De Dapper's results, Anderson (1990: 28) inferred dense forest throughout the Holocene and suggested a last glacial maximum antiquity for the savanna interlude. Adams (1998) similarly reconstructs a change from monsoon forest at 18,000 B.P. to tropical rain forest, by 8,000 B.P. and thereafter, across the entire present-day extent of the Malay Peninsula. To extend this model of environmental change back to before the last glacial maximum, Holocene climatic conditions (including dense forest cover) would have prevailed at those times when the sea level lay above today's 50-m sea-depth contour (fig. 4.3). Evidence of relatively moist conditions prior to the last glacial maximum is provided by the 5f-level pollen sample recovered from a context older than 26,000 B.P. in the Moh Khiew rock shelter. The sample is dominated by *Brownlowia,* whose habitat includes freshwater swamp, evergreen forest, and scrub (Wattanasak, 1994: 396–397). As a last note to the model on paleoenvironmental change summarized in table 4.2, anthropogenic forest disturbance was evidently an increasing factor after 3000 B.P., though early to middle Holocene incidences are also suggested (Maloney, 1999; Bowdery, 1999: 164–167).

Late Pleistocene Hunter-Gatherers of the Lowlands

Kota Tampan, Moh Khiew, and Lang Rongrien are the three sites in the Malay Peninsula with chronometric dates older than the last glacial maximum (fig. 4.3). All lie less than 100 m above median present sea level and reflect occupation of the lowlands near the west coast. Higher sites, such as Gua Gunung Runtuh (150 m above sea level), already formed as a rock shelter by 30,000 B.P. (Mokhtar and Tjia, 1994), do not record traces of habitation till the very end of the Pleistocene. Lack of cultural materials dated from approximately 25,000 B.P. to 15,000 B.P. suggests that the inhabitants relocated to the exposed continental shelf when the sea retreated during the last glacial maximum. These now-submerged lands would have supported tropical rain forest along the west coast and tropical savanna in the east (Adams, 1998).

Kota Tampan, an open site, is sealed by a volcanic ash with a fission-track date of 31,000 ± 3000 B.P. (Stauffer et al., 1980). Problematically, the ash is ascribed to the Toba eruption (Stauffer et al., 1980; Zuraina and Tjia, 1988), which other sources date to around 70,000 B.P. (Ambrose, 1996). An antiquity in excess of 75,000 B.P. would bring the coastline at the time close to Kota Tampan (table 4.2). Level 1, the basal cultural layer at Moh Khiew, could also reach far back into the past. It extends 1.5 m beneath the 25,800 ± 600 B.P. radiocarbon date at the top of the layer (Pookajorn, 1994b). It may well be coeval with units 8 to 10 at Lang Rongrien; those units have produced four charcoal dates, in stratigraphic succession, between 27,110 ± 615 B.P. near the top of unit 8 and 37,000 ± 1780 B.P. in unit 9 (Anderson, 1990: 21). A small lens of charcoal, recovered from the meter-thick layer of limestone rubble that overlies unit 8, is dated to greater than 43,000 B.P., and so is interpreted as an inclusion of older material (Anderson, 1990). Both of the oldest Lang Rongrien dates lie near the 40,000 B.P. "barrier," where radiocarbon determinations should be treated as minimum estimates. Hence there is some chronometric evidence for occupation of the Malaya lowlands in excess of 40,000 years ago.

Zuraina (1998b) and Mokhtar (1998a,b) report three new assemblages near Kota Tampan. These sites have been described too recently to qualify as confirmed discoveries, unlike Kota Tampan, whose credentials seem to have been generally accepted, given the lack of contrary debate in the literature. Bukit Jawa and Kampung Temelong, along with Kota Tampan, are interpreted as sites along the same ancient lake (72–76 m above sea level). Lawin is considered to have been deposited on the shores of another lake that reached 92–122 m above sea level. Bukit Jawa is suspected to date very approximately to 100,000 B.P., based on terrace geomorphology and its large "crude" artifacts, while Kampung Temelong and Lawin are considered to illustrate the transition to the industry at Kota Tampan (i.e., more than 70,000 B.P.?). All four sites have assemblages centered on knapping waterworn quartz or quartzite pebbles, presumably collected from the foreshore.

Lang Kamnan, which lies at 110 m above sea level at the northern border

Table 4.3. Uncalibrated Lang Kamnan Radiocarbon Determinations

	From probable cultural contexts		No cited evidence of a cultural context	
Layer	Date (B.P.)	Material	Date (B.P.)	Material
2	150 ± 60	Charcoal		
	7540 ± 180	Land snail from shell midden	7740 ± 140	Land snail
	8305 ± 90	Land snail from feature	7990 ± 100	Land snail
	10,030 ± 110	Riverine shell[a,b] (≡ 8000–7000 B.P.)		
	15,150 ± 70	Charcoal[a]		
	16,170 ± 175	Riverine shell[b] (≡ 14,000–13,000 B.P.)		
3	6110 ± 60	Burnt clay & charcoal	6680 ± 150	Land snail
	15,170 ± 70	Carbonized wood	18,280 ± 320	Land snail
	15,345 ± 190	Ash	20,020 ± 240	Land snail
4	15,640 ± 150	Land snail from shell midden	17,130 ± 220	Land snail
	23,165 ± 330	Land snail from shell midden	21,120 ± 460	Land snail
	30,880 ± 760	Land snail from shell midden	26,920 ± 210	Land snail
			27,110 ± 500	Land snail

SOURCE: Data from Shoocongdej (2000: table 4).
[a]Claimed by the excavator to be from a disturbed context.
[b]Probably 2000–3000 years too old (cf. Adi, 2000: 111).

of the peninsula, is claimed to have been occupied from about 27,000 B.P. onwards (Shoocongdej, 2000). However, the only dates which fall between approximately 23,000 B.P. and 16,000 B.P. come from land snails, which could be natural inclusions, as they lack cited evidence of a cultural context (table 4.3). The dates, which do probably derive from cultural contexts (excluding, for the moment, recent and pre–last glacial maximum dates), cluster between circa 8000–6000 B.P. and circa 16,000–14,000 B.P., and both clusters cross the recognized stratigraphic layers. These points suggest repeated abandonment of the shelter for up to six millennia at a time. Further, the dates on (presumably) natural land snail also chronologically confound stratigraphic layer 2 with layer 3, and layer 3 with layer 4 (table 4.3). This chronometric confusion cautions against treating the reported contents of the layers (Shoocongdej, 2000) as a chronological sequence. Nonetheless, the dates themselves are consistent with Pleistocene occupation in excess of 23,000 B.P., followed by abandonment until approximately 15,000 B.P., in accord with Pleistocene sites to the south. Reoccupation of Lang Kamnan toward the end of the last glacial maximum would strike a particular chord with another lowland shelter, Gua Sagu, with its basal charcoal date of 14,410 ± 180 B.P. (Zuraina, 1998a: 246). As both sites would have lain far inland at the time (fig. 4.3), they represent colonization of the region's deep interior during the late Pleistocene, though probably not throughout the last glacial maximum.

Four assemblages of flaked lithics that probably predate the last glacial maximum have been described in a form amenable to comparison with later assemblages (table 4.4). Their contrasts with each other and with Hoabinhian assemblages can be attributed to a combination of site function differences and unreliable data. The interpretation of Kampung Temelong and Kota Tampan as flaking floors tallies with their incidence of debitage ("other flakes," about 95%), higher than in any other assemblage. In stark contrast, a remarkable incidence of flake tools (60–64%) has been identified in the Pleistocene levels at Lang Rongrien and Moh Khiew. In the case of Lang Rongrien, the sample size may be too small to justify Anderson's (1990) claim of a shift from flake tools in the late Pleistocene to pebble tools in the Holocene. As regards Moh Khiew (Pookajorn, 1994a, 1996), percentages of "flake tools" in excess of 57% supposedly characterize its post–last glacial maximum assemblages as well, despite a vacillating focus between shale and chert (table 4.4). Contrast these figures with the 0–20% incidence of flake tools observed in other post–last glacial maximum assemblages of flaked lithics (table 4.4) or with the 6.8% incidence recorded on the excavated lithics from Lang Kamnan (Shoocongdej, 2000: table 3). Either retouching and utilization of flakes were extraordinarily intensive at Moh Khiew or Pookajorn's classification scheme is incommensurate with that of other observers.

The Malay Peninsula apparently witnessed a switch from unifacial and bifacial cores before the last glacial maximum to unifacial and bifacial pebble tools (and utilized pebbles) after the last glacial maximum. Only one bifacially flaked pebble (from Lang Rongrien) has been positively identified in a context earlier than the last glacial maximum. Recognizable cores seem to fall in frequency relative to flaked pebbles in post–last glacial maximum assemblages (table 4.4). Utilized and hematite-stained pebbles (as opposed to hammer stones and anvils, used in flaking) also appear to be a common Holocene occurrence (table 4.5). The focus on flaked and utilized pebbles further marks the Hoabinhian of northern Sumatra (Bellwood, 1997: 170), in contrast with the majority of coeval lithic assemblages from Lang Kamnan northwards, which are characterized as core-tool industries (Anderson, 1990: 72; see also Shoocongdej, 2000: 28–30). Assemblages linking Malaya's more core-based pre–last glacial maximum industries and its pebble-based Hoabinhian tradition may lie in the submerged continental shelf that borders the peninsula.

Published evidence on pre–last glacial maximum faunal resources is available only from level 1 at Moh Khiew. Its sparse refuse includes remains of water buffalo (Pookajorn, 1994a: 341), pig, barking deer and other

Table 4.4. Late Pleistocene and Holocene Flaked Lithic Assemblages

Assemblage	Flaked pebbles (%)		Cores (%)		Flaked pieces (%)		Flakes (%)	
Pre-LGM	Unifacial flaking	Bifacial flaking	Unifacial	Bifacial	All unifaces	All bifaces	Flake tools	Other
Kampung Temelong (n = 4291)[a]	2.9	0.0	0.0	0.5	2.9	0.5	1.0	95.5
Kota Tampan (n = 1726)[b]	1.3	0.0	0.0	1.2	1.3	1.2	3.0	94.6
Lang Rongrien Units 8–10 (n = 36)[c]	2.8	2.8	8.3	5.6	11.1	8.3	63.9	16.7
Moh Khiew Level 1 (n = 1190)[d]	8.6		8.6		8.6		60.6	30.8
Pre-LGM Preceramic	Unifacial flaking	Bifacial flaking	Unifacial	Bifacial	All unifaces	All bifaces	Flake tools	Other
Gua Sagu spits 7–13[e] (n = 30)[f]	10.0	20.0	0.0	0.0	10.0	20.0	0.0	70.0
Gua Tenggek (n = 117)[g]	5.1	9.4	0.0	0.0	5.1	9.4	0.0	85.5
Moh Khiew Levels 2–3 (n = 12,441)[h]	1.1		1.1		1.1		68.1	30.8
Gua Guntung Runtuh (n = 397)[i]	20.4	6.3	1.6		>20.4	>6.3	1.4	55.4
Gua Teluk Kelawar B (n = 71)[j]	0.0	0.0	0.0	0.0	0.0	0.0	7.0	93.0
Gua Teluk Kelawar (n = 1465)[j]	5.9	2.3	0.3		>5.9	>2.3	2.5	88.9
Lang Rongrien Units 5–6 (n = 182)[k]	8.8	1.6	1.6	2.2	10.4	7.7	14.3	67.6
Gua Cha Layers 3–4 (n = 544)[l]	4.4	8.1	5.5		>4.4	>8.1	3.7	78.3
Gua Batu Tukang (n = 86)[l]	3.5	1.2	0.0	0.0	3.5	1.2	5.8	89.5
Gua Ngaum (n = 91)[j]	8.8	4.4	0.0	0.0	8.8	4.4	19.8	67.0

Holocene Ceramic	Unifacial flaking	Bifacial flaking	Unifacial	Bifacial	All unifaces	All bifaces	Flake tools	Other
Moh Khiew Levels 4–5 (n = 5920)[m]		±0.4			±0.4		57.6	42.0
Gua Kecil 14–34 inches deep (n = 62)[n]	3.2	4.8	0.0	0.0	3.2	4.8	17.7	74.2
Gua Sagu spits 1–6 (n = 121)[g]	15.7	14.0	0.0	0.0	15.7	14.0	2.5	67.8

NOTE: Pebbles have a waterworn cortex and lack the platform preparation characterizing cores. Pebbles with unifacial flaking include tabular pieces variously called a "hand-adze" (Anderson, 1990), "paleoadzes" (Zuraina et al., 1994) and "paleo-beliung" (Zolkurnian, 1998). Cores include core tools. Flaked pieces include pebbles, cores, and seven biface fragments from Lang Rongrien. Flake tools are restricted to utilized flakes in the cases of Gua Cha and Gua Kecil. Flakes include broken flakes, flake fragments, and debris.

[a]Mokhtar (1998a). All alat-alat pebbles are treated as having unifacial flaking. Puingan are placed under "other flakes." Predominantly quartz and quartzite.

[b]Zuraina (1990). Pebbles with unifacial flaking include seven "failed cores." Debitage is placed under "other flakes." Higher counts are now available for at least some artifact classes (Mokhtar, 1998a). Predominantly quartzite and quartz.

[c]Anderson (1990: 56–58). Excludes six core fragments and one possible artifact. Mainly chert.

[d]Pookajorn (1994a, 1996). The pieces classified as "unifacial and bifacial core tools" (no further divided) are described as Hoabinhian-like core tools similar to the unifacially flaked pebbles from Kota Tampan (Pookajorn, 1994a: 320–324), suggesting that some at least would be classified as pebble tools by other workers. Predominantly siliceous shale.

[e]These hoabinhian assemblages lie to the east of the main range.

[f]Zuraina et al. (1998). "Miscellaneous" in this assemblage are pebbles with unifacial flaking. Mainly slate, quartzite, and basalt.

[g]Zuraina et al. (1998). Excludes seven "miscellaneous." Mainly sandstone, slate, and basalt.

[h]Pookajorn (1994a, 1996). See note d above. Predominantly chert and jasper.

[i]Zuraina et al. (1994). "Choppers," "perimeter flaked pebbles," and miscellaneous pebbles are treated as pebbles with unifacial flaking. Mainly quartzite.

[j]Zolkarnian (1998). Mainly quartzite and quartz.

[k]Anderson (1990: 37–46). Excludes four core fragments. Mainly shale.

[l]Adi (1981: 72). Truncated pebble tools and pebbles with one end flaked are classified as pebbles with unifacial flaking. Highly variable lithology including chert, shale, and quartz.

[m]Pookajorn (1994a, 1996). Axes and adzes, often polished, took over from Hoabinhian-like core tools in these levels. Predominantly siliceous shale.

[n]Dunn (1964: 114–115). Excludes three "fragments of . . . chipped and flaked tools."

Table 4.5. Prehistoric Assemblages of Non-flaked Lithics

Assemblage	Hammers (%)	Anvils (%)	Utilized and stained pebbles (%)	Mortars (%)	Polished stone (%)
Pre-LGM					
Kampung Temelong (n = 368)	84.8	15.2	0.0	0.0	0.0
Kota Tampan (n = 33)	54.5	45.5	0.0	0.0	0.0
Lang Rongrien Units 8–10 (n = 3)	0.0	33.3	66.7	0.0	0.0
Post-LGM Preceramic					
Gua Sagu spits 7–13 (n = 12)	33.3	8.3	58.3	0.0	0.0
Gua Tenggek (n = 23)	60.9	13.0	26.1	0.0	0.0
Gua Gunung Runtuh (n = 101)	85.1	14.9	0.0	0.0	0.0
Gua Peraling (n = 324)	61.4	5.6	33.0	0.0	0.0
Gue Teluk Kelawar B (n = 12)	33.3	0.0	66.7	0.0	0.0
Gua Teluk Kelawar (n = 61)	86.9	13.1	0.0	0.0	0.0
Lang Rongrien Units 5–6 (n = 14)	50.0	0.0	35.7	0.0	14.3
Gua Cha Layers 3–4 (n = 118)	0.0	0.0	100.0	0.0	0.0
Gua Ngaum (n = 54)	51.9	0.0	48.1	0.0	0.0
Gua Batu Tukang (n = 20)	50.0	0.0	50.0	0.0	0.0
Post-LGM Ceramic					
Moh Khiew Levels 4–5 (n = ±23)	0.0	0.0	0.0	0.0	100.0
Gua Kecil 4–22 inches deep (n = 10)	0.0	0.0	0.0	0.0	100.0
Gua Sagu spits 1–6 (n = 33)	12.1	6.1	72.7	9.1	0.0

NOTE: Data from sources given in table 4.4, except Gua Peraling, for which data on debitage are unavailable. The grindstones and/or pounders from Gua Peraling (Adi, 2000: tables 4.5 and 4.6) are treated here as utilized and stained pebbles.

deer, otter, macaque, langur, flying lemur, porcupine, bamboo rat, and fish (Chaimanee, 1994). This assemblage would not seem to substantiate Pookajorn's (1994a: 343) reconstruction of grassland rather than forest in the site's vicinity. Remains of freshwater shellfish were not recorded, in contrast to their abundance in almost every Holocene rock-shelter deposit, including Moh Khiew (Waiyasadamrong, 1994). Hence the pre–last glacial maximum hunter-gatherers had probably not yet shifted to the broad-spectrum economy, typical of Hoabinhian sites in the peninsula, which included dependable, labor-intensive resources such as shellfish.

Shell and Its Dating Implications

Rising sea levels greatly increase the availability of littoral and estuarine shellfish, as the continental shelf and lower drainages are flooded by shallow expanses of saline water (Oyuela-Caycedo, 1996). The increasing abundance of marine shellfish when the sea rapidly advanced between 14,000 B.P. and 6000 B.P. (table 4.2), combined with sea's closer proximity to rock shelter sites during the mid-Holocene, suggests that most marine shell in the shelters should date broadly to the middle Holocene. Indeed, marine shell was not recorded in the terminal Pleistocene to early Holocene levels at Gua Sagu and Gua Tenggek, but it occurs in those sites' upper spits (Zuraina et al., 1994). At Moh Khiew, the substantial presence of sea and mangrove shell in level 2, which is undated radiometrically (Waiyasadamrong, 1994), suggests that the initial Holocene dates in the overlying level 3 (Pookajorn, 1996: 206) may be fractionally too old. On balance, levels 2 and 3 would appear to be contemporaneous, as supported by their similarities in lithic technology and vertebrate refuse (see Chaimanee, 1994; Pookajorn, 1994a: 343–345), and both of early Holocene age (table 4.4). Marine shellfish also help to date Gua Kerbau, whose cultural deposits plunged 5.6 m beneath the surface (figs. 4.4 and 4.5). Shellfish occurred only in the upper 3 m, along with late Holocene artifacts such as glazed ceramics and earthen sherds, but its presence here confirms other indications of substantial mixing of Hoabinhian and later materials (see van Stein Callenfels and Evans, 1928; Evans, 1928).

The middle Holocene encroachment of the sea beyond the present coastline left a legacy of huge shell middens stranded well inland. Northeast Sumatra used to have numerous middens of cockle shells buried beneath the alluvium. When quarried for lime, the middens were often found to exceed 100 m in diameter and 12 m in height. A charcoal sample salvaged from a level at about one-third of the height of the Sukajadi Pasar III midden clocked in at 7340 ± 360 B.P. (Edwards McKinnon, 1991). The base of the aceramic shell midden at Bukit Kerang (fig. 4.2) is dated to 5970 ± 50 B.P. (Adi, 2000: 157; Comparative Study, 1987). The Guar Kepah "heaps of cockle shells" (fig. 4.2), excavated in 1860 and 1934, originally stood 5 m high. The 1934 season recovered unifacial pebble tools, pounders, grinding stones and slabs, waisted axes, and potsherds from all levels of all three excavated middens (Matthews, 1961: 26–28). The artifacts and sea-level comparisons together suggest a 4000–5000 B.P. dating for Guar Kepah.

Freshwater shell, so abundant in the Holocene deposits in limestone shelters, is a tempting material to submit for radiocarbon dates. Unfortunately, the determinations will be at least 1500 years too old, owing to the radiometrically dead calcium carbonate that the shells ingest (Spriggs, 1989). Adi has been able to obtain parallel sequences of radiocarbon dates on charcoal and freshwater shell from Gua Chawas and Gua Peraling in order to gauge the requisite correction factor, aided by the impressive agreement between increasing age and depth within any of his sequences (table 4.6).

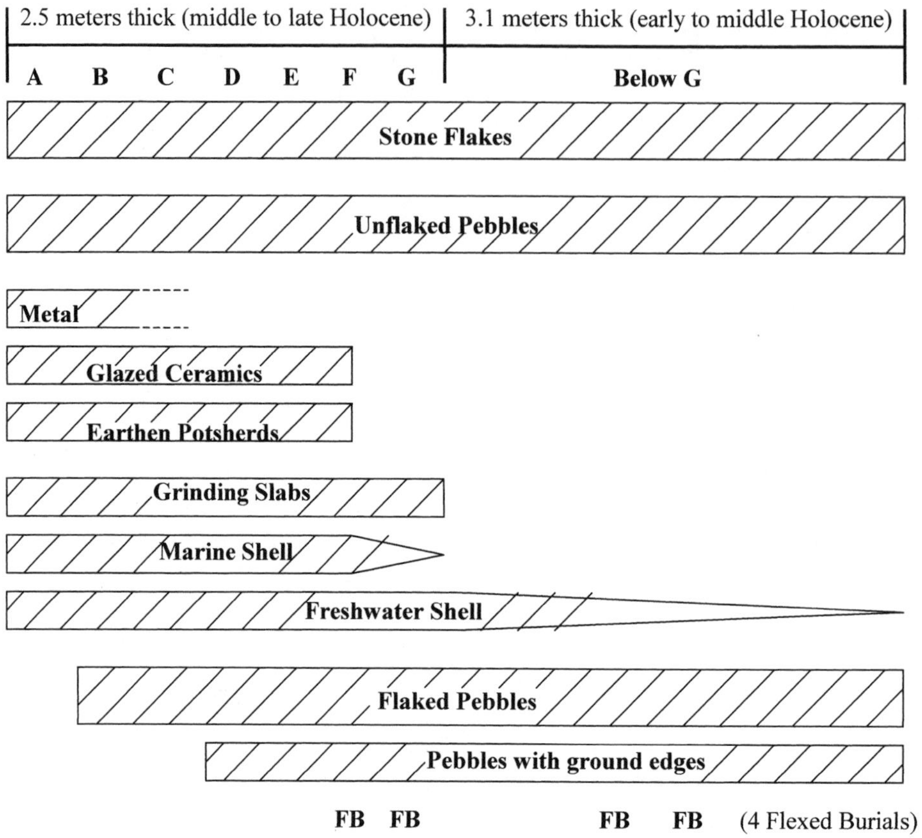

Fig. 4.4. Schematic representation of the finds from layers A to G and below G at Gua Kerbau. Summarized from the accounts in van Stein Callenfels and Evans (1928) and Evans (1928).

The paired shell and charcoal dates consistently differ by 1500 to 5000 years, recommending that Adi's (2000: 111) preferred correction, which is to subtract 2000–3000 years from any date on freshwater shell, be adopted as a conservative measure. Accordingly, radiocarbon determinations on karstic freshwater shell from Malayan sites can be represented as lozenges covering the period 2000–3000 years younger than the laboratory date (fig. 4.6). The potential error can be enlarged to 0–3000 years when the material supplying the determination is undisclosed, but could be freshwater shell, as at Gua Bukit Taat (Shuhaimi et al., 1990).

These error allowances are independent of any standard error computed by the laboratory in estimating the date. Hence, for the sake of consistency, radiocarbon dates on charcoal will be represented as single points corresponding to the median determination. Also, the large potential error

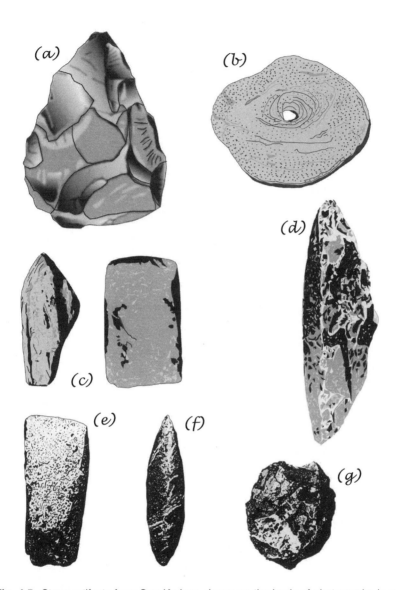

Fig. 4.5. Stone artifacts from Gua Kerbau, drawn on the basis of photographs in van Stein Callenfels and Evans (1928) and Evans (1928): *a,* pointed flake, dorsal view (layer B, late Holocene); *b,* grinding slab (layer D, middle-late Holocene); *c,* "chisel" with polished sides as well as edge polishing (layer E/F, middle-late Holocene); *d,* largest of the unifacially flaked pebbles or "sumatraliths" (layer G, middle-late Holocene); *e,* axe-shaped pebble tool; *f,* "pick" with edge ground to a point on three sides; *g,* edge-modified flake or "scraper," dorsal view; *e–g,* all from beneath layer G (early-middle Holocene). No scale provided in original source.

Table 4.6. Comparisons of Charcoal and Freshwater Shell Radiocarbon Determinations (in Years B.P.) from Gua Chawas and Gua Peraling

Site	Trench/ layer	Spit	Charcoal date	Shell date	Artifactual associations
Gua Chawas	Trench 2	1	Modern		Iron, bone needles
	Trench 2	3	820 ± 50		Glass bead, celadon sherd, Buddhist votive tablets
		4	400 ± 60		
	Trench 2	5	2200 ± 70	3620 ± 80	Pot sherds, polished stone (including adze preform and bracelet)
	Trench 1	7	1840 ± 70		
	Trench 2	10	1770 ± 80		
	Trench 1	26	4390 ± 80	9270 ± 100	Preceramic Hoabinhian
	Trench 1	29	4560 ± 160		
	Trench 2	29		12,210 ± 90	
	Trench 1	40	6100 ± 60		
	Trench 1	48		10,410 ± 90	
	Trench 2	50–51		12,550 ± 100	
	Trench 1	65		10,770 ± 90	

Gua Peraling					
Layer 3	1	Modern		Glass beads, stoneware sherds, sparse pottery, stone adze, Hoabinhian lithics
	5	5720 ± 210		
Layer 5	1	5330 ± 100		Preceramic Hoabinhian (including edge-ground pebble tools)
	7	6250 ± 80		
	14	5850 ± 310	9200 ± 90	
	15		9590 ± 100	
	17		6910 ± 250	
Layer 6	7		9730 ± 90	Preceramic Hoabinhian
Layer 7	3		9750 ± 70	Preceramic Hoabinhian
	12		11,930 ± 100	
	20		11,770 ± 90	

SOURCE: Information extracted from Adi (2000).

NOTE: Adi (2000) characterizes layer 3 (and the archaeologically sterile layer 4) as Neolithic, even though less than 200 potsherds (p. 118) were recovered from an excavated area of around 30 m^2 and Hoabinhian lithics tended to occur at their densest concentrations in the uppermost spits (tables 4.5–4.6). These points, along with the ~5700 B.P. date in layer 3, suggest that most of the cultural contents from layer 4 upwards are of middle Holocene age, even as small amounts of later materials (including one or two human burials) have infiltrated or been buried into these deposits.

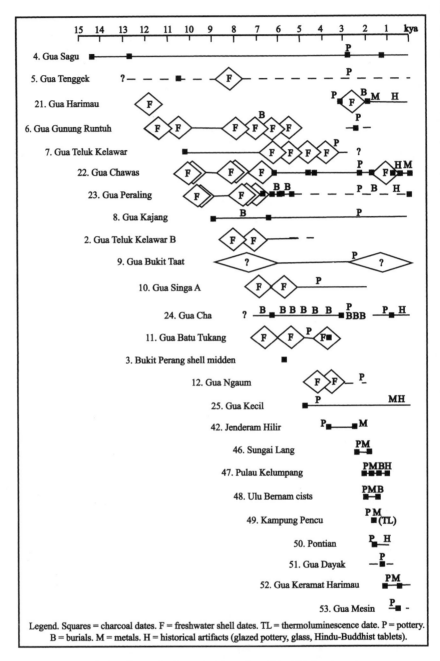

Fig. 4.6. Post-LGM radiometric determinations from West Malaysian sites: *squares* = charcoal dates; F = freshwater shell dates; TL = thermoluminescence date; P = pottery; B = burials; M = metals; H = historical artifacts (glazed pottery, glass, Hindu-Buddhist tablets). *Sources:* Adi, 1981, 2000; Azman, 1998; Bellwood, 1997: 278–286; Bulbeck, 2000; Dunn, 1964; Evans, 1927; Henson, 1989; Kamaludin, 1991; Leong, 1989, 1991a, 2000; Matthews, 1961; Shuhaimi et al., 1990; Zuraina, 1998a; Zolkurnian, 1998.

associated with radiocarbon dates on freshwater shells rules out calibrating them. The peninsula's archaeological chronology will accordingly be discussed in terms of uncalibrated dates. This facilitates the correlation with sea-level changes, which Tjia (1991) reconstructed in terms of uncalibrated radiocarbon dates. Finally, separate dates can be linked into a sequence of occupation if they fall close together or are bridged by archaeological evidence of continuous habitation (fig. 4.6). This is not always the case; for instance, an isolated date on freshwater shell from Gua Harimau at 14,140 ± 795 B.P. (Zuraina, 1998a: 244) is one of its few indications of pre-Neolithic occupation (Zolkurnian, 1998).

Post–Last Glacial Maximum Artifacts and Their Chronological Associations

West Malaysia's archaeologists have been very active in augmenting the inventory of excavated sites and obtaining radiometric dates (table 4.7). The available database suggests a steady increase in the occupation of rock shelters throughout the terminal Pleistocene and early Holocene (fig. 4. 6). At the middle Holocene, some rock shelters were effectively abandoned (Gua Teluk Kelawar B, Gua Batu Tukang) while others registered their first signs of habitation (Gua Ngaum, Gua Kecil). During the late Holocene, the use of rock shelters seems to have remained steady, or even increased. Open sites appear to have increased in number from the middle to the late Holocene, consistent with a scenario of continuous population growth. The circumstances of the use of rock shelters must have changed significantly as farmers increasingly claimed the more fertile and readily cleared patches of land, and hamlets and larger settlements sprung up along the coast and valleys.

The local Hoabinhian focus on flaked and utilized pebble tools is already present at transitional Pleistocene-Holocene sites such as Gua Sagu and Gua Tenggek (tables 4.4 and 4.5). Most variation evidently relates to whether the sites lie to the east or the west of the Malay Peninsula's main range. The assemblages to the east (asterisked in table 4.4) show a lower ratio of unifacially to bifacially flaked pebbles tools (0.5:1.12) than the assemblages in the west (2.0:5.5). The western lowland assemblages link up with their northern Sumatran counterparts, across the Melaka Strait, where bifacial tools are similarly rare and sumatraliths (flaked all over one face: fig. 4.5d) and other unifacially flaked pebbles hold sway (Bellwood 1997: 170). As regards temporal change, Adi (2000: 114) observes that pebble tools with ground edges appeared as a new element at Gua Peraling at about 6000 B.P., even if Zuraina (1990: 83) noted one pebble with a ground, beveled edge at the Pleistocene site of Kota Tampan. Adi's observation suggests that an

Table 4.7. Holocene Prehistoric and Protohistoric Sites of West Malaysia

Preceramic Hoabinhian only
1, Gua Chawan; 2, Gua Teluk Kelawar B; 3, Bukit Kerang shell midden
(3 sites, 2 radiometrically dated; dating range ~8.5−6.0 B.P.)

Hoabinhian (preceramic) to Neolithic (ceramic)
4, Gua Sagu; 5, Gua Tenggek; 6, Gua Gunung Runtuh; 7, Gua Teluk Kelawar; 8, Gua Kajang;
9, Gua Bukit Taat; 10, Gua Singa A; 11, Gua Batu Tukang; 12, Gua Ngaum; 13, Gua Putih; 14,
Kota Tongkat; 15, Kota Balai; 16, Gua Kolam Tujuh; 17, Gua Serendah; 18, Gua Madu; 19, Bukit
Chintamani; 20, Guar Kepah
(17 sites, 9 radiometrically dated; dating range ~14.4−1.2 B.P.)

Hoabinhian (preceramic) to Metal phase (metal, glass and stone beads, glazed pottery, Buddhist tablets)
21, Gua Harimau; 22, Gua Chawas; 23, Gua Peraling; 24, Gua Cha; 25, Gua Kecil; 26, Gua
Kerbau; 27, Gua Baik; 28, Gua Kelawar; 29, Gua Tok Long; 30, Gua Bintong; 31, Gua Berhala;
32, Gua Debu
(12 sites, 4 radiometrically dated; dating range ~12.2−0.1 B.P. [Kitchen Qing ceramics])

Neolithic (ceramic)
33, Gua Singa B; 34, Gua Musang; 35, Gua Jaya; 36, Gua Pasir; 37, Gua Taufan; 38, Gua Gergasi; 39, Kampung Chinggong; 40, Kampung Dusun Raja; 41, Bukit Pulai
(9 sites, none radiometrically dated)

Neolithic (ceramic) to Iron Age (metals, imported pottery)
42, Jenderam Hilir; 43, Bukit Tengku Lembu; 44, Nyong (Tembeling); 45, Kampung Pagi
(4 sites, 1 radiometrically dated; dating range ~3.7−0.4 B.P. [Ming ceramics])

Iron Age
46, Sungai Lang; 47, Pulau Kelumpang (Kuala Selinsing); 48, Ulu Bernam cists; 49, Kampung
Pencu; 50, Pontian; 51, Gua Dayak; 52, Gua Keramat Harimau; 53, Gua Mesin (see fig. 4.6);
54, Sungkai; 55, Klah; 56, Slim; 57, Kerling (all sites with cist graves); 58, Klang; 59, Bukit Kuda
(bronze bells of Kampung Pencu type); 60, Batu Pasir Garam; 61, Kuala Terengganu; 62, Kampung Gaung (all sites with Dong Son drum); 63, Pengkalan Bujang (Hindu-Buddhist architecture);
64, Sungai Mas; 65, Bukit Meriam; 66, Sungai Muda; 67, Bukit Choras; 68, Cerok Tokun (all sites
with 5th−6th-century A.D. inscriptions); 69, Kinta (6th−8th-century A.D. bronze Buddha statue); 70,
Bidor (7th−10th-century A.D. bronze Buddha statue); 71, Bukit Berhala; 72, Gua Kurong Batang
(Buddhist tablets); 73, Jeram Koi; 74, Bukit Komel (includes Ming porcelain)
(29 sites, 8 radiometrically dated, dating range ~2.5−0.4 B.P. [Ming ceramics])

SOURCES: Adi, 1988, 1989, 1991, 1993, 2000; Adi and Jafaar, 1990; Comparative Study, 1987; Bulbeck,
1997a; Callenfels, 1939; Evans, 1931a,b, 1932; Shuhaimi et al., 1988; Shuhaimi et al., 1993; Shuhaimi
et al., 1998; Leong, 1991b; Wisseman Christie, 1990.

early-middle Holocene dating is the oldest age that should be entertained
for any part of the Gua Kerbau sequence, as pebble tools with ground edges
were excavated from the site's deepest levels (figs. 4.4 and 4.5).

Pottery appears at several sites in contexts securely dated between circa
4000 B.P. and 5000 B.P. (fig. 4.6). Examples include Gua Batu Tukang (where

almost all sherds lie beneath the 3600 B.P. determination on charcoal), Gua Kecil, Gua Teluk Kelawar, and Jenderam Hilir. Pottery may have been of marginal utility to hunter-gatherers, explaining its absence from Gua Cha until circa 3000 B.P. and its similarly late appearance at some other sites. Artifacts of polished stone would also appear to have been adopted on a piecemeal basis. Polished Neolithic tools are reported in only 6 of the 18 rock shelters summarized by Matthews (1961), including Gua Kerbau (fig. 4.5c), and none of the 12 shelters discussed by Zuraina (1998a). Gua Cha is unique in its evidence of a workshop for flaking adze pre-forms (Sieveking, 1954: 86). Other "Neolithic" artifacts, such as the mortars from the upper spits at Gua Sagu (table 4.5) and the polished adzes and bracelets from Gua Kecil (Dunn, 1964: 115) and Gua Chawas (table 4.6), are late Holocene. Polishing stone to produce formal artifacts probably represents an intrusive technological tradition unrelated to the edge-ground Hoabinhian pebble tools (Bellwood, 1993, 1997).

Iron fragments are often reported from the upper layers of rock shelters, e.g. Gua Kerbau, Gua Baik (Matthews, 1961), Gua Kecil (Dunn, 1964), Gua Kelawar (Adi and Jafaar, 1990), Gua Chawas (Adi, 2000: 79, 86), Gua Harimau (Zolkurnian, 1998: 247) and Gua Keramat Harimau (Zuraina, 1998a: 248). The extreme susceptibility of iron to corrosion would explain why iron goes unreported in many rock shelters with signs of late occupation (fig. 4.6), notably Gua Cha and Gua Peraling, which contain sherds of glazed ceramics, rice remains, and other campsite debris dating to the last millennium (Sieveking, 1954; Adi, 1981; Adi, 2000: 62, 117). Ethnographically, iron is ubiquitous among the Semang, yet they lack pottery or tools of polished or flaked stone. In Rambo's (1979) summary of the use of stone among the Semang, three of the seven functions specifically involve iron (anvils and hammer stones for working iron, whetstones for sharpening iron knives and arrowheads, and quartzite strike-a-lights, struck with steel hammers). The minimal Hoabinhian-Neolithic overlap observed by Bellwood (1997: 165–168) in several West Malaysian sites would, I suspect, largely stem from iron's tendency to replace flaked and polished stone tools as soon as iron became available.

Burial Chronology and Body-Size Reduction in the Malay Peninsula

Human interments, by definition, are placed in deposits older than the moment of burial. Where the top of the grave cutting has not been documented, the burial's maximum possible age, as represented by material at the same level, becomes the relevant stratigraphic anchor. Hence the freshwater shell of 9460 B.P. from the same level as the flexed inhumation at Gua Gunung Runtuh (Zuraina, 1994: 30) supplies the maximum age of the specimen. Moh Khiew burials 1 and 2 (Pookajorn, 1994a), excavated in level 2 and level 3, respectively, also appear to be primary flexed inhumations; loss of the leg bones from burial 1 does not make it a secondary burial (pace Pookajorn, 1994a: 339). While these three cases of flexed burials are possibly early Holocene, the great majority found in the Malay Peninsula would be middle Holocene. These include the flexed Hoabinhian burials at Gua Cha (fig. 4.6), the four burials from Gua Kerbau (fig. 4.4), and the possibly flexed burial from Gua Peraling (Adi, 2000: 222).

So-called secondary burials, some of which would be redeposited primary burials, are evidently at least as old as the flexed inhumations at Gua Cha (Bulbeck, n.d.) and Gua Peraling (Adi, 2000: 113, 222) , i.e., broadly middle Holocene. The secondary burials in the middle levels of Gua Baik, overlying a flexed inhumation at the base (Snell, 1949), would probably also date to the middle Holocene. A similar age is likely for the large number of secondary burials excavated in the Guar Kepah middens (see Matthews, 1961: 27). The latest prehistoric mortuary practice, present in rock shelters as well as open sites, involved supine inhumations, with skeletons lying on their back. The Neolithic examples at Gua Cha date to circa 3000 B.P. (Bulbeck, 2000), and the Iron Age cases at Gua Harimau (Zolkurnian, 1998) and Pulau Kelumpang (Bulbeck, 1998) would date to between 2500 B.P. and 1000 B.P. A late Holocene antiquity would also apply to the single, extended burials in the pottery-bearing layers near the top of Gua Peraling (Adi, 2000: 221) and Gua Baik (Snell, 1949), so completing the chronological scheme laid out in table 4.8.

The useful record of human burials in the Malay Peninsula enables a diachronic perspective on body size. My analysis will employ limb-bone lengths (as defined by Martin, 1957) rather than stature for two reasons. Many skeletons are represented by a single limb bone, so estimating stature via regression formulae introduces potential error without improving on the original data. Second, the Gua Gunung Runtuh individual had scoliosis of the vertebral column, a pathological condition which significantly shortened the length of his backbone (Jamaludin, 1994).

A decrease in the lengths of limb bones clearly occurred in the hinterland during the Holocene. Not one early Holocene value falls within the recorded range of the Semang and Senoi (table 4.8). When limb-bone lengths are expressed as a percentage of the Orang Asli average, all Orang Asli limb bones fall in the 87–107% range, whereas the early Holocene limb bones range between 111% and 118% (fig. 4.7). Assuming no change in body proportions, a reduction in body size of approximately 15% would have occurred in the peninsula's hinterland since the early Holocene. This diminution in body size would appear to have been gradual: the middle Holocene values overlap with the upper segment of the Orang Asli range, while the hinterland Neolithic and early Metal phase values range between the Orang Asli and early Holocene midpoints.

The remains from Guar Kepah and Pulau Kelumpang, however, suggest stasis in body size among coastal populations (fig. 4.7). Ethnographically, the predominantly coastal Malays are substantially taller than the Semang and Senoi: e.g., average male stature between 162 and 163 cm (Martin, 1957: 768), compared to 153–154 cm for male Semang and 155–158 cm for male Senoi (Bulbeck, 1996). Though the sample size of coastal archaeological specimens is very small, the available data suggest that the current coastal-hinterland difference in body size was established millennia ago. Body-size reduction among the Holocene inhabitants of Malaya's hinterland rain forests may be a biological response to life beneath the canopy and/or to dietary restrictions. Small size affords ease of movement through the jungle and potential thermoregulatory benefits, while also allowing a lean diet to be eked out (Bulbeck, 1996, 1999a; Shea and Bailey, 1996). Whether the genetically mediated, multigenerational adaptation or process of exaptation has yet led to genetically scaled-down growth potential among the Semang is unclear; a study which attempts to prove their low concentrations of insulin-like growth factor type 1 (Ishida et al., 1998) is best described as inconclusive.

Nonetheless, the evidence is commensurate with a scenario in which coastally focused hunter-gatherers, who had conceivably accessed the hinterland on a sporadic basis, have a place in the ancestry of the Holocene rain forest denizens. The hinterland Hoabinhian men (approximate stature of 162–175 cm) and women (155–160 cm) would appear to have been full-sized people (Bulbeck, 1996, 1997b). Taller than any ethnographically recorded rain forest hunter-gatherers (see Martin, 1957: 778–780; Bulbeck, 1999a), they probably survived in a way reminiscent of dwellers of a more open landscape. They could have maintained a network of well-cleared tracks through the rain forest, linking up favorable resource patches. These patches would have included encounter locations with game, similar to the bearded pig migration routes recorded in southern West Malaysia (Hislop, 1954), and foci for sessile foodstuffs, such as freshwater shellfish beds. The smaller, more agile band members would still have benefited from being less bound to the tracks, allowing access to the full range of rain forest resources, especially arboreal game (cf. Endicott, 1979: 23). Exaptation of the resultant advantages may help to explain why body size in the hinterland reduced over time.

Table 4.8. Malayan Archaeological and Orang Asli Limb Bone Lengths (mm)

Males	Femur length (mm)		Tibia length (mm)	Humerus length (mm)	Radius length (mm)	Ulna length (mm)
	Maximum (M1)	Physiological (M2)	Maximum (M1a)	Maximum (M1)	Maximum (M1)	Maximum (M1)
Early Holocene: Gua Gunung Runtuh [a]	451.5 RL	448.5 RL 467.5 R	395 R	324 RL	259 RL	275 RL
Middle Holocene: Gua Kerbau [b]						
Middle Holocene: Gua Cha As.33.5.4 [c]	±488 R (449–528)					
Middle Holocene: Gua Cha 1 [c]	±425 RL (417– 424 R, 428–432 L)		±355 L (approximate estimate)	±309 RL (304– 307 R, R 313 L)	248 R	266.5
Middle Holocene: Guar Kepah C82 [c]				±340 R (330–350)		
Middle Holocene: Guar Kepah B289 [c]	±420 L (396–444)					
Neolithic: Gua Baik [d]					217 R	
Iron Age: Gua Harimau [c]	±458 L (455–460)			±313 L		
Iron Age: Pulau Kelumpang 4 [e]			±381R (366–397)	317.3 RL		
Iron Age: Pulau Kelumpang 5 [e]	±452 R (449–455)			±315 R		
Menriq (Semang) [f]	411.5 RL	408 RL	345 RL	302 L	237 L	252 L
Jehai (Semang) [f]	432 L	426 L	369 L	306.5 RL		
Kensiu (Semang) [f]	425.5 RL			302.5 RL	231 RL	249.5 RL
Semang [f]					239 L	253 L
Senoi [f]	368 R	365 R	299 R	249.5 RL	202 RL	222 L
Batang Labu Senoi [f]	389 RL	386 RL	323 RL	277.5 RL	217.5 RL	234 RL
Orang Asli average	405.2	396.3	334.0	287.6	225.3	242.1

	Femur length (mm)	Tibia length (mm)	Humerus length (mm)	Radius length (mm)	Ulna length (mm)
Females	Maximum (M1)	Total (M1)	Maximum (M1)	Maximum (M1)	Maximum (M1)
Early Holocene: Moh Khiew 1 [g]			±308 R	±247 R	±271 R
Early Holocene: Moh Khiew 2 [g]			±293 L (289–297)		
Middle Holocene: Gua Peraling [h]			±288 L		
Neolithic: Gua Cha As.33.5.8 [i]	384 R	327 RL	266.5 RL		248.5 RL
Neolithic: Gua Cha As.33.5.3 [j]	373.4 RL	305 R	265 L	211 R	
Jehai No. 7 (Semang) [f]	389 RL		267 RL		
Jehai No. 4 (Semang) [f]	395 RL	325.5 RL	268.5 RL	219.5 RL	234.5 RL
Semang No. 3 [f]	389 RL	312 RL	269 RL		
Semang No. 1 [f]	383 R	295 RL	269 RL	197.5 RL	211.5 RL
Mai Darat (Senoi) [f]			270.5 RL	214 RL	232 RL
Batang Labu Senoi [f]	393.5 RL	309.5 RL			
Orang Asli average	387.2	309.4	268.2	210.8	226.9

NOTE: Where both the left (L) and right (R) members of the same limb bone are present on an individual, the average (signified RL) is taken.
[a] Measurements of complete limb bones (Matsumara and Zuraina, 1999).
[b] Complete femur (Duckworth, 1934).
[c] Previously unreported data, including estimates from measurements on the extant fragments, using the regression formulae in Steele (1970) and Steele and McKern (1969). These sources give different regression formulae for Mesoamericans, American Blacks and American Whites, and it is not always clear which formulae are most suitable, so I note the range of estimates and treat the average as the best estimate.
[d] Complete radius (Snell, 1949).
[e] Includes measurements estimated from extant fragments (Bulbeck, 1998).
[f] Martin (1905); Schebesta and Lebzelter (1926).
[g] Estimated from the scaled drawings in Pookajorn (1994a: figs. 12 and 15) as no metrical data appear to have been published.
[h] Estimated from measurements on the extant fragment (Bulbeck, 1997b). I maintain the view that the specimen is more probably female than male, notwithstanding reports to the contrary in Adi (2000: 223).
[i] Previously unreported data, employing the Amerindian regression formula.
[j] Measurements of complete limb bones (Trevor and Brothwell, 1962).

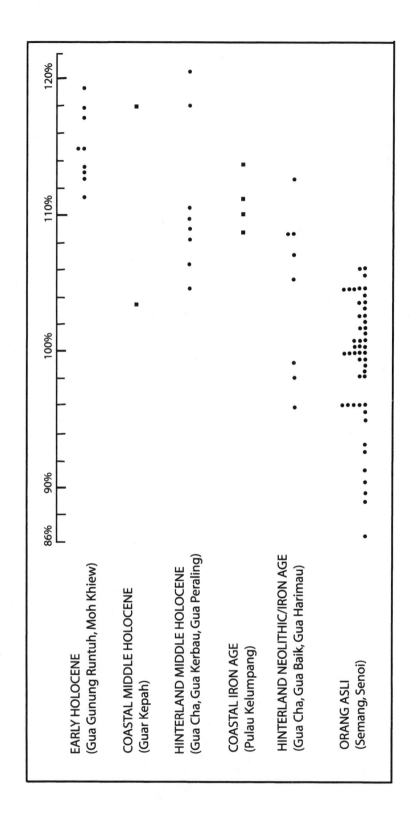

Holocene Landscape Archaeology
and Exploited Resources

Archaeological evidence indicates a rapid expansion by hunter-gatherers across the Malay Peninsula at the Pleistocene-Holocene junction. The first sites inhabited after the last glacial maximum—Lang Kamnan; Gua Sagu; and Gua Tenggek, where about 30% of the artifacts lie beneath the approximately 10,500 B.P. level (Zuraina et al., 1998)—lie at 110 m or less above sea level. The next sites to be occupied lie at slightly higher elevations: Gua Gunung Runtuh (150 m above sea level), Gua Chawas (124 m above sea level), Gua Peraling (195 m above sea level), and the Harimau and Teluk Kelawar shelters along the western foothills of the Lenggong (Zuraina, 1998a). Sakai Cave, lying at about 200 m above sea level, slightly to the south of Moh Khiew, registers its initial occupation at circa 9300 B.P. (Pookajorn, 1994a,b). At some point after 9000 B.P., occupation of the really high land is evidenced at Gua Bukit Taat, which lies at about 1500 m above sea level in an isolated mountain block. The site's location well to the east of the ethnographically recorded Orang Asli distribution (fig. 4.2) indeed hints at more extensive hunter-gatherer occupation during much of the Holocene, more extensive than in recent times.

Population pressure on the coast and adjacent lowlands may well have spurred the evident establishment of residential populations in the deep hinterland. The post–last glacial maximum rise in sea levels incurred a rapid loss of land (fig. 4.3), notwithstanding the bonanza of marine shellfish for those Hoabinhians who stayed near the coast. Several subcoastal rock shelters were intensively occupied during the early Holocene. Levels 2 and 3 at Moh Khiew together yielded 12,441 flaked-stone artifacts (Pookajorn, 1996: 204–205). All the published Holocene charcoal dates from Lang Rongrien fall between 9700 B.P. and 7500 B.P. (Anderson, 1990), comparable to the series of radiocarbon dates from 8000 B.P. to 6000 B.P. at Lang Kamnan (table 4.3). Similar evidence of intensive occupation during the middle Holocene can then be found at more interior sites, such as Gua Kerbau (fig. 4.4), Gua Cha (Adi, 1981), and Gua Peraling (table 4.6). These data suggest a process of increasing forays into the interior as population pressure mounted, followed by the establishment of hinterland communities which, in turn, explored and penetrated ever more remote habitats.

Prey species ranged from the largest to the minute. Bones of elephant, bear, rhinoceros, and deer occur in the Tamiang shell midden in Sumatra

Fig. 4.7. Limb bone lengths expressed as a percentage of the average Orang Asli limb bone length

(Edwards McKinnon, 1991: 132). Every assemblage with more than 21 mammalian identifications in table 4.9 (which arranges these assemblages in approximate chronological order) includes at least one perissodactyl or large carnivore. Yet those assemblages all include rodents, accounting for up to 31% of identifications. Tortoises occur with all of the assemblages in table 4.9 except Gua Ngaum and Teluk Kelawar B. Lizard remains are reported from Gua Sagu, Gua Peraling, Moh Khiew (levels 2–3), and all four Lenggong Valley sites. Fish, undoubtedly underrepresented owing to the small size of their bones, have been identified in Gua Peraling, Gua Ngaum, and Moh Khiew (levels 2–3). Freshwater shell is consistently abundant.

At around 4000 B.P. to 3000 B.P., however, a shift evidently occurred from large terrestrial game to arboreal game (table 4.9). Hunting in coordinated groups and ambushing the relatively dangerous ungulates apparently gave way to very small hunting parties less likely to disturb the ever-watchful arboreal prey. (The two Gua Sagu assemblages will be excluded from the following generalizations, owing to their small sample size, but they follow the same trend.) The early and middle Holocene assemblages tend to be dominated by artiodactyls, whose identifications account for 31% to 89% of the total. Monkeys and the occasional gibbon are always present but at a lower frequency than the artiodactyls. This focus on artiodactyls was maintained despite the peak in sea level (and thus rainfall?) at about 6000 B.P., as Gua Cha, Gua Peraling, and Gua Ngaum demonstrate. The pattern then reverses in Moh Khiew (levels 4–5) and Gua Harimau. Artiodactyls drop to 10% to 15%, while monkeys and gibbons account for over 50% of the identifications, and bats and flying lemurs appear at Moh Khiew.

The Batek Semang have evidently lopped off both extremes from the broad-spectrum Hoabinhian economy. Karen Endicott's (1979) account of their subsistence strategies does not mention pigs or other large game. Instead, blowpipe hunting, in pursuit of monkeys as the major quarry, dominates time spent hunting. Kirk Endicott (1984: 35) notes that the Batek rarely pursue pigs as the chase is considered too bothersome and dangerous. The late Holocene trend toward arboreal game evidently intensified among the direct ancestors of the Semang. At the other extreme, contemporary Orang Asli rarely harvest shellfish (Adi, 2000: 86), just as Karen Endicott (1979) makes no reference to shellfish collection.

The late Holocene trends can be explained by the dual processes of progressive loss of land and increased opportunities for trade, as farmers began to make serious inroads into the hinterland. The farmers presumably targeted the more open terrain, where artiodactyl numbers would have been higher. Moreover, the blowpipe, the ideal weapon in the forest undergrowth, was probably adopted by the Semang only during the Iron Age or slightly earlier (Bellwood, 1997: 150–152). Karen Endicott (1979) stresses

Table 4.9. Proportions of Mammals (NISPs, MNIs, or Occurrences) in Archaeological Sites

Assemblage	Artiodactyls	Perissodactyls	Bear, cats, Cuon dog	Small carnivores	Gibbons, monkeys	Flying lemurs	Bats	Rodents
Early Holocene								
Gua Sagu, spits 5–13	8 (80%)	0	1 (10%)	0	0	0	0	1 (10%)
Moh Khiew, levels 2–3	15 (31%)	0	3 (6%)	5 (10%)	11 (22%)	0	0	15 (31%)
Gua Gunung Runtuh	41 (51%)	1 (1%)	4 (5%)	1 (1%)	29 (36%)	0	1 (1%)	3 (4%)
Lang Rongrien	160 (62%)	2 (1%)	0	0	50 (19%)	0	0	47 (18%)
Gua Teluk Kelawar B	10 (56%)	0	0	0	8 (44%)	0	0	0
Gua Peraling, layers 5–7	105 (54%)	15 (8%)	1 (1%)	7 (4%)	46 (24%)	0	0	21 (11%)
Middle Holocene								
Gua Cha	22 (48%)	0	1 (2%)	3 (7%)	10 (22%)	0	2 (4%)	8 (17%)
Gua Peraling, layers 3–4	47 (89%)	1 (2%)	0	0	4 (8%)	0	0	1 (2%)
Gua Ngaum	11 (52%)	0	0	0	5 (24%)	0	0	5 (24%)
Late Holocene								
Moh Khiew, levels 4–5	6 (15%)	1 (3%)	1 (3%)	1 (3%)	21 (54%)	1 (3%)	2 (5%)	6 (15%)
Gua Harimau	2 (11%)	0	1 (6%)	0	12 (67%)	0	0	3 (17%)
Gua Sagu spits	5 (50%)	0	0	0	2 (20%)	0	0	3 (30%)

SOURCE: Data from Zuraina et al. (1998) for Gua Sagu, Chaimanee (1994) for Moh Khiew, Davison (1994) for Gua Gunung Runtuh, Kinigam (1990) for Lang Rongrien, Zolkurnian (1998) for Gua Teluk Kelawar B, Gua Ngaum, and Gua Harimau, Adi (1981) for Gua Cha preceramic, and Adi (2000: table 3.15) for Gua Peraling.

NOTE: For a discussion of the probable middle Holocene age of the faunal remains from layers 3–4 at Gua Peraling, see the note to table 4.6. Artiodactyls include boars, deer, the seladang bovid, and the Capricornis sumatraensis goat, in decreasing order of frequency. Perissodactyls include rhinoceros and tapir.

the importance, among the Batek, of fishing , often with hooks and nets obtained through trade, which would have obviated the need to collect shellfish. Land loss and trade growth, along with the threat of slaving, emerge as dominant concerns for Malaya's late Holocene forest foragers, as the next section of this chapter explores.

Plants must have always been critical to Malaya's rain forest inhabitants, not only as a raw material for tools, utensils, and poisons, but also as foodstuffs (Bowdery, 1999). According to Karen Endicott (1979), the Batek may subsist mainly on honey and fruits for nearly half the year. Outside of the fruit and honey seasons, easily grubbed tubers supply well over half of the calories obtained through foraging. The only known archaeological echo of these foodstuffs is the unusually high incidence of caries and abscesses, by hunter-gatherer standards, in both the Hoabinhian and the Neolithic teeth at Gua Cha (Bulbeck, 1999b). Fruit and honey, being widely dispersed forest resources (Endicott, 1983, 1984), probably became increasingly important during the Holocene, while tubers would have been a dominant staple as communities settled down to their increasingly bounded patches of forest.

The Semang: Traders Who Forage or Foragers Who Trade?

Gua Cha epitomizes the leap in material prosperity that began with the Neolithic. Jewelry of marine shell, presumably traded in from the east coast, was one among numerous wares (fine pottery, polished adzes, bark cloth beaters) interred in the Neolithic burials. Conceivably, all these goods were obtained by trade, and probably in exchange for forest produce. Rattan, fine timber such as gharu, resins, honey, thatch, *petai* pods, and hornbill beaks are the main exports of the Batek today (Endicott, 1979). These items would have tapped a ready market as soon as sedentary communities arose, focused on their crops and local forest resources. Settled communities would likewise have traded rice and other tasty domesticates, as well as pottery and other wares, to their neighbors in the jungle. The almost universal occurrence of potsherds in rock shelters across Malaya testifies to the complementary relationship between foragers and farmers during the late Holocene. The widespread occurrence of bark cloth beaters across central West Malaysia (fig. 4.2) may also reflect the spread of material wealth (bark cloth) as much as the expansion of farming groups.

West Malaysia is renowned for its circa 2000-year-old bronzes. They include axes, whose distribution follows the bark cloth beaters (fig. 4.2); six Dong Son drums, usually found in coastal locations (sites 46 and 59–62); and clapperless bells from the lowlands (see fig. 4.2 and table 4.7 for sites). West Malaysia was also home to a thriving and possibly unique industry of shafted iron tools, often found in association with the early bronzes and slabbed cist graves (Sieveking, 1956b). The Sanskrit inscriptions located between site 67 and site 68 (fig. 4.2), dated to the fifth and sixth centuries A.D., are among the oldest in Southeast Asia (Wisseman Christie, 1990). They are associated with the Hindu-Buddhist Pengkalan Bujang state (site 63), based near the northwestern border of the ethnographic Semang distribution. Two Buddha statues from the first millennium A.D. have been recovered a short distance to the south (sites 69 and 70). Pulau Kelumpang (site 47) appears to have been a major center for manufacturing jewelry of glass and semiprecious stone during the first millennium A.D.

These developments in luxury goods and regalia, and early Buddhist civilization, must have been associated with the peninsula's abundance of tin and other natural resources and its key location on the international trade routes through Southeast Asia (Allen, 1991). They accordingly complement the persistence of mobile foragers across much of the interior until modern times, as centers of industry and high culture emerged along the west coast without the need for an agrarian base (Allen, 1991). Items useful for survival in the forests, notably iron, moved inland in exchange for valuable forest produce coming to the coast. One attendant ill of civilization, the demand for slaves, imperiled social reproduction among any Orang Asli whose relations with outsiders were not adequately guarded (Endicott, 1983). The retreat of the highland Senoi to remote locations greatly reduced the likelihood of slave raids, while the Semang tendency to forage in nuclear family units facilitated their flight into the jungle at the first scent of danger (Rambo, 1984; Benjamin, 1985, 1987).

The late Holocene shift from hunting along maintained tracks to dispersed foraging in the forest, as hypothesized in this chapter, would have served as a useful pre-adaptation against the risk of enslavement. Lack of trails would have minimized the scope for slave raiders to ambush or pursue the Semang, while the blowpipe–arboreal game complex would have suited small, virtually undetectable hunting parties. Rain forest hunter-gatherers who could access iron and rice without losing family members to human predation would have enjoyed substantial reproductive benefits compared to those who could not. Thus, the rapid expansion of a North Aslian–speaking Semang complex, across the lowlands that lie at a comfortable remove from the western coast (fig. 4.2), emerges as a distinctly feasible scenario.

Conclusions

Late Pleistocene sites in the Malay Peninsula include several whose initial occupation possibly antedate the effective limit of radiocarbon dating (40,000 years ago). However, no cultural materials have yet been dated to the height of the last glacial maximum, when the sea underwent its farthest retreat from the present-day landmass. Late Pleistocene hunter-gatherers were apparently concentrated on the lowlands up to approximately 150 m above the prevailing sea level. During the early and middle Holocene, the littoral zone and adjacent hinterland probably stayed as the prime habitat, but by this time hunter-gatherers were dispersing across the peninsula and occupying every available niche.

Solid empirical results to emerge from this study include the focus of the local Hoabinhian on utilized and flaked pebble tools, an east-west distinction in the ratio of bifacial to unifacial examples (tables 4.4 and 4.5), reduced body size in the interior during the course of the Holocene (table 4.8), and the shift from artiodactyls to arboreal game after 4000 b.p. (table 4.9). To explain the last two observations, this chapter hypothesizes that the pattern of long-distance mobility through the rain forest along well-cleared trails, a pattern which prevailed until the middle Holocene, shifted to foraging in the undergrowth within more restricted territories during the late Holocene.

This hypothesis further portrays the Semang as successful, expansive foragers adapting to new social circumstances increasingly dominated by non-forager networks. The late Holocene spread of farming and, subsequently, organized long-distance trade would have curtailed the territories of those communities whose members continued to forage and would have threatened them with slave raiding. On the other hand, these developments would also have augmented the value of forest produce and introduced novel goods and foodstuffs to the peninsula. There is every reason to suspect that the major features of the Semang tradition—foraging in nuclear family units; the focus on yams, fruit, and honey; and the collection of forest produce to trade for iron and rice—originated when farming communities spread across the peninsula and intensified with the rise of early states during protohistoric and historical times. This conclusion does not in the least deny Endicott and Bellwood's (1991) view of the Semang as having continuously occupied, for more than 10 millennia, an equatorial rain forest uncommonly favorable to hunter-gatherers. Rather, the Malay Peninsula's wide portfolio of rain forest resources provided the ecological basis for its hunter-gatherers to successfully respond to the dramatic changes of the last millennia.

Acknowledgments

The fieldwork to inspect the West Malaysian human remains was funded by a small Australian Research Council grant. My thanks to Adi Haji Taha (Kuala Lumpur), Zuraina Majid (Penang), Hamid Moho Isa (Taiping), and John de Vos (Leiden), as well as Robert Foley and Margaret Bellatti (Cambridge), for their hospitality and

assistance. The arguments presented in this chapter have benefited from dis-
cussions with Adi Haji Taha (who also supplied the altitudinal data from his sites),
Peter Bellwood, Geoffrey Benjamin, Mohammad Mokhtar Saidin, Mohammad
Mahfuz Nordin, Zulkifli Jaafar, and Zuraina Majid. Keith Mitchell (Cartography,
Australian National University) drafted figure 4.1. Thanks also to Julio Mercader
and an anonymous reviewer for their constructive and informative comments.

References

Adams, J. (1998): Global atlas of palaeovegetation since the last glacial maximum.
Http://www.esd.ornl.gov/ern/qen/adams1.html.

Adi Haji Taha (1981): The re-excavation of the rockshelter of Gua Cha, Ulu Kelantan, West Malaysia. M.A. thesis, Australian National University.

———. (1988): Penyelidikan arkeologi di Kota Balai, Jerantut, Pahang-laporan
awal. *Jurnal Arkeologi Malaysia* 1:1–11.

———. (1989): Archaeological, prehistoric, protohistoric, and historic study of
the Tembeling Valley, Pahang West Malaysia. *Jurnal Arkeologi Malaysia* 2:47–69.

———. (1991): Archaeological discoveries in Peninsular Malaysia (1987–1990).
Journal of the Malaysian Branch of the Royal Asiatic Society 64:75–96.

———. (1993): Archaeological discoveries in Peninsular Malaysia (1991–1993).
Journal of the Malaysian Branch of the Royal Asiatic Society 66:67–83.

———. (2000). Archaeological excavations in Ulu Kelantan, Peninsular Malaysia. Ph.D. diss., School of Archaeology and Anthropology, Australian National
University.

Adi Haji Taha and Zulkifli Jafaar (1990): A preliminary report on archaeological
research and excavation at Gua Kelawar, Sungai Siput, Perak. *Jurnal Arkeologi
Malaysia* 3:111–124.

Allen, J. (1991): Trade and site distribution in early historic-period Kedah: Geoarchaeological, historic, and locational evidence. *Bulletin of the Indo-Pacific Prehistory Association* 10:307–319.

Ambrose, S. H. (1996): Late Pleistocene human population bottlenecks, volcanic
winter, and differentiation of modern humans. *Journal of Human Evolution* 34:
623–651.

Anderson, D. D. (1990): *Lang Rongrien rockshelter: A Pleistocene–early Holocene site
from Krabi, southwestern Thailand.* Philadelphia: University of Pennsylvania.

Azman Mohd Noh (1998): Ekskavasi Gua Singa dan Gua Putih di Gerik, Perak.
Malaysia Museums Journal 34:221–239.

Bailey, R. C., G. Head, M. Jenike, B. Owen, R. Rechtman, and E. Zechenter (1989):
Hunting and gathering in tropical rain forest: Is it possible? *American Anthropologist* 91:59–82.

Bellwood, P. (1993): Cultural and biological differentiation in Peninsular Malaysia: The last 10,000 years. *Asian Perspectives* 32:37–60.

———. (1997): *Prehistory of the Indo-Malaysian Archipelago.* Rev. ed. Honolulu: University of Hawai'i Press.

Benjamin, G. (1985): In the long term: Three themes in Malayan cultural ecology.

In *Cultural values and human ecology in Southeast Asia,* edited by K. A. Hutterer, A. T. Rambo, and G. Lovelace, 219–271. Ann Arbor: University of Michigan Papers on South and Southeast Asia.

———. (1987): Ethnohistorical perspectives on Kelantan's prehistory. In *Kelantan zaman awal: Kajian arkeologi dan sejarah di Malaysia,* edited by Nik Hassan Shuhaimi bin Nik Abdul Rahman, 108–153. Kota Bharu: Perbadanan Muzium Negeri Kelantan Istana Jahar.

———. (1996): Issues in the ethnohistory of Pahang. In *Seminar Arkeologi Pahang: Kearah pembangunan arkeo-pelancungan,* edited by Nik Hassan Shuhaimi bin Nik Abdul Rahman and Ahmad Hakimi bin Khairuddin, 82–121. Kota Bahru: Perbadanan Muzium Negeri Kelantan.

Bowdery, D. (1999): Phytoliths from tropical sediments: Reports from Southeast Asia and Papua New Guinea. *Bulletin of the Indo-Pacific Prehistory Association* 18: 159–168.

Bulbeck D. (n.d.): The Gua Cha burials—concordance, chronology, demography. Report submitted to the Duckworth Laboratory, Cambridge University, England.

———. (1996): Holocene biological evolution of the Malay Peninsula Aborigines *(Orang Asli). Perspectives in Human Biology* 2:37–61.

———. (1997a): Description and preliminary analysis of the human remains from Gua Tok Long, Pahang, Malaysia. Report to the Department of Museums and Antiquity, Kuala Lumpur, Malaysia.

———. (1997b): Description and preliminary analysis of the human remains from Gua Peraling, Kelantan, Malaysia. Report to the Department of Museums and Antiquity, Kuala Lumpur, Malaysia.

———. (1998): Description and preliminary analysis of the human remains from Pulau Kelumpang (Kuala Selinsing), Perak, Malaysia. Report to the Department of Museums and Antiquity, Kuala Lumpur, Malaysia.

———. (1999a): Current biological anthropological research on Southeast Asia's Negritos. *SPAFA Journal* 9(2):14–22.

———. (1999b): The Hoabinhian to Neolithic foraging transition at Gua Cha, West Malaysia. Paper presented at the 13th Annual Conference of the Australasian Society for Human Biology, Sydney, Australia, 6–9 December.

———. (2000): Dental morphology at Gua Cha, West Malaysia, and the implications for "Sundadonty." *Bulletin of the Indo-Pacific Prehistory Association* 19:17–40.

Bulbeck, D., and A. R. Kadir (1999): Brief summary (preliminary report) of research findings: Study into Malay Peninsula Aborigines' dental differentiation; and the Orang Asli project, 1999. Report to the Prime Minister's Department, Malaysia.

Chaimanee, Y. (1994): Mammalian fauna from archaeological excavations at Moh Khiew cave, Krabi Province, and Sakai cave, Trang Province, Southern Thailand. In *Final report of excavations at Moh Khiew Cave, Krabi Province; Sakai Cave, Trang Province; and ethnoarchaeological research of hunter-gatherer group, socall Mani or Sakai or Orang Asli at Trang Province,* edited by Surin Pookajorn, A. Waiyasadamrong, S. Sinsakul, M. Wattanasak, and Y. Chaimanee, vol. 2, *The Hoabinhian Research Project in Thailand*, 405–418. Bangkok: Faculty of Archaeology, Silka-porn University.

Chappell, J. A., A. Omura, T. Esat, M. McCulloch, J. Pandolfi, Y. Ota, and B. Pillans (1996): Reconciliation of late Quaternary sea levels derived from coral terraces at Huon Peninsula with deep-sea oxygen isotope records. *Earth and Planetary Science Letters* 141:227–237.

Comparative study. (1987): In *Final Report: Seminar in Prehistory of Southeast Asia (T-WII)*, 19–22. Project in Archaeology and Fine Arts. Bangkok: Southeast Asian Ministers of Education and Organization.

Davison, G.W.H. (1994): Some remarks on vertebrate remains from the excavations at Gua Gunung Runtuh, Perak. In *The excavation of Gua Gunung Runtuh and the discovery of Perak Man in Malaysia*, edited by Zuraina Majid, 141–148. Kuala Lumpur: Department of Museums and Antiquity Malaysia.

Duckworth, W.L.H. (1934): Human remains from rock-shelters and caves in Perak, Pahang, and Perlis and from Selinsing. *Journal of the Malayan Branch of the Royal Asiatic Society* 37:149–167.

Dunn, F. L. (1964): Excavations at Gua Kechil, Pahang. *Journal of the Malaysian Branch of the Royal Asiatic Society* 37:87–124.

Edwards McKinnon, E. (1991): The Hoabinhian in the Wampu/Lau Biang Valley of northeastern Sumatra: An update. *Bulletin of the Indo-Pacific Prehistory Association* 10:132–142.

Endicott, Karen (1979): Batek Negrito sex roles. M.A. thesis, Australian National University.

Endicott, Kirk (1983): The effects of slave raiding on the Aborigines of the Malay Peninsula. In *Slavery, bondage, and dependency in Southeast Asia*, edited by A. Reid and J. Brewster, 216–241. Brisbane: University of Queensland Press.

———. (1984): The economy of the Batek of Malaysia: Annual and historical perspectives. *Research in Economic Anthropology* 6:29–52.

Endicott, Kirk, and P. Bellwood (1991): The possibility of independent foraging in the rain forest of Peninsular Malaysia. *Human Ecology* 19:151–185.

Evans, I.H.N. (1927): Notes on the remains of an old boat from Pontian, Pahang. *Journal of the Federated Malay States Museums* 12:93–96.

———. (1928): Further excavations at Gunong Pondok. *Journal of the Federated Malay States Museums* 12:161–162.

———. (1931a): A search for antiquities in Kedah and Perlis. *Journal of the Federated Malay States Museums* 15:43–50.

———. (1931b): Excavations at Nyong, Tembeling River, Pahang. *Journal of the Federated Malay States Museums* 15:51–62.

———. (1932): Buddhist bronzes from Kinta, Perak. *Journal of the Federated Malay States Museums* 15:135–136.

Fix, A. G. (1999): *Migration and colonization in human microevolution*. Cambridge: Cambridge University Press.

Henson, F. G. (1989): The Butuan Balangay site: Earliest evidence of formative civilization in the Philippines. In *Development of Pacific region agriculture, domestication and emergence of formative civilizations: Circum-Pacific prehistory conference.* Vol. 4. Seattle: Washington Centennial.

Hislop, J. A. (1954): Notes on the migration of bearded pig. *Malayan Historical Journal* 1–2:134–136.

Hope, G. (in press): The physical geography of Southeast Asia. In *The Quaternary in Southeast Asia,* edited by Avijit Gupta. Oxford: Oxford University Press.

Ishida, Takafumi, Juri Suzuki, Phaibool Duangchan, and Wannapa Settheetham-Ishida (1998): Preliminary report on the short stature of Southeast Asian forest dwellers, the Manni, in southern Thailand: Lack of an adolescent spurt in plasma IGF-1 concentration. *Southeast Asian Journal of Tropical Medicine and Public Health* 29:62–65.

Jamaludin, Mohamad (1994): The Gua Gunung Runtuh human skeleton: First impressions of an orthopaedic surgeon. In *The excavation of Gua Gunung Runtuh and the discovery of Perak Man in Malaysia,* edited by Zuraina Majid, 70–76. Kuala Lumpur: Department of Museums and Antiquity Malaysia.

Kamaludin bin Hassan (1991): Quaternary geological investigation at Pulau Kelumpang archaeological site, Perak, Peninsular Malaysia. *Jurnal Arkeologi Malaysia* 4:74–94.

Kinjgam, A. (1990): Identification of faunal remains from stratigraphic units 5 and 6, Lang Rongrien rockshelter, Thailand. In *Lang Rongrien rockshelter: A Pleistocene–early Holocene site from Krabi, southwestern Thailand,* otherwise written by D. D. Anderson, 76. Philadelphia: University of Pennsylvania.

Leong Sau Heng (1989): Satu perbincangan mengenai peninggalan objek-objek gangsa kecil dari zaman pra-sejarah Malaysia. *Jurnal Arkeologi Malaysia* 2:1–8.

———. (1991a): Jenderam Hilir and the mid-Holocene prehistory of the west coast plain of Peninsular Malaysia. *Bulletin of the Indo-Pacific Prehistory Association* 10:150–160.

———. (1991b): Kubur kepingan dan alat besi bersoket: Satu perbincangan kronologinya. *Jurnal Arkeologi Malaysia* 4:103–115.

———. (2000): The chronology of the Bernam cist graves in Peninsular Malaysia. *Bulletin of the Indo-Pacific Prehistory Association* 19:65–72.

Loewenstein, Prince J. (1956): The origin of the Malayan Metal Age. *Journal of the Malayan Branch of the Royal Asiatic Society* 29(2):5–78.

Maloney, B. K. (1999): A 10,600-year pollen record from Nong Thale Song Hong, Trang Province, South Thailand. *Bulletin of the Indo-Pacific Prehistory Association* 18:129–137.

Martin, R. (1905): *Die Inlandstämme der Malayischen Halbinsel: Wissenschaftliche Ergebnisse einer Reise durch die Vereinigten Malayischen Staaten.* Stuttgart: Gustav Fischer Verlag.

———. (1957): *Lehrbuch der Anthropologie.* Stuttgart: Gustav Fischer Verlag.

Matsumara Hirofumi and Zuraina Majid (1999): Metric analysis of an early Holocene human skeleton from Gua Gunung Runtuh, Malaysia. *American Journal of Physical Anthropology* 109:327–340.

Matthews, J. (1961): *A check-list of "Hoabinhian" sites excavated in Malaya, 1860–1939.* Singapore: Eastern Universities Press Ltd.

Mokhtar Saidin, M. (1998a): Kota Tampan dan Kampung Temelong: Kajian perbandingan tapak-tapak Paleolitik di Lembah Lenggong. *Malaysia Museums Journal* 34: 131–153.

———. (1998b): Palaeoenvironmental reconstruction of Palaeolithic sites in Lenggong and Tingkayu, Malaya. Paper presented at the International Conference Sangiran: Manusia, Budaya dan Lingkungan pada Kala Plestosen, 21–25 September, Solo, Indonesia.

Mokhtar Saidin, M., and H. D. Tjia (1994): Gua Gunung Runtuh: The cave. In *The excavation of Gua Gunung Runtuh and the discovery of Perak Man in Malaysia,* edited by Zuraina Majid, 9–21. Kuala Lumpur: Department of Museums and Antiquity Malaysia.

Oyuela-Caycedo, A. (1996): The study of collector variability in the transition to sedentary food producers in northern Columbia. *Journal of World Prehistory* 10: 49–90.

Pookajorn, S. (1994a): The geology setting and excavation report of Moh Khiew Cave, In *Final report of excavations at Moh Khiew Cave, Krabi Province; Sakai Cave, Trang Province; and ethnoarchaeological research of hunter-gatherer group, socall Mani or Sakai or Orang Asli at Trang Province,* edited by S. Pookajorn, A. Waiyasadamrong, S. Sinsakul, M. Wattanasak, and Y. Chaimanee, vol. 2, *The Hoabinhian Research Project in Thailand,* 306–347. Bangkok: Faculty of Archaeology, Silkaporn University.

———. (1994b): Interpretation and conclusion of archaeological context from Moh Khiew and Sakai Caves. In *Final report of excavations at Moh Khiew Cave, Krabi Province; Sakai Cave, Trang Province; and ethnoarchaeological research of hunter-gatherer group, socall Mani or Sakai or Orang Asli at Trang Province,* edited by S. Pookajorn, A. Waiyasadamrong, S. Sinsakul, M. Wattanasak, and Y. Chaimanee, vol. 2, *The Hoabinhian Research Project in Thailand,* 419–442. Bangkok: Faculty of Archaeology, Silkaporn University.

———. (1996): Human activities and environmental changes during the late Pleistocene to middle Holocene in southern Thailand and Southeast Asia. In *Humans and the end of the Ice Age: The archaeology of the Pleistocene-Holocene transition,* edited by L. G. Straus, B. V. Eriksen, J. M. Erlandson, and D. R. Yesner, 201–213. New York: Plenum Press.

Rambo, A. T. (1979): A note on stone tool use by the Orang Asli (Aborigines) of Peninsular Malaysia. *Asian Perspectives* 22: 113–119.

———. (1984): Why are the Semang? In *Ethnic diversity and the control of natural resources in Southeast Asia,* edited by A. T. Rambo, K. Gillogly, and K. L. Hutterer, 19–35. Ann Arbor: University of Michigan Papers on South and Southeast Asia.

Schebesta, P. (1928): *Among the forest dwarfs of Malaya.* London: Hutchinson.

Schebesta, P., and V. Lebzelter (1926): Schädel und Skelettreste von drei Semang-Individuen. *Anthropos* 21: 959–990.

————. (1928): Anthropological measurements in Semangs and Sakais in Malaya (Malacca). *Anthropologie* 6:183–251.

Shea, B. T., and R. C. Bailey. (1996): Allometry and adaptation of body proportions and stature in African pygmies. *American Journal of Physical Anthropology* 100: 311–340.

Shoocongdej, R. (2000): Forager mobility organization in seasonal tropical environments of western Thailand. *World Archaeology* 32:14–40.

Shuhaimi bin Nik Abdul Rahman, Nik Hassan; Ariffin, Abdul Latib; and Mat Kasa, Abdul Jamil (1988): Penemuan tapak arkeologi pra sejarah di Pergunungan Kenderong dan Kerunai, Grik, Perak. *Jurnal Arkeologi Malaysia* 1:36–49.

Shuhaimi bin Nik Abdul Rahman, Nik Hassan; Rahman, Mohd. Kamaruzaman Abd.; and Abdullah, Mohd. Yusof (1990): Tapak prasejarah Gua Bukit Taat Hulu Terengganu (8920 ± 120 B.P.–2630 ± 80 B.P.). *Jurnal Arkeologi Malaysia* 3:1–9.

Shuhaimi bin Nik Abdul Rahman, Nik Hassan; and Kamaruddin bin Zakaria (1993): Recent archaeological discoveries in Sungai Mas, Kuala Muda, Kedah. *Journal of the Malaysian Branch of the Royal Asiatic Society* 66:73–80.

Shuhaimi bin Nik Abdul Rahman, Nik Hassan, and Che Muhamad Azmi Ngah (1998): Galicara arkeologitapak gendang Dong Son di Kampung Gaung, Jertah, Terengganu. *Jurnal Arkeologi Malaysia* 11:61–89.

Sieveking, G. de G. (1954): Excavations at Gua Cha, Kelantan, 1954. *Malayan Historical Journal* 1–2:75–134.

————. (1956a): The distribution of stone bark cloth beaters in prehistoric times. *Journal of the Malayan Branch of the Royal Asiatic Society* 29(3):78–85.

————. (1956b): The Iron Age collections of Malaya. *Journal of the Malayan Branch of the Royal Asiatic Society* 29(2):79–138.

Sinsakul, S. (1994): Environmental geology of the prehistoric archaeological site at Muang Krabi District, Krabi Province. In *Final report of excavations at Moh Khiew Cave, Krabi Province; Sakai Cave, Trang Province; and ethnoarchaeological research of hunter-gatherer group, socall Mani or Sakai or Orang Asli at Trang Province*, edited by S. Pookajorn, A. Waiyasadamrong, S. Sinsakul, M. Wattanasak, and Y. Chaimanee, vol. 2, *The Hoabinhian Research Project in Thailand*, 293–305. Bangkok: Faculty of Archaeology, Silkaporn University.

Snell, C.A.R.D. (1949): Human skeletal remains from Gol Ba'it, Sungai Siput, Perak, Malay Peninsula. *Acta Neerlandica Morphologica Normalis et Pathologicae* 6: 353–377.

Spriggs, M. (1989): The dating of the Island Southeast Asian Neolithic: An attempt at chronometric hygiene and linguistic correlation. *Antiquity* 63:587–613.

Stauffer, P. H., S. Nishimura, and B. C. Batchelor (1980): Volcanic ash in Malaya from a catastrophic eruption of Toba, Sumatra, 30,000 years ago. In *Physical geology of Indonesian island arcs*, edited by S. Nishimura, 156–164. Kyoto: Kyoto University.

Steele, D. G. (1970): Estimation of stature from fragments of long limb bones. In *Personal identification in mass disasters*, edited by T. D. Stewart, 85–96. Washington, D.C.: Smithsonian Institute.

Steele, D. G., and T. W. McKern (1969): A method for assessment of maximum long bone lengths and living stature from fragmentary limb bones. *American Journal of Physical Anthropology* 31:215–228.

Tjia, H. D. (1987): Ancient shorelines of Peninsular Malaysia. In *Final Report: Seminar in Prehistory of Southeast Asia (T-W11),* 239–257. Project in Archaeology and Fine Arts. Bangkok: Southeast Asian Ministers of Education and Organization.

———. (1991): Pertukaran garisan tepilaut Perak: Selama sembilan ribu tahun terakhir. *Jurnal Arkeologi Malaysia* 4:1–15.

Trevor, J. C., and D. R. Brothwell (1962): The human remains of Mesolithic and Neolithic date from Gua Cha, Kelantan. *Federations Museum Journal* 7:6–22.

van Stein Callenfels, P. (1939): An interesting Buddhistic bronze statue from Bidor, Perak. *Journal of the Federated Malay States Museums* 15:175–179.

van Stein Callenfels, P., and I.H.N. Evans (1928): Report on cave excavations in Perak. *Journal of the Federated Malay States Museums* 12:145–159.

Waiyasadamrong, A. (1994): Shell samples analysis of Moh Khiew and Sakai Cave. In *Final report of excavations at Moh Khiew Cave, Krabi Province; Sakai Cave, Trang Province; and ethnoarchaeological research of hunter-gatherer group, socall Mani or Sakai or Orang Asli at Trang Province,* edited by S. Pookajorn, A. Waiyasadamrong, S. Sinsakul, M. Wattanasak, and Y. Chaimanee, vol. 2, *The Hoabinhian Research Project in Thailand,* 400–404. Bangkok: Faculty of Archaeology, Silkaporn University.

Wattanasak, M. (1994): Archaeopalynology of Moh Khiew and Sakai Cave, Southern Thailand. In *Final report of excavations at Moh Khiew Cave, Krabi Province; Sakai Cave, Trang Province; and ethnoarchaeological research of hunter-gatherer group, socall Mani or Sakai or Orang Asli at Trang Province,* edited by S. Pookajorn, A. Waiyasadamrong, S. Sinsakul, M. Wattanasak, and Y. Chaimanee, vol. 2, *The Hoabinhian Research Project in Thailand,* 393–399. Bangkok: Faculty of Archaeology, Silkaporn University.

Wisseman, Christie, J. (1990): The Sanskrit inscription recently discovered in Kedah, Malaysia. *Modern Quaternary Research in Southeast Asia* 11:39–53.

Zolkurnian, H. (1998): Urutan kebudayaan prasejarah Lembah Lenggong, Hulu Perak, Perak pada zaman Holosen. Sarjana (M.A.) thesis, Centre for Archaeological Research Malaysia, Universiti Sains Malaysia.

Zuraina Majid (1990): The Tampanian problem resolved: Archaeological evidence of a late Pleistocene lithic workshop. *Modern Quaternary Research in Southeast Asia* 11:71–96.

———. (1994): The excavation of Perak man, an epi-Paleolithic burial at Gua Gunung Runtuh. In *The excavation of Gua Gunung Runtuh and the discovery of Perak Man in Malaysia,* edited by Zuraina Majid, 23–43. Kuala Lumpur: Department of Museums and Antiquity Malaysia.

———. (1998a): Radiocarbon dates and culture sequence in the Lenggong Valley and beyond. *Malaysia Museums Journal* 34:241–249.

———. (1998b): Man and technology in late Pleistocene Malaysia. Paper presented at the International Conference Sangiran: Manusia, Budaya dan Lingkungan pada Kala Plestosen, 21–25 September, Solo, Indonesia.Zuraina Majid and H. D. Tjia (1988): Kota Tampan, Perak: The geological and archaeological evidence for a late Pleistocene site. *Journal of the Malaysian Branch of the Royal Asiatic Society* 56:123–134.

Zuraina Majid; Saidin, Mohd. Mokhtar; Soon, Stephen Chia Ming; and Zolkurnian, Hasan (1994): Artifacts from the Gua Gunung Runtuh excavation. In *The excavation of Gua Gunung Runtuh and the discovery of Perak Man in Malaysia*, edited by Zuraina Majid, 149–168. Kuala Lumpur: Department of Museums and Antiquity Malaysia.

Zuraina Majid, Ang Bee Huat, and J. Ignatius (1998): Late Pleistocene–Holocene sites in Pahang: Excavations of Gua Sagu and Gua Tenggek. *Malaysia Museums Journal* 34:65–110.

CHAPTER 5

More Than a Million Years of Human Occupation in Insular Southeast Asia

The Early Archaeology of Eastern and Central Java

François Sémah, Anne-Marie Sémah, and Truman Simanjuntak

Java, birthplace of the Pithecanthropus (the Javanese form of *Homo erectus*), holds an important place in world prehistory. The archaeological interest of the central and eastern parts of the island was noticed more than a century ago, before the discovery of well-known sites such as Trinil, Wajak, and Sangiran (see, for instance, Vrolik, 1850; Schmülling, 1864; Dubois, 1894, 1920). The famous series of lower and middle Pleistocene hominid finds with several dozens of specimens suggests that archaic humans arrived in central and eastern Java between 1.7 and 1.2 million years ago (see Sémah, 1986; Kadar and Watanabe, 1985; Swisher et al., 1994; Sémah et al., 2000). These hominids underwent conspicuous evolution from the most archaic forms represented by the Sangiran 4 or Sangiran 31 fossils (von Koenigswald, 1936; Sartono and Grimaud-Hervé, 1983) to the most progressive forms, such as the "Solo Man" (Santa-Luca, 1980), which could date to the late Pleistocene (Swisher et al., 1996). It is difficult to correlate morphological and behavioral development among these Pleistocene hominids. Indeed, in spite of long-standing paleoanthropological evidence indicative of sustained occupation of Java for more than 1 million years, there is no conclusive behavioral evidence associated with human remains to shed light on relevant cultural variables such as technology and human ecology among the oldest islanders of the world.

Until recently, the stone industries that could be considered part of the tool kit produced by *Homo erectus* consisted of isolated and poorly dated stone pieces, like those found on the Solo terraces (van Heekeren, 1972), and the so-called "Sangiran flakes" (see von Koenigswald and Gosh, 1973). New research, however, has expanded the available archaeological evidence (presumably produced by *Homo erectus*) from the Pleistocene period to include the industries from Sambungmacan (Jacob et al., 1978) and Miri (Djubiantono, 1992; Djubiantono et al., 1992) and the new finds from Sangiran (Sémah et al., 1992; Widianto et al., 1998).

161

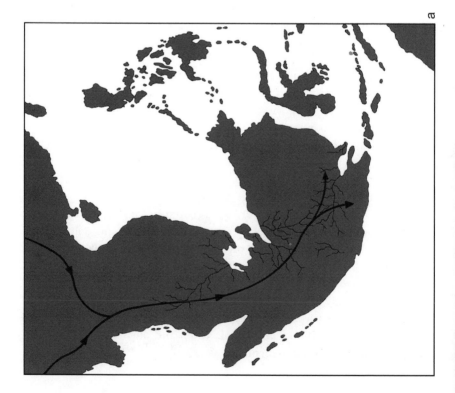

The reader should note, however, that the occupation sequence of Java comprises more than just the famous *Homo erectus*–bearing sites. In the area outside Sangiran, an Indonesian-French team has worked in the Southern Mountains of Java documenting a long sequence of occupation and adaptation to wet tropical environments from the late middle Pleistocene to the Holocene. In the Punung area, along the banks of the Baksoka River, archaeological remains have been known since the pre–World War II days, when the lithic assemblages of the Patjitanian (von Koenigswald, 1936; see also Teilhard de Chardin, 1937; Movius, 1944; van Heekeren, 1955) and the Punung mammalian fauna (Badoux, 1959) were first discovered. Karstic caves and rock shelters often contain lengthy sedimentary records, some of which are currently being investigated. These sites were previously identified with the Mesolithic, e.g., at Goa Lawa cave, near Sampung, in eastern Java (van Stein Callenfels, 1932; van Heekeren, 1972), but actually represent a much longer record that precedes that period. Rock shelters and open-air Neolithic sites were also documented near Punung (Sutikno, 1999; Simanjuntak, 1999).

Early hominid adaptations to the Southeast Asian forest environment are almost unknown and are the source of lively academic debate. This problem is of utmost importance. It is known that hominid settlement of Indonesia started with *Homo erectus* more than 1 million years ago (Swisher et al., 1994; F. Sémah et al., 2000) and that, for these archaic humans to reach the southernmost part of the Sunda shelf, where Java is located, they would have had to use land bridges, depending on the changing paleogeography of the Sunda shelf throughout the Pleistocene (see fig. 5.1). This equatorial region was, to some extent, covered by tropical forests and underwent drastic environmental changes during the climatic oscillations that characterize the Quaternary period. Although drier climatic conditions during glacial periods resulted in more open botanical formations (e.g., grasslands [Caratini and Tissot, 1985] or monsoon forests), it seems that some tropical rain forests could have persisted in riverine galleries (Hooghiemstra, 1997) and highlands (Stuijts, 1993). The rain forest certainly prevented some animals from reaching Java at some points during the Pleistocene, resulting in the development of isolated mammalian faunas (de Vos et al., 1982), and may have played a significant role in cultural evolution on the island. Thus, the diverse geographical and environmental mosaic to which hominids adapted could explain the appearance of highly specific stone and wood industries among the early occupants of Java. Some workers (Bartstra, 1982, 1983; Bartstra and Basoeki, 1989) assumed that the use of small flake tools by *Homo*

Fig. 5.1. Quaternary land bridges from mainland Southeast Asia to the Indonesian Archipelago: *a,* during glacial times, and *b,* during interglacial times.

erectus could predate the use of the larger pebble tools by *Homo sapiens*. Others have emphasized the possibility that the specific rain-forest environment of Java could have triggered a significant reliance on bamboo implements rather than stone (see, for example, Pope, 1985).

In this chapter we will review the available archaeological and paleoecological data on the Pleistocene and early Holocene of central and eastern Java to correlate hominid and human developments with their paleoecological frame in order to determine the extent to which local prehistoric groups adapted to the ever-changing environments of Java.

Geography, Climate, and Vegetation

The island of Java reached a shape somewhat similar to the present one only in the early middle Pleistocene (Djubiantono, 1992). Yet the coastal morphology experienced major modifications during sea regressions and transgressions through the late Pleistocene. Java's relief, with ridges and depressions oriented east to west (van Bemmelen, 1949; also see fig. 5.2a), is

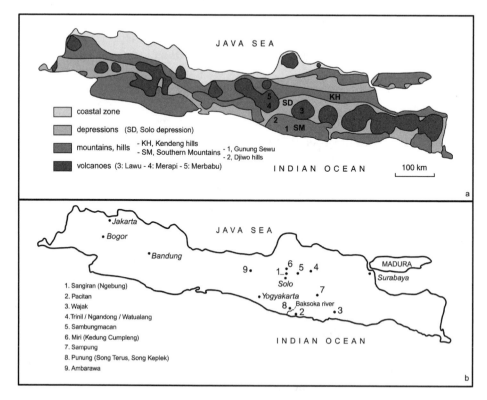

Fig. 5.2. Java: *a,* geological map of Java (after van Bemmelen, 1949); *b,* sites cited in the text.

determined by geotectonics. These relief lines follow the Sunda arc, which resulted from the subduction of the Indian Ocean tectonic plate beneath the Eurasian plate (Katili, 1974). Java is part of a volcanic system. The study area covers part of a large basin, called the Solo depression, in central Java and the Southern Mountains of the island. The Solo basin is filled with Quaternary alluvium and volcanic debris from volcanoes such as Merapi, Merbabu, and Lawu, while the Southern Mountains yield Miocene andesites and reef limestones. It is worthwhile to note that much of the preferred raw material, such as chalcedony, chert, quartz, and silicified wood, used by prehistoric humans to make their stone artifacts originates from this hill range.

Notably, the study area lies in a transitional climatic and an ecological boundary. The region falls between the seasonal eastern Indonesian domain and the hyperhumid evergreen forest domain that covers western Java and Sumatra. Climatic patterns in eastern Java are characterized by the existence of a marked dry season *(musim kemarau)*, and, therefore, they contrast with those of the almost always wet western part of the island (Estienne and Godard, 1970). Precipitation near Surabaya, in eastern Java, reaches 1500 mm or less per year, while rainfall at Bogor (western Java) is 4400 mm. In addition to a pronounced, distinct dry season, eastern and central Javanese sites located near the southern coast support strong daily temperature fluctuations because of open exposure to trade winds. Thus, it is possible that (1) ancient climatic shifts (even slight El Niño Southern Oscillation-like events) could be recorded in local sites, (2) these shifts had an impact on the geographical extent of the forest cover in the study area, and (3) human groups had to adapt to such a shifting landscape.

Because of anthropogenic impact, many areas in eastern Java presently have botanical formations that do not represent the original vegetation cover, even in the recent past. In areas where primary vegetation still persists, there is a seasonal and open semi-deciduous forest with *Albizzia* and many trees of the Leguminosae, resembling the so-called monsoon forest. In the highlands, especially on the slopes of high volcanoes, wetter conditions support tropical rain forest (Backer and Bakhuizen van den Brink, 1965; Whitemore, 1975). Dense forest formations may not have been geographically extensive in the lowlands, especially in the light of a complex coastal geography and common volcanic disruptions. During the dry, glacial stages of the Pleistocene, humans would have to face open environments dominated by grasslands (Caratini and Tissot, 1985), open woodlands, and gallery forests (Hooghiemstra, 1997). In wet periods, some sort of forest existed, but, in central and eastern Java, this forest probably was part of a matrix that enclosed both closed and open terrain. In this respect, our findings are not indicative of closed forest occupation, but of the various types of ecosystems that exist in this part of Java.

Stratigraphy, Fauna, and Vegetation from the Early Pleistocene to the Late Holocene

We will restrict our comments to three sites in central and eastern Java, from a stratigraphical, paleontological, and palynological perspective. Fossil-bearing locales have not always proven to be suitable for a paleoecological interpretation, in which case we will present off-site pollen sources in order to reconstruct ancient landscapes. Our sites include (1) the Sangiran dome, with a lengthy sequence that includes diverse sedimentary facies ranging from marine to palustrine, lacustrine, and fluviatile: (2) Song Terus, a cave sequence that covers the last 200,000 years; and (3) the Ambarawa basin, which yields an accurate picture of paleoenvironmental evolution for the last 20,000 years.

Early and Middle Pleistocene: The Sangiran Dome

At Sangiran, the stratigraphical series (fig. 5.3) starts in the late Pliocene, when open marine *Globigerina* marls where replaced by blue clays deposited in a shallow sea (the so-called lower and upper Kalibeng beds; see Marks, 1957). The installation of regressive conditions can be linked with the oldest eustatic drops in sea level, but also with the volcanic activity of the extant Sunda arc (Sémah, 1986; Djubiantono, 1992). The weathering of the abundant volcanic ashes resulted in abundant clays. At this time, there was a tropical rain forest in the area, while mangrove forests colonized shorelines. It is possible that temperatures could have dropped, as suggested by pollen from highland taxa such as *Dacrycarpus* (A.-M. Sémah, 1982, 1984, 1986).

Coastal facies indicate the extension of emerged land near the Pliocene-Pleistocene boundary. For example, *Balanus* limestone (van Es, 1931; von Koenigswald, 1940) and pollen (A.-M. Sémah, 1982) indicate the presence of shoreline and the development of mangrove forest (fig. 5.4). Ostracods confirm a shallow sea-level (McKenzie and Sudijono, 1981).

Laharic flows accumulated in the Solo basin at 1.6 to 1.8 million years ago (Sémah, 1986; Swisher et al., 1994; Sémah et al., 2000). G.H.R. von Koenigswald (1940), in fact, found the oldest continental mammals (cervid teeth) from Sangiran precisely in these flows. It seems that the vegetation was repetitively disturbed during long periods of volcanic activity. Pollen data show open grasslands with taxa such as *Imperata cylindrica* and *Casuarina* trees (pioneer taxa adapted to quick recolonization after major disruptions; van Steenis, 1972).

After this stage, clayey sedimentation resumed in a swampy basin that was occasionally invaded by the sea. These series are called Pucangan, and the series' lower parts contain a poor mammal fauna (de Vos et al., 1982;

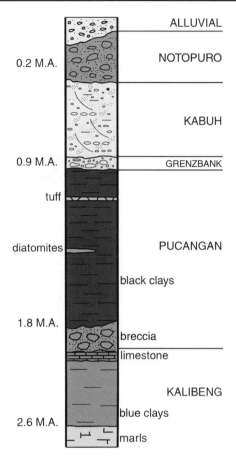

Fig. 5.3. Schematic stratigraphic of the Sangiran dome (not at scale). M.A. = million years.

Sondaar, 1984) with *Tetralophodon, Hexaprotodon,* and cervids. At present, it is unclear whether certain archaic Javanese fossils (Sangiran 31, Sangiran 27) would be contemporaneous with this mammalian fauna. Hominids could have arrived with more recent invaders such as *Stegodon trigonocephalus* and the first carnivora (Sondaar, 1984). Yet it is clear that hominids had already arrived in Sangiran between circa 1.7 million years ago (age of the first Pucangan beds) and circa 1.2 million years ago (age of a tuff within the Pucangan layers; Sémah, 1986; Kadar and Watanabe, 1985; Swisher et al., 1994; Sémah et al., 2000; Saleki, 1997).

Throughout this long period represented by the Pucangan series, *Homo erectus* lived in various environments, with different types of lowland forest ecosystems, such as mangrove and swamp forests. These forest ecosystems

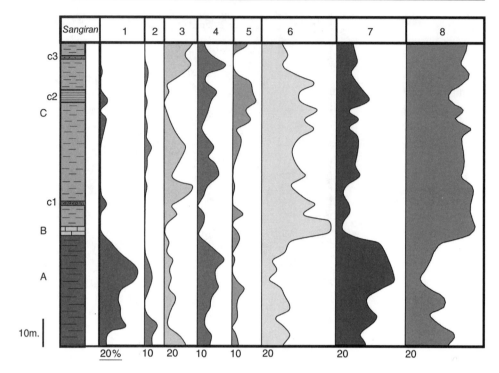

Fig. 5.4. Schematic pollen diagram of the late Pliocene and lower Pleistocene series at Sangiran: *A*, late Pliocene shallow marine blue clays; *B, Balanus* limestone; *C,* Pucangan clays; *c1,* lower breccia; *c2,* diatomitic beds; *c3,* tuff; *1,* mangrove forest; *2,* swamp forest; *3,* Cyperaceae; *4,* rain forest; *5,* open forest; *6,* Poaceae; *7,* arboreal pollen; and *8,* non-arboreal pollen.

may not have been closed forests. They were often disrupted and disturbed by both volcanic activity and marine transgressions. Thus, the Pucangan layers comprise brackish diatomitic incursions (Lizon-Sureau, 1979) as well as volcanic tuffaceous beds.

With regards to neighboring slopes and hills, it seems that these were covered by a rain forest with *Podocarpus,* Fagaceae, Dipterocarpaceae, and *Altingia* during interglacial periods, and by open-forest formations during glacial periods. The pollen analysis indicates that, in some parts of the landscape, the tropical forest never disappeared: significantly, the pollen record shows a sustained presence over time of tree taxa such as *Podocarpus, Castanopsis, Quercus, Myrica, Casuarina,* which appear as both mature forest taxa and pioneer taxa after volcanic disruptions (A.-M. Sémah, 1996a; van Zeist, 1984). The top series of the Pucangan layers comes with a more evolved form of *Homo erectus,* as seen in the Hanoman 1 skull (Widianto et al., 1994) that resembles the well-known Trinil skullcap (Dubois, 1894) and is dated to circa 1 million years ago.

The uplifting of the Kendeng hills to the north and the Southern Mountains contributed to the complete refilling of the Solo basin and the disappearance of marine influence. This is the time when fluviatile sedimentation begins with a synorogenic deposit that mixes marine and continental elements. This episode was called Grenzbank (boundary bed) by G.H.R. von Koenigswald (1940). It is first characterized by a pollen-and-spores assemblage reflecting volcanic disruptions, then by a diversification of the flora, with many more tree species, in which the still high percentage of non-arboreal taxa indicates an open vegetation.

The Grenzbank episode was followed by the deposition of clay, sand, and tuffaceous cross-bedded alluvia, called the Kabuh beds. This part of the Sangiran series, dated circa 0.8 million years ago (Jacob, 1975; Saleki, 1997), yielded the majority of the hominid fossils. Pollen analysis documents a somewhat drier climate and open vegetation with Poaceae, Asteraceae, Fabaceae, and Mimosaceae. Nevertheless, especially at the base of the Kabuh layers, there still are humid forest taxa, with Fagaceae, *Altingia excelsa*, *Engelhardia* and *Podocarpus*. Antelopes and *Elephas* appeared for the first time (Sondaar, 1984). Atop, there is the Notopuro breccia and lahar cap, dated to the end of the middle Pleistocene (Saleki, 1997).

Late Pleistocene to Holocene: The Song Terus Cave at Punung

The geological configuration of the Punung area (fig. 5.5) differs from that of the Solo depression. Here, the Miocene volcanic arc yields the "Old Andesite" formation (Marks, 1957; Saint-Marc et al., 1977). During the Miocene, some type of emersion occurred, as shown by silicified fossil wood found in ancient river beds. Then, the area stayed at sea level (van Bemmelen, 1949) and was subjected to volcanic tuff deposition and formation of reef limestone. The age of Quaternary uplifting is not known, but it may be contemporaneous with the flux of clastic materials that originated here but are found in the Solo depression and are dated 0.8–0.9 million years ago. Subsequent karstic activity was very intense, resulting in the conspicuous "Thousand Hills" landscape and the formation of caves. In the valleys, river terraces were heavily dissected (Sartono, 1964; Teilhard de Chardin, 1937; de Terra, 1943; Bartstra, 1976; van Heekeren, 1955; Sartono et al., 1978), making difficult any attempt to study the chronology of their deposits.

In the Song Terus cave we exposed a 15 m long stratigraphy (figs. 5.6 and 5.7) The basal part of the section contains block falls, thick clays, and several fluviatile sandy lenses. From 4 to 12 m down, clays were overlaid by

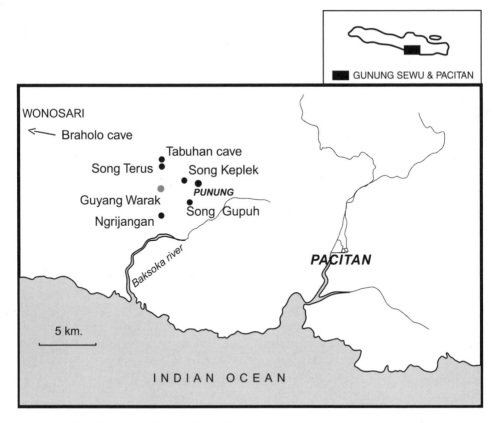

Fig. 5.5. Archaeological sites in the Punung area

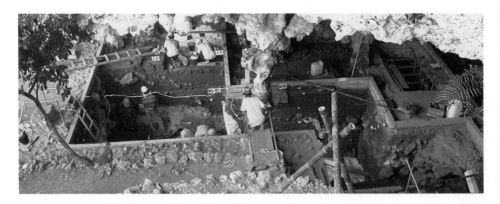

Fig. 5.6. The Song Terus excavation. Courtesy of the Semenanjung research group.

Fig. 5.7. (Opposite) The Song Terus stratigraphy (K1 pit), showing the uppermost 12 m of the sequence

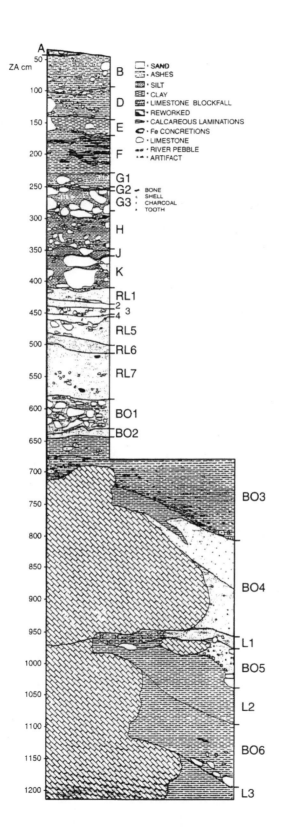

coarse alluvial deposits with sands, gravels, pebbles, and iron nodules. These deposits were occasionally interrupted by the deposition of thin layers with weathered volcanic ashes. Even though these deposits were laid in a cave setting, they are thought to represent a fossil alluvial terrace whose upper tread is currently located 12 m above the present-day riverbed. Fossils in this stratigraphical unit are scarce and pollen was not preserved. The fauna shows taxa such as *Tapirus,* some carnivores, *Axis lydekkeri,* Rhinoceros, and bovids. Mineralized charcoal also appears here. Uranium-series dating on bones and teeth (Falguères et al., 1998; Hameau, 1999) has provided ages for this layer from 187,000 ± 20,000 B.P. to 79,000 ± 3000 B.P.

The cave filling deposited between a depth of 1.5 and 4.0 m has been dated to around 39,000 B.P. in its lower part. From 25,000 B.P. onwards, volcanic ash lenses were deposited, including a layer which provided a bovid bone with an age of 18,100 ± 600 B.P. Pollen from this upper unit indicates a somewhat dry vegetation, with Poaceae, Arecaceae, *Casuarina,* and Moraceae/Urticaceae. Cervids (including Axis) dominate and are followed by bovids. Rare instances of *Elephas* and two carnivora *(Panthera tigris)* were also found. This drier climate and open environment may explain the existence of aeolian sediment in the cave (Gallet, 1999). This type of observation is in consonance with similar paleoenvironmental observations made in the Southern Mountains (Urushibara-Yoshino and Yoshino, 1997).

The top part of the Tabuhan sequence is marked by a recrudescence of karstic activities and is assumed to represent the Pleistocene to Holocene boundary. This part of the sequence comprises significant amounts of rain forest pollen *(Quercus, Podocarpus,* Dipterocarpaceae, *Castanopsis)* and fern spores.

The Holocene sequence from this site is called "Keplek," and is represented in all the local caves (Simanjuntak, 1999). It contains huge block-falls, probably due to earthquakes, at its lower part. Almost everywhere, the uppermost, unprotected part of the Keplek layers well above the blocks has been washed out, leaving only relicts on the walls of the caves. The fauna is diverse, including *Cuon javanicus, Paradoxorus, Felis,* a large felid, two Mustelidae, *Sus, Tragulus, Axis, Cervus timorensis, Bubalus, Bos, Rhinoceros sondaicus,* and *Elephas* (Moigne et al., 2001). The faunal assemblage is dominated by *Presbytis* and *Macaca.* Charcoal carbon-14 ages from the Keplek units at Song Terus range from 8340 ± 310 B.P. to 5770 ± 60 B.P. Alternate rain forest–open forest ecosystems were noticed besides an increased number of *Barringtonia,* a tree that usually requires beach conditions.

Lastly, off-site records, such as that from the Guyang Warak swamp, near Punung, with a 9-m-deep column (Chacornac, 2000), allowed us to study environmental conditions during the last 3000 years B.P., showing the first modification of the forest environment through extensive use of fire.

Off-site Paleoecological Records from the Late Pleistocene and Holocene: Ambarawa

The majority of available pollen data used to reconstruct vegetation and climate changes in insular Southeast Asia comes from highland cores (Flenley, 1979a,b, 1985; Morley, 1982; Haberle et al., 1991; Haberle, 1998; Stuijts, 1984). Verstappen (1997) has emphasized that while temperature drop was the most influential climatic parameter in the highlands, precipitation was the key climatological factor in the ecological configuration of the lowlands. Yet direct data from lowland cores are very scarce (van der Kaars, 1998, 2001). Our data come from the Ambarawa lowlands, in central Java (fig. 5.2b), and therefore are directly relevant to the paleoclimatological debate (A.-M. Sémah et al., 1992; A.-M. Sémah, 1996b).

At 460 m above sea level, the Ambarawa depression contains a shallow lake presently closed by a dam. Two cores were extracted from the lake: (1) Pojoksari G core, 40 m deep, in the center of the basin; and (2) Rowoboni Kebumen core, 10 m deep, in the southern part, closer to volcanic slopes. Carbon-14 chronology indicates a stable sedimentation rate of about 2 mm a year (A.-M. Sémah et al., 1992).

The Pojoksari G core may be divided in two main zones (fig. 5.8). Zone 1 comprises the basal part of the core and extends to a depth of about 26 m. This zone yields clay and silt facies with several sandy layers. The rest of the sequence, or zone 2, contains peat-like deposits. The boundary between both zones has been placed at 12,000 B.P. The carbon content (fig. 5.8) shows a drastic increase near the boundary between the two zones. Layers with high smectite proportions can be used to detect the existence of climates with a marked dry season, while high proportions of halloysite suggest ever-wet climates. In the Pojoksari G core the smectite-halloysite ratio decreases sharply from the bottom of the sequence to the boundary with zone 2, reflecting the appearance of ever-wet conditions of weathering around the basin after 12,000 B.P. (Moudrikah, 1992). The late Pleistocene was dominated by a fluviatile sedimentation with repeated floods, indicating an important erosion of the poorly protected neighboring slopes. During the Pleistocene-Holocene boundary and onwards, a swampy environment developed in the basin.

The pollen record from this core (A.-M. Sémah, 1996b) shows the following features. Below 26 m, pollen taxa are indicative of a downward shift of altitudinal vegetation belts, as suggested by the presence of taxa such as *Dacrycarpus, Altingia,* and *Engelhardia,* adapted to cool conditions. Poaceae are abundant. This evidence indicates open and herbaceous landscapes with patches of rain forest during drier and cooler periods of the late Pleistocene. Between 26 and 21 m, there is an increase in *Typha,* suggestive of

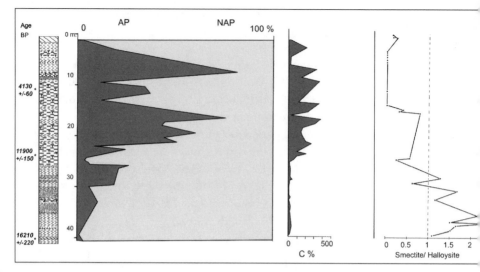

Fig. 5.8. Stratigraphic section, pollen diagram, carbon content, and clay mineralogy from the Pojoksari G core, Ambarawa basin. AP = arboreal pollen; NAP= non-arboreal pollen.

swampy conditions in the depression. More abundant arboreal pollen includes seasonal, open-forest taxa and points to marked seasonal climates at the Pleistocene to Holocene boundary. Above 21 m, the arboreal pollen dominates the sequence, with local swamp forest trees and taxa indicative of rain forest landscape on the neighboring relief.

The Rowoboni Kebumen core (A.-M. Sémah et al., 1992) shows four phases of rain forest regression. The first two regressions took place at 4000 B.P. and 2800 B.P. and indicate periods with a long dry season. The second set of forest regressions took place at 1500 B.P. and A.D. 1300 and are associated with the building of the oldest Hindu monuments, and with local farming practices.

Pleistocene Archaeological Materials

Early and Middle Pleistocene: Did Homo erectus Make Stone Tools?

This question has proven elusive throughout many decades of research. G.H.R. von Koenigswald was the first person interested in the stone tools made by early prehistoric humans; he worked for the Geological Survey of Bandung at Ngandong, Sangiran, and Mojokerto (von Koenigswald, 1956). The first finds were inconclusive and consisted of surface pieces from the Watualang alluvial deposits with bone artifacts (van Heekeren, 1972) as well as the famous andesitic bolas of unclear cultural agency. The quest for *Homo erectus* tools has indeed continued until recent times. Thus, in 1975, a chopper made on a big flake and a smaller retouched flake were found at the site where the Sambungmacan skull was discovered (Jacob et al., 1978), but it is suspected that these presumed tools come from a layer older than the bed in which the skull was found. In 1984, several limestone artifacts were discovered in a conglomeratic layer dating back to the lower Pleistocene at Kedung Cumpleng, near Miri (Djubiantono, 1992; Djubiantono et al., 1992); but, although stratigraphically in situ in a deltaic deposit, these implements had suffered extensive redeposition.

As early as 1936, Pierre Teilhard de Chardin mentioned two possible industries that could be assigned to *Homo erectus* (Teilhard de Chardin, 1937) based on the discoveries by von Koenigswald. These industries included the Sangiran flakes and the Patjitanian. A third industrial complex has been added to the list and comes from the lowermost layers at Song Terus cave (the so-called Terus industry; Simanjuntak and Sémah, 1996).

The Sangiran flakes were initially discovered by G.H.R. von Koenigswald near Ngebung, in the northwest part of the Sangiran dome. These are small, rolled flakes and cores, some of which were scraper-like retouched pieces (fig. 5.9). This industry also yields borers (von Koenigswald and Gosh, 1973).

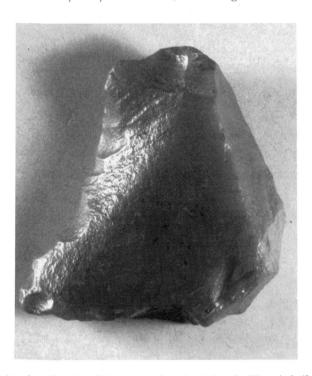

Fig. 5.9. Retouched Sangiran flake scraper (maximum length, 55 mm). Artifact from the collection of the Institut de Paléontologie Humaine, France. Courtesy of the Semenanjung research group.

The Sangiran industry is made of exotic raw materials from the Southern Mountains of Java, with materials such as quartz, chalcedony, jasper, or hardly silicified tuff. Von Koenigswald conducted several excavations and found Sangiran flakes together with fossil mammals that appeared to belong to the so-called Trinil fauna, usually associated with *Homo erectus* (von Koenigswald, 1936). His point of view was contradicted by G. J. Bartstra (1985), who noticed that the Sangiran flakes actually came from the uppermost units of Ngebung, and above a geological unconformity that separates the hominid-bearing deposits from later materials. However, recent research by the authors has confirmed von Koenigswald's interpretation regarding the antiquity of at least part of the industrial complex known as Sangiran flakes. These recent finds include the new excavations in the Kabuh layers at Ngebung (see Simanjuntak and Sémah, 1996; Sémah et al., 1992), and the work by Widianto et al. (1998), who reported several chalcedony artifacts from the Sangiran dome, within the conglomeratic layer that marks the Grenzbank episode (mentioned earlier). In the light of these recent finds, we believe that, regardless of the secondary nature of the deposits in which many of these flake industries are found, the oldest Sangiran flakes are coeval with the classical Javanese *Homo erectus*.

THE SITE OF NGEBUNG

Prior to our work at this site, R. P. Soejono (1982) and G. J. Bartstra (1985) had mentioned the existence of massive artifacts, including bolas. The Kabuh layers are dated to circa 0.8 million years ago at Ngebung (fig. 5.10), by Argon methods (see Saleki, 1997) on pumices, then onto single grains taken from homogeneous tuff layers that overlay the excavated horizon. The results are consistent with the dates recorded in many places of the Sangiran dome on such layers (see also Langbroek and Roebroeks, 2000). The deposition began with thick grayish clays (documented also in the sections published by Kadar and Watanabe, 1985) that underwent erosion and were covered by the typical cross-bedded Kabuh sands. The contact between the clayey and the sandy series forms an erosional surface that could be interpreted as a fossil riverbank. The sandy and clayey matrix also trapped large fractions, including stones and fossil bones (Sémah et al., 1992). We excavated this sandy and clayey matrix (called ensemble A; see fig. 5.11), which yielded many instances of "granulometrically abnormal" fractions, stone tools, and mammal remains. The preservation conditions seem exceptional. For example, this layer yielded prints of bamboo, bark, and leaves. The freshly broken bones, as well as the artifacts, seem to exist in an almost undisturbed occupation floor (fig. 5.12). The influence of fluvial forces on the

Fig. 5.10. Archaeological excavations at Ngebung, in the Sangiran dome. Courtesy of the Semenanjung research group.

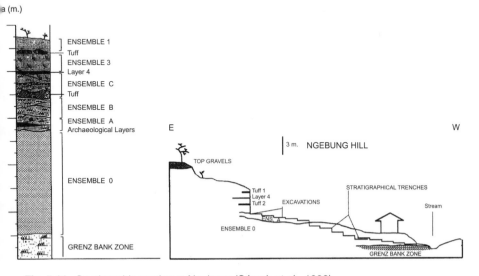

Fig. 5.11. Stratigraphic section at Ngebung (Sémah et al., 1992)

Fig. 5.12. Archeological excavations at Ngebung: association between cleaver, cervid, and bovid antlers.

sedimentological configuration of this layer was limited, as shown by our granulometric study. However, the sediment itself did not produce much pollen (except for Mimosaceae and Poaceae).

The Ngebung faunal list can be compared with that published by J. de Vos et al. (1982). Although carnivores are almost lacking (only a Mustelidae mandible was recovered), about 2000 remains of large herbivores were discovered at Ngebung (65% of which could be determined to the genus or species level). Large bovids are most common *(Bibos palaeosondaicus* and *Bubalus palaeokerabau)*, followed by *Axis lydekkeri* (an archaic cervid), *Cervus hippelaphus,* and *Stegodon trigonocephalus.* Less frequent taxa comprise *Duboisia, Hexaprotodon sivalensis, Sus brachygnathus* and *Rhinoceros palaeosondaicus* (Moigne et al., 2001). Closed rain forest taxa such as *Muntiacus* or monkeys are missing. The study of the differential preservation of skeletal parts shows that skeletons were already dismembered, the skulls opened, and the mandibles broken before their deposition in the riverbank. Bones are often grouped, but were never found in anatomical connection. These bones were buried quickly, as shown by lack of evidence suggestive of transport or carnivore activity. Yet the bones were weathered by water percolation, as indicated by their surface aspect and by abundant, though thin, iron and manganese-oxide layers within ensemble A. Human activity could be represented by intentional fragmentation of fresh bones. Impacts and cut marks are visible on long bones from adult bovids as well as on *Stegodon* tusks. Lithics from Ngebung include in situ Sangiran flakes, and big "bola" stones in association with the above-mentioned faunal remains. These stones can be grouped in two categories. The first one shows various artifacts made of fine-grained andesite, with choppers, horse-hoof scrapers, and a cleaver (fig. 5.13). A big quartz hammer stone is included in this category. This industrial group is not dominant. The second category includes round-shaped artifacts made on coarse-grained andesite. Some of these tools are perfect bolas. Yet others are no more than unretouched pebbles, trimmed pebbles, polyhedral tools, or spheroids (fig. 5.14). This industrial group is dominant. It seems that if these industries indeed belong to the industrial repertoire produced by *Homo erectus,* their typological distribution was determined by the availability of strictly local raw material. Coarse-grained pebbles emerge from the erosion of laharic formations and are quite common in the region, whereas it is much more difficult to find higher-quality fine-grained andesitic materials; and one has to go further south to find big quartz pebbles. There is little evidence for an intentional shaping of bones, except for the *Stegodon* tusk fragment cited above.

THE PATJITANIAN AND TERUS LITHIC INDUSTRIES

The so-called Patjitanian industry was found along the banks of the Baksoka River, near Punung. This is a crude and heavily patinated industry made of chert, chalcedony, jasper, and silicified wood (fig. 5.15) and originally attributed to *Homo erectus* by G.H.R. von Koenigswald. It is not our purpose here to recapitulate the Patjitanian typology, and we refer to H. L. Movius (1944), H. R. van Heekeren (1955; 1972) and G. J. Bartstra (1976) for further reference. The Patjitanian includes a significant proportion of pebble tools and large flake tools. Its relationship with *Homo erectus* has been a source of debate until recently, since the alluvial deposits in which it was discovered are hardly suitable for dating (see, for instance, Bartstra, 1984).

The Terus industry comes from the basal units of Song Terus and was deposited under fluvial conditions, and part of it could date back to the middle Pleistocene (see above). The Terus industry is lighter than the Patjitanian

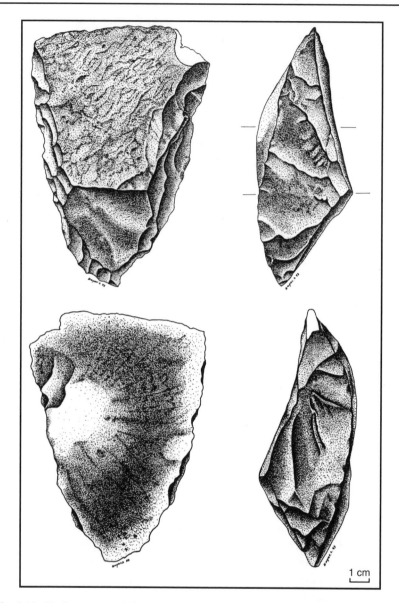

Fig. 5.13. Ngebung stone industry: cleaver. Drawing by Dayat Hidayat.

and comprises abundant flakes and no pebble tools. We retrieved four small chopping tools among more than 300 retouched tools from an assemblage with more than 800 artifacts. The flakes included different sizes, and 40% of them had been retouched. The dominant tool types include scrapers and denticulates (fig. 5.16).

Fig. 5.14. Stone artifacts from Ngebung: *a and b,* cleavers (see fig. 5.13; length, 130 mm); *c,* polyhedric tool (diameter, 90 mm); *d,* "bola" (diameter, 90 mm).

Late Pleistocene to Holocene: The Tabuhan and Keplek Industries

The Tabuhan layers date to the late Pleistocene in the Song Terus and Guwo Tabuhan caves. To our surprise, these layers did not yield many artifacts (unlike the underlying Terus layers), but they represent the oldest systematic cave occupation around 45,000 B.P. The lithic repertoire shows cores and chert and limestone flakes. It also contains freshly broken faunal remains. Reddened limestone is indicative of proximity to hearths.

The Keplek industry is fully Holocene in age and is extensively represented in the upper layers of the caves and rock shelters of the Punung area and also further west, in the Wonosari karstic zone (Simanjuntak, 1999; Widiasmoro, 1999). The main cultural innovations that appear with the

Fig. 5.15. Patjitanian hand-axe. Artifact from the collection of the Institut de Paléontologie Humaine, France. Courtesy of the Semenanjung research group.

Keplek horizon include a bone industry, hunting and gathering of freshwater turtles and shellfish, and a small collection of marine shells used as tools. Physical anthropological evidence allows us to attribute the Holocene layers under discussion to *Homo sapiens* (fig. 5.17; see Widianto, 1999). The stone industry from the Keplek horizon was studied at the type site of Song Keplek cave (Forestier, 1998; Simanjuntak and Forestier, 1998). The industry has

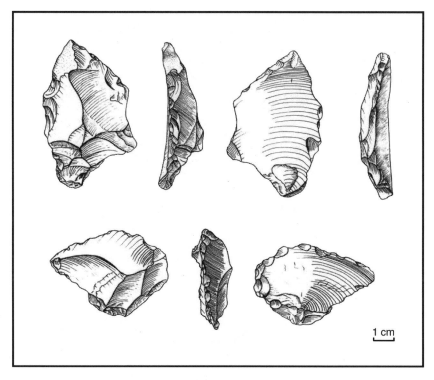

Fig. 5.16. The Terus industry from Song Terus: denticulated flake artifacts. Drawing by Dayat Hidayat.

been characterized as a "mousteroid" industry, with a great variety of flake tools, scrapers, denticulated scrapers, notches, backed knives, and borers (fig. 5.18).

Conclusions

Homo erectus was adapted to life in the seasonal tropics of eastern and central Java and lived in what probably was a diverse group of ecosystems that included several types of variably open forests. Archaic humans *(Homo erectus)*, such as those represented by hominid remains from the Pucangan layers of the Sangiran dome, inhabited the hills around the Solo depression during the lower Pleistocene, as suggested by the discovery of several stone tools attributable to this hominid in the Kedung Cumpleng deltaic conglomerate at Miri, as well as by the Sangiran flakes found in the Grenzbank. The 800,000-year-old Ngebung industry sheds additional light on the Pithecanthropus way of life, indicating that these hominids relied mostly on local raw materials and used simple reduction sequences to produce bolas,

Fig. 5.17. The Song Keplek burial. Photograph by H. T. Simanjuntak.

pebble tools, and flake implements made of andesite. *Homo erectus* also occupied the Southern Mountains of eastern Java during the late middle Pleistocene, as indicated by the Song Terus sequence and the Patjitan industry.
Homo sapiens occupied the cave sites of the Punung area around 45,000 years
ago and continued to inhabit the changing environments of central and
eastern Java during the entire Holocene.

Acknowledgments

The Ngebung excavations were supported by the French Ministère des Affaires
Etrangères and the Leakey Foundation. The Song Terus and Tabuhan excavations
are supported by the French Ministère des Affaires Etrangères. We wish to thank
Julio Mercader and two anonymous readers for their comments.

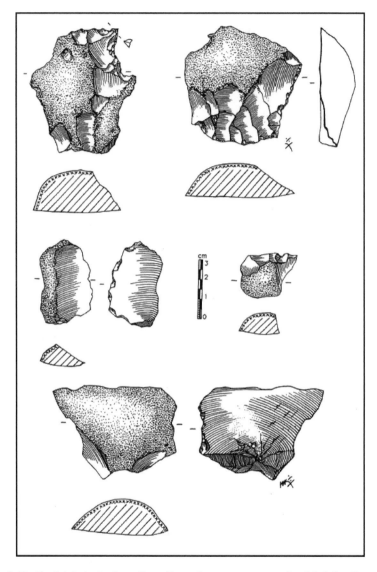

Fig. 5.18. Keplek industry from Song Terus: borer, scraper, and notch (after Forestier, 1998).

References

Backer, C. A., and R. C. Bakhuizen van den Brink Jr. (1965): *Flora of Java*. Vol. 2. Groningen: Noordhoff éditeur.

Badoux, D. M. (1959): *Fossil mammals from two fissure deposits at Punung (Java)*. Utrecht: Kemink en Zoon NV ed.

Bartstra, G. J. (1976): *Contributions to the study of the Paleolithic Patjitan culture, Java, Indonesia: Part I.* Leiden: E. J. Brill éditeur.

―――. (1982): *Homo erectus erectus:* The search for his artifacts. *Current Anthropology* 23(3):318–320.

―――. (1983): Some remarks upon fossil man from Java, his age, and his tools. *Bijdragen tot de Taal-, Land en Volkenkunde* 139(4):421–434.

―――. (1984): Dating the Pacitanian: Some thoughts. *Courier Forschungs Institut Senckenberg* 69:253–258.

―――. (1985): Sangiran, the stone implements of Ngebung, and the Paleolithic of Java. *Modern Quaternary Research in Southeast Asia* 9:99–113.

Bartstra, G. J., and Basoeki (1989): Recent work on the Pleistocene and the Palaeolithic of Java. *Current Anthropology* 30(2):241–244.

Caratini, C., and C. Tissot (1985): *Le sondage MISEDOR. Etude palynologique.* Etudes de géographie tropicale, C.E.G.E.T./C.N.R.S., Talence, n°3.

Chacornac, M. (2000): Etude palynologique du marais de Guyang Warak (Java Central, Indonésie): Mémoire de D.E.A.—Quaternaire: Géologie, Paléontologie Humaine, Préhistoire., M.N.H.N., Paris.

de Terra, H. (1943): Pleistocene geology and early man in Java. *Transactions of the American Philosophical Society* 32(3):437–464.

de Vos, J., S. Sartono, S. Hardjasasmita, and P.-Y. Sondaar (1982): The fauna from Trinil, type locality of *Homo erectus*, a re-interpretation. *Geologie en Mijnbouw* 61:207–211.

Djubiantono, T. (1992): Les derniers dépôts marins de la dépression de Solo (Java Central, Indonésie)—Stratigraphie et paléogéographie. Thèse de doctorat du Muséum National d'Histoire Naturelle, Paris.

Djubiantono, T., F. Sémah, and A.-M. Sémah (1992): Chronology and palaeoenvironment of Plio-Pleistocene deposits in the Solo depression (Central Java): The Kaliuter area and its relations with the ancient Javanese settlements. Paper presented at the Comm. 6th National Archaeological Congress Indonesian Archaeology, Malang, July.

Dubois, E. (1894): *Pithecanthropus erectus, eine menschenähnliche Uebergangsform aus Java, Batavia.* Monograph, 40 pp.

―――. (1920): De proto-Australische fossiele mensch van Wadjak, Java. *Verslag Koninklijke Ned. Akademie van Wetenschappen Te Amsterdam* 29:88–105, 866–887.

Estienne, P., and A. Godard (1970): *Climatologie.* Paris: Armand Colin ed.

Falguères, C., F. Sémah, A.-M. Sémah, and H. T. Simanjuntak (1998): The geochronological frame of the cave deposits in the Southern Mountains karst, central and east Java, Indonesia: First results. Comm. 1st IPPA Congress. Bulletin of the Indo-Pacific Prehistory Association 17(1):37.

Flenley, J. R. (1979a): The late Quaternary vegetational history of the equatorial mountains. *Progress in Physical Geography* 3:488–509.

―――. (1979b): *The equatorial rain forest: A geological history.* London and Boston: Butterworths ed.

―――. (1985): Quaternary vegetational and climatic history of island Southeast Asia. *Modern Quaternary Research in Southeast Asia* 9:55–63.

Forestier, H. (1998): Technologie et typologie de la pierre taillée de deux sites holocènes des Montagnes du Sud de Java (Indonésie). Thèse de doctorat du Muséum National d'Histoire Naturelle, Paris.

Gallet, X. (1999): *Dynamique de la sédimentation dans les grottes des Montagnes du Sud de Java: Exemple de Song Terus (Java Est, Indonésie).* Mémoire de D.E.A.—Quaternaire: Géologie, Paléontologie Humaine, Préhistoire., M.N.H.N., Paris, 60 pp.

Haberle, S. G. (1998): Late Quaternary vegetation change in the Tari Basin, Papua New Guinea. *Palaeogeography, Palaeoclimatology, Palaeoecology* 137:1–24.

Haberle, S. G., G. S. Hope, and Y. de Fretes (1991): Environmental change in the Baliem Valley, montane Irian Jaya, Republic of Indonesia. *Journal of Biogeography* 18:25–40.

Hameau, S. (1999): *L'âge de l'occupation des grottes dans les Montagnes du Sud de Java (Indonésie): Datation Uranium-Thorium des sites de Song Terus et Guwo Tabuhan.* Mémoire de D.E.A.—Quaternaire: Géologie, Paléontologie Humaine, Préhistoire., M.N.H.N., Paris, 33 pp.

Hooghiemstra, H. (1997): Tropical rain forest versus savanna: Two sides of a precious medal? NWO/Huygenslezing (series of lectures open to the public), 31–43. Den Haag: Netherlands Organization for Scientific Research NWO.

Jacob, T. (1975): L'Homme de Java. *La Recherche* 6:1027–1032.

Jacob, T., R. P. Soejono, L. G. Freeman, and F. H. Brown (1978): Stone tools from mid-Pleistocene sediments in Java. *Science* 202:885–887.

Kadar, D., and N. Watanabe, eds. (1985): *Geology of hominid-bearing formations of Java.* Bandung: Geological Research and Development Centre.

Katili, J. A. (1974): Geological environment of the Indonesian mineral deposits: A plate tectonic approach. *Direktorat Geologi Bandung, Publikasi Teknik, Seri Geologi Ekonomi* 7:18.

Langbroek, M., and W. Roebroeks (2000): Extraterrestrial evidence on the age of the hominids from Java. *Journal of Human Evolution* 38:595–600.

Lizon-Sureau, B. (1979): Essai de reconstitution du paléoenvironnement du Pithécanthrope de Java, au Pléistocène moyen, dans la région de Sangiran (Java central), par les méthodes d'analyse en sédimentologie classique et moderne. Thèse de Doctorat de Spécialité, Paris, Université Paris VI.

Marks, P. (1957): Stratigraphic lexicon of Indonesia. *Pusat Jawatan Geologi Bandung*, Seri Geologi, n°31.

McKenzie, K. G., and Sudijono (1981): Plio-Pleistocene Ostracoda from Sangiran, Java. *Geol. Res. Dev. Center*, Paleontological series, n°1:29–51.

Moigne, A.-M., R. Due Awe, F. Sémah, and A.-M. Sémah (2001): The cervids from the Ngebung site("Kabuh" series, Sangiran Dome, central Java) and their biostratigraphical significance. *Modern Quaternary Research in Southeast Asia* (Rotterdam) 18.

Morley, R. J. (1982): A palaeoecological interpretation of a 10,000-years pollen record from Danau Padang, Central Sumatra, Indonesia. *Journal of Biogeography* 9:151–190.

Moudrikah, R. (1992): *Premières analyses concernant la sédimentation dans la dépression d'Ambarawa (Java Central) durant les quinze derniers millénaires.* Mémoire de D.E.A.— Quaternaire: Géologie, Paléontologie Humaine, Préhistoire, M.N.H.N., Paris.

Movius, H. L. (1944): *Early man and Pleistocene stratigraphy in southern and eastern Asia.* Papers of the Peabody Museum of American Archaeology and Ethnology, 19, n°3. Harvard: Harvard University.

Pope, G. G. (1985): Taxonomy, dating, and paleoenvironment: The paleoecology of the early Far Eastern hominids. *Modern Quaternary Research in Southeast Asia* 9:65–80.

Saint-Marc, P., F. Paltrinieri, and B. Situmorang (1977): Le Cénozoïque d'Indonésie occidentale. *Bulletin Société Géol. Fr.,* 7° series, 19(1):125–133.

Saleki, H. (1997): Apport d'une intercomparaison des méthodes nucléaires (230Th/234U, ESR et 40Ar/39Ar) à la datation de couches fossilifères pléistocènes dans le dôme de Sangiran (Java, Indonésie): Thèse de doctorat du Muséum National d'Histoire Naturelle, Paris.

Santa-Luca, A. P. (1980): *The Ngandong fossil hominids: A comparative study of a Far Eastern* Homo erectus *group.* New Haven: Yale University Publications in Anthropology.

Sartono, S. (1964): *Stratigraphy and sedimentation of the easternmost part of Gunung Sewu (East Java).* Direktorat Geologi Bandung, Publik. Tek. Seri Geol Umum, 1.

Sartono, S., and D. Grimaud-Hervé (1983): Les pariétaux de l'Hominidé Sangiran 31. *L'Anthropologie* 87(4):465–468.

Sartono, S., H. Syarif, J. Zaim, U. P. Nababan, and T. Djubiantono (1978): Undak sungai Baksoko. *Proyek Penelitian dan Penggalian Purbakala,* #19B, Jakarta, p. 23–52.

Schmülling, P.E.C. (1864): Mededeeling over fossiele wervelidierresten. *Natuur Tijdschift van Nederlandsch Indië* 7:399.

Sémah, A.-M. (1982): A preliminary report on a Sangiran pollen diagram. *Modern Quaternary Research in Southeast Asia* 7:165–170.

⸻. (1984): Palynology and Javanese Pithecanthropus environment. *Courier Forschungs Institut Senckenberg* 69:237–243.

⸻. (1986): Le milieu naturel lors du premier peuplement de Java. Vol. 2. Thèse de Doctorat d'Etat es Sciences, Université de Provence.

⸻. (1996a): Pleistocene and Holocene environmental changes. In *Indonesian Heritage,* edited by J. Miksic, vol. 1, 20–21. Singapore: Didier Millet ed.

⸻. (1996b): Evolution of the vegetation in the Ambarawa depression (central Java) during the last 16,000 B.P. Paper presented at the 9th IPC Meeting, Houston, Texas, 23–28 June.

Sémah, A.-M., F. Sémah, C. Guillot, T. Djubiantono, and M. Fournier (1992): Etude de la sédimentation pollinique durant les quatre derniers millénaires dans le bassin d'Ambarawa (Java Central, Indonésie)—Mise en évidence de premiers défrichements. *C.R. Academy of Science* (Paris) 315(2):903–908.

Sémah, F. (1986): Le peuplement ancien de Java—Ebauche d'un cadre chronologique. *L'Anthropologie* 90(3):359–400.

Sémah, F., A.-M. Sémah, T. Djubiantono, and H. T. Simanjuntak (1992): Did they also make stone tools? *Journal of Human Evolution* 23:439–446.

Sémah, F., G. Féraud, H. Saleki, C. Falguères, and T. Djubiantono (2000): Did early man reach Java during the late Pliocene? *Journal of Archeological Science* 27:763–769.

Simanjuntak, H. T. (1998): New discoveries at Braholo Cave, Western Gunung Sewu. Paper presented at 7th Eurasean Conference, Berlin, 31 August–3 September.

Simanjuntak, H. T., ed. (1999): *Gunung Sewu: Exploitation in Holocene.* Jakarta: National Center of Archaeology.

Simanjuntak, H. T., and F. Sémah (1996): A new insight in the Sangiran flake industry. *Bulletin of the Indo-Pacific Prehistory Association* 14(1):22–26.

Simanjuntak, H. T., and H. Forestier (1998): Lithic typo-technology of Holocene Song Keplek cave, East Java. Comm. 16th IPPA Congress. *Bulletin of the Indo-Pacific Prehistory Association* 17(1):69.

Soejono, R. P. (1982): New data on the Paleolithic industry in Indonesia. l'Homo erectus et la place de l'homme de Tautavel parmi les hominidés fossiles, 2, 578–592. Paper presented at the I° Congrès International de Paléontologie Humaine, Nice, Colloque International du C.N.R.S.

Sondaar, P. Y. (1984): Faunal evolution and the mammalian biostratigraphy of Java. *Courier Forschungs Institut Senckenberg* 69:219–235.

Stuijts, I. (1984): Palynological study of Situ Bayongbong, West Java. *Modern Quaternary Research in Southeast Asia* 8:17–27.

———. (1993): Late Pleistocene and Holocene vegetation of West Java, Indonesia. *Modern Quaternary Research in Southeast Asia* 12:173.

Sutikno, T. (1999): Situs Song Gupuh di Punung, Pacitan: Profil gua payung hunian berciri neolitik. Paper presented at the Comm. 8th I.P.P.A. meeting, Yogyakarta, February.

Swisher C. C., III, G. H. Curtis, T. Jacob, A. G. Getty, A. Suprijo, and Widiasmoro (1994): Age of the earliest known hominids in Java, Indonesia. *Science* 263:1118–1121.

Swisher C. C., III, W. J. Rink, S. C. Anton, H. P. Schwarcz, G. H. Curtis, A. Suprijo, and Widiasmoro (1996): Latest *Homo erectus* of Java: Potential contemporaneity with *Homo sapiens* in Southeast Asia. *Science* 274:1870–1874.

Teilhard de Chardin, P. (1937): Notes sur la paléontologie humaine en Asie méridionale. *L'Anthropologie* 47:23–33.

Urushibara-Yoshino, K., and M. Yoshino (1997): Palaeoenvironmental change in Java island and its surrounding areas. *Journal of Quaternary Science* 12(5):435–442.

van Bemmelen, R. W. (1949): *The geology of Indonesia and adjacent archipelagoes.* Den Haag: Martinus Nijhoff éditeur.

van der Kaars, S. (1998): Marine and terrestrial pollen records of the last glacial cycle from the Indonesian region: Bandung Basin and Banda Sea. *Palaeoclimates* 3(1–3):209–219.

van der Kaars, S., D. Penny, J. Tibby, J. Fluin, R.A.C. Dam, and P. Suparan (2001):

Late Quaternary palaeoecology, palynology, and palaeolimnology of a tropical lowland swamp: Rawa Danau, West-Java, Indonesia. *Palaeogeography, Palaeoclimatology, Palaeoecology* 171(3–4):185–212.

van Es, L.J.C. (1931): *The age of Pithecanthropus.* Den Haag: Martinus Nijhoff éditeur.

van Heekeren, H. R. (1955): *New investigations on the lower Palaeolithic Patjitan culture in Java.* Berita Dinas Purbakala—Bulletin of the Archaeological Service of the Republic of Indonesia. Jakarta, 1.

———. (1972): *The Stone Age of Indonesia.* Verhandelingen van het Kon. Inst. Voor Taal-, Land-, en Volkenkunde, n°61, Den Haag: Martinus Nijhoff éditeur.

van Steenis, G.G.G.J. (1972): *The mountain flora of Java.* Leiden: E. J. Brill ed.

van Stein Callenfels, P. V. (1932): Note préliminaire sur les fouilles dans l'abri sous roche du Guwa Lawa à Sampung. In *Hommage du Service Archéologique des Indes Néerlandaises au 1er Congrès des préhistoriens d'Extrême-Orient à Hanoi,* 25–29. Albrecht: Batavia.

van Zeist, W. (1984): The prospects of palynology for the study of prehistoric man in Southeast Asia. *Modern Quaternary Research in Southeast Asia* 8:1–15.

Verstappen, H. Th. (1997): The effect of climatic change on Southeast Asian geomorphology. *Journal of Quaternary Science* 12(5):413–418.

von Koenigswald, G.H.R. (1936): Early Palaeolithic stone implements from Java. *Bulletin of the Raffles Museum Singapore* 1:52–60.

———. (1940): Neue Pithecanthropus-Funde 1936–1938. *Dienst van den Mijnbouw in Nederlandsch Indië Wetenschappelijke Mededeelingen* 28.

———. (1956): *Les premiers hommes sur la terre.* Paris: Denoël ed.

von Koenigswald, G.H.R., and A. K. Gosh (1973): Stone implements from the Trinil beds of Sangiran, Central Java, I. *Proceedings Koninklijke Ned. Akademie van Wetenschappen,* series B, 76(1):1–34.

Vrolik, W. (1850): Berigt over zes steenen wiggen op Java gevonden. *Tijdschrift Voor de Wis- en Natuurk Wetenschappen* 3:99–102.

Whitemore, T. C. (1975): *Tropical rain forests of the Far East.* London: Clarendon Press.

Widianto, H. (1999): Manusia penghuni gua Gunung Sewu. In *Gunung Sewu: Exploitation in Holocene,* edited by T. Simanjuntak. Jakarta: National Center of Archaeology.

Widianto, H., A.-M. Sémah, T. Djubiantono, and F. Sémah (1994): A tentative reconstruction of the cranial human remains of Hanoman 1 from Bukuran (central Java). *Courier Forschungs Institut Senckenberg* 171:47–59.

Widianto, H., B. Toha, and T. Simanjuntak (1998): The discovery of stone implements in the Grenzbank: New insight to the Sangiran flake industry's chronology. Paper presented at the 16th IPPA Congress, Melaka, July.

Widiasmoro (1999): Genesa gua Braholo dari proses sedimentasi deposit arkeologi di Gunung Kidul, Yogyakarta. Paper presented at the 8th I.P.P.A. meeting, Yogyakarta, February.

CHAPTER 6

An Archaeological Assessment of Rain Forest Occupation in Northeast Queensland, Australia

Brit Asmussen

This chapter reviews the archaeology and research potential of the tropical rain forest region of northeast Queensland. Comparatively, little work has been done in this region, with only two sites excavated and analyzed (Horsfall, 1987b; see also Cosgrove, 1996). At present there is no direct archaeological evidence for a Pleistocene occupation of the rain forests of northeast Queensland, although this is unsurprising given the limited amount of archaeological research carried out to date. Ancient occupation is possible, as hunter-gatherers were present throughout Australia before the last 30,000 years B.P. (Roberts et al., 1990; Smith and Sharp, 1993; Fullagar et al., 1996; O'Connell and Allen, 1998; Thorne et al., 1999; Cosgrove, 1999; Turney et al., 2001) and occupied the forest-bordering region (Rosenfeld et el., 1981; David, 1994; Morwood and Hobbs, 1995) and Southeast Asia during the late Pleistocene (Bowdler, 1993).

The current Australian models claiming a distinct occupation of the northeast Queensland rain forests during the last 5000 years have been based on ethnography and, while intriguing, cannot be supported by unequivocal archaeological data. In order to understand the nature of rain forest occupation we need to start from the ground up and develop our analytical methodologies. If this can be achieved, the Australian evidence will have the potential to shed light on key issues facing tropical rain forest archaeology throughout the world.

The Tropical Rain Forests of North Queensland

The northeast Queensland tropical rain forests are relatively small by international standards, covering less than 0.1% of the area of the continent (Stocker and Unwin, 1989; see fig. 6.1). Rain forests of varying spatial, structural, and floristic diversity occur along the northeastern Queensland coastline from Cooktown to Ingham and have been classified into different structural and floristic types (Webb and Tracey, 1981; Tracey, 1982). The rain forests of the area vary from complex multi-storied vine forests on the

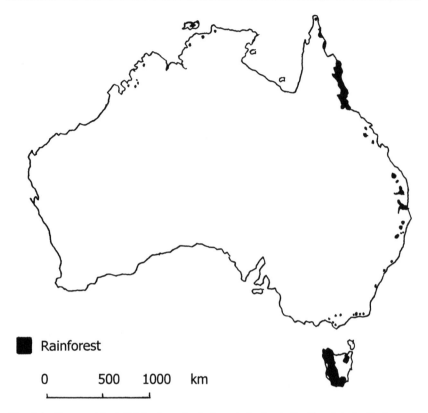

Fig. 6.1. Geographical distribution of rain forests in Australia. Reprinted with the permission of Cambridge University Press.

richer soils in warmer and wetter areas to simple vine forests on poor soil and in drier areas and fern forests at higher, wetter, altitudes (Webb and Tracey, 1981; Bell et al., 1987; Werren and Kershaw, 1991).

The most highly developed and extensive area of wet lowland tropical rain forest is located between Cairns and Innisfail. The lowland rain forest receives an average rainfall of 2500–4300 mm per annum, with 75% delivered seasonally over the four-month summer period (Werren and Kershaw, 1991: 7). These rain forests are surrounded by tall, open eucalyptus forests to the west, north, and south; these eucalyptus forests also divide parts of the lowland rain forest itself. This feature is claimed to be unique in Australia (Werren and Kershaw, 1991: 12). Extensive mangrove communities, several large river systems, estuaries, lakes, and swamps are distributed amongst the rain forests of the coastal lowlands and provide a diverse range of resources.

The structure and floristic assemblages in the lowland rain forests of northeast Queensland are different from those in other tropical rain forests. These forests bear the closest resemblance to the humid tropical lowland forests of Southeast Asia (Werren and Kershaw, 1991: 12). In contrast to the African and Asian forests (Whitmore, 1989: 198; Hamilton, 1989: 164), the upper canopy of the northeast Queensland rain forest is usually less than 45 m in height, with several semi-open, closely spaced lower canopies of various heights that allow light to reach the forest floor (Werren and Kershaw, 1991: 12). Undergrowth taxa include several endemic species such as ferns, cycads, palms, and tubers (Werren and Kershaw, 1991), although the forest lacks the annual herbs common in American Neotropical forests (Prance, 1989: 110). Together, this range of plants supplies several different sources of carbohydrates, with several nut species with high oil and fat content available for several months of the year (Boland et al., 1984; Harris, 1987; Miller et al., 1993; Hiddins, 1999). Rain forests are not unchanging ecosystems, and frequent cyclones, dry winters, droughts, floods, landslides, fires, and frosts result in frequent alterations to the rain forest (Stocker, 1981; Catling and Newsome, 1981; Ash, 1988; Stocker and Unwin, 1989; Gill et al., 1990; Bowman, 2000).

Northeast Queensland rain forests are high in species richness and comparatively rich by Australian standards, but low in comparison to other tropical rain forests (Crome, 1990: 54; Adam, 1992: 181–182). The wet tropical rain forests have the richest fauna in Australia and contain 30% of the Australian marsupial species, 60% of the bat species, 30% of the frog species, 23% of the reptile species, 18% of the bird species, and 54 vertebrate animals unique to the area (Werren and Kershaw, 1991: 21). Mammalian fauna includes a diverse range of rain forest specialists and also a number of open- and closed-forest generalists (Adam, 1992: 181–183). Many animals in the area show a graded ability to use a range of vegetation types from closed to open forest (Crome, 1990: 54). Today some species are differentiated spatially and altitudinally (Strahn, 1983; Kikkawa, 1990). A diverse range of animal species from swamp and riverine environments are also present due to the environmental heterogeneity of the area. When the juxtaposition of the rain forest with other environmental types in the region is considered, this area of northern Queensland becomes Australia's richest faunal region (Crome, 1990: 54).

Paleoenvironmental Evidence

Our ability to answer questions about the nature, antiquity, and continuity of rain forest use depends on developing good paleoenvironmental data. It is generally accepted by Australian archaeologists that occupation of the continent began in the last 50,000 years with the arrival of *Homo sapiens* from the north (White and O'Connell, 1982; Smith et al., 1993; Lourandos, 1997). Some researchers have suggested even earlier dates for occupation (Roberts et al., 1990; Bowdler, 1992; Kershaw et al., 1993; add Fullagar et al., 1996; Thorne, 1999; refer also to White, 1994; Hope, 1994; Anderson, 1994). During this period there have been dramatic and complex climatic changes in northeast Queensland, changes which have impacted on the extent and composition of rain forest communities.

Paleoenvironmental evidence for the Pleistocene period is largely based on pollen data from Lynch's Crater, on the Atherton Tablelands (Kershaw, 1974, 1975, 1976, 1978, 1983, 1985, 1986, 1995), 50 km from the lowland rain forest environment. More information is available for vegetation changes during the Holocene period from the same area (Kershaw, 1970, 1971; Chen, 1986, 1988; Walker and Chen, 1987). However, the Atherton Tablelands are considered to be climatically sensitive (Butler, 1998), so it is not clear whether the patterns of change on the tablelands are representative of changes in the lowlands (Butler, 1998; Walker, 1991).

The Pleistocene evidence suggests the following changes around Lynch's Crater. A dry type of rain forest was present around 40,000 B.P. (Kershaw, 1983). There is evidence of large-scale fires, in the form of a dramatic influx of charcoal in the palynological core, around 38,000 B.P. Frequent but less intense fires continued until 26,000 B.P., and during this time the vegetation changed from being equal proportions of rain forest and schlerophyl taxa to being dominated by schlerophyl vegetation (Kershaw, 1978, 1986; Harrison and Dodson, 1993). It remained dominant until 9000 B.P. (Kershaw, 1974, 1986: 48–49).

There is considerable uncertainty about the extent, distribution, composition, and structure of rain forests from 26,000 B.P. to 10,000 B.P. Low, open schlerophyl savanna woodlands are thought to have been at their greatest extent and covered much of northern Queensland, including the tablelands and significant areas of the highlands (Ash, 1988; Hopkins et al., 1990; Hopkins et al., 1993; Allen and Kershaw, 1996: 177) and lowlands (Kershaw, 1975: 185 and 1995: 664). Significantly, no continuous palynological data are available for the lowlands during this period.

There are two different refugia models concerning the fate of the rain forest during this period. Some palynologists have suggested that small isolated patches of rain forest retreated to moist, protected areas such as stream gullies (Kershaw, 1974: 222), swampy lake margins (Bohte and Kershaw, 1999), mountain valleys, and coastal ranges (Kershaw, 1975). Other researchers suggest that substantial areas of rain forest vegetation were present on the lowlands, including the exposed continental shelf (Walker and Chen, 1986; Ash 1988: 627). D. Walker and Y. Chen have drawn from global data regarding rain forest changes to suggest that if rain forests were present on the lowlands, they may have had greater niche differentiation and more heterogeneous habitats distributed less continuously and in different locations prior to 10,000 B.P. (Walker and Chen, 1987: 82–83). There is more data for the history of the tableland rain forests during the Holocene, with six palynological sites within a 15-km radius of the Atherton Tablelands. Once again, the exact implications of highland data for coastal highland and lowland rain forest dynamics are uncertain. The data still provide some important information on rates of rain forest colonization, diversity of floristic composition, and the impact of fire (Hopkins et al., 1993; Walker and Chen, 1987: 83).

This palynological data indicates a complex and variable process of rain forest colonization of the tablelands. Rain forest was established at the various palynological sites over a period of 3000 years, between 10,000 B.P. and 7000 B.P. It then took from 1000 to 1500 years to become fully established (Walker and Chen, 1986). Rain forests reached their maximum geographical extent throughout the region between 6000 B.P. and 3500 B.P. (Allen and Kershaw, 1996: 179). The pattern at any one site appears to have been substantially affected by site-specific factors, including variations in soil and rainfall conditions but also the return rate of fires and proximity to frequently burned schlerophyl vegetation (Hopkins et al., 1993; Walker and Chen, 1987: 84). Around 3500 B.P., there was a fragmentation of the rain forest. Pollen data is also available for a restricted time period (7000 B.P. to 6000 B.P.) from the lowlands at Mulgrave River. This indicates that wet and dry types of rain forest taxa were present by 7000 B.P. (Crowley et al., 1990).

Archaeological Evidence

Only two sites have been excavated from within the wet rain forests: Jiyer Cave and Mulgrave River 2 (fig. 6.2). Both were excavated by Horsfall as part of her doctoral research (Horsfall, 1983, 1984, 1987a,b, 1996). Several excavations have been conducted in regions around the rain forest during the last 50 years. Excavated archaeological sites around Chillagoe and Laura do not show the use of specifically rain forest resources (see Woolston, 1965; Rosenfeld et al., 1981; David 1994; Morwood and Hobbs, 1995). The use of a range of environments, including open forest, marine, and mangrove areas, is indicated at Bare Hill. Faunal remains from this site have been interpreted as indicating the use of rain forest resources; however, the site has not been dated and no other details are available (Horsfall 1996: 176). Analysis of faunal remains from excavations at Kennedy, Jourama, and Herveys Range suggests that animal resources were derived from open schlerophyl and savanna communities (Archer and Brayshaw, 1978), while shell species from nearby mangrove communities were also used (Brayshaw,

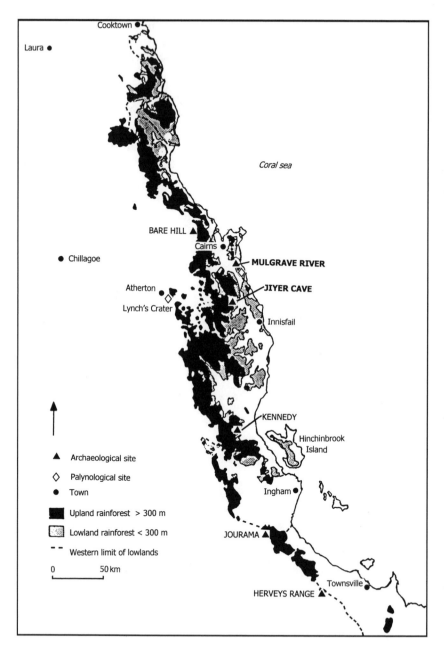

Fig. 6.2. Queensland rain forests, showing archaeological and palynological sites mentioned in the text

1995). Shell middens and fish traps are generally restricted to the southern part of the region, around Hinchinbrook Island and the adjacent coast, and are considered to provide information on coastal economies, not rain forest use per se (Campbell, 1982; Horsfall, 1987b: 109).

There is a wide range of site types (fig. 6.3) and the majority of them fall along the coast or along the western periphery of the present rain forest block. There are over 100 sites known between Cairns and Ingham, although most have not been archaeologically investigated (Horsfall, 1987b: 312–318; Horsfall and Furay, 1988: 42). The range of site types known to exist in rain forests are like those across the rest of the continent and include rock shelters, with or without rock art, and a variety of open sites such as shell middens, shell and artifact scatters, isolated artifact finds, quarries, carved and scarred trees, engravings, fish traps, axe-grinding grooves, stone arrangements, ceremonial grounds, and walking tracks (Horsfall and Furay, 1988: 33–36).

Jiyer Cave is a small rock shelter located near the bank of the Russell River (Horsfall, 1987b: 119; and see fig. 6.4). The total floor area is 324 m^2, although much of this is covered with roof fall (Horsfall, 1996: 176). Water seeps through the shelter from the roof, producing an inner drip-line, and water runs across the shelter floor. The Russell River has been known to rise to the level of the floor of the shelter when in flood (Horsfall, 1987b). The date obtained from the lower excavated level, at a depth of 110–133 cm, was 5130 ± 40 B.P. (fig. 6.5). Horsfall has suggested that periodic flooding of the site is likely to have occurred between 5000 B.P. and 3000 B.P. The site was excavated in 5-cm units. There are three stratigraphic units: an upper, loose sandy layer, 10–12 cm thick; a middle layer of sand-clay mixture, the bottom of which contains increasing quantities of coarse sand at an approximate depth of 140 cm; and the lowest layer, which is an archaeologically sterile clay (Horsfall, 1987b, 1996). Some dating inversions are evident in the site.

A range of archaeological materials were excavated from the site, including nutshells, bone, shell, grindstones, quartz artifacts, and metal and glass obtained directly or indirectly in the colonization period. Dating problems make it difficult to clearly establish when increases in deposition occurred. However, it appears that the majority of materials were deposited in the last 200 to 400 years. The exception to this rule is the deposition of ochre at the site. There are several peaks in discard of ochre, between about 130 cm to 140 cm, at about 110 cm, and one coinciding with an increase in the deposition of charcoal between 30 cm and 40 cm. The majority of the cultural materials increase in deposition in the top 20 cm.

The majority of nutshell remains are deposited in the top 20 cm, the largest peak of nutshell remains is 16.3 grams at 16 cm (fig. 6.6). Nutshell

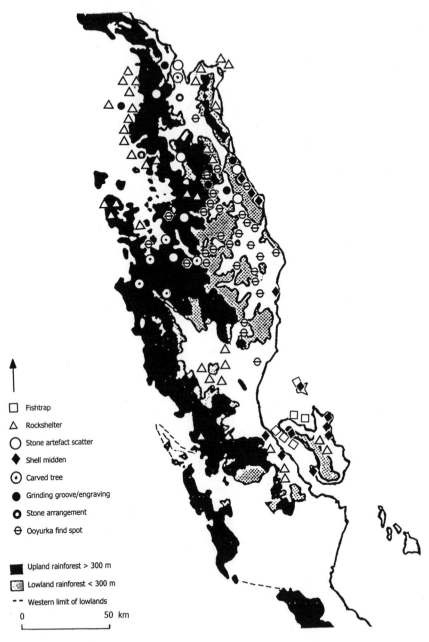

Fig. 6.3. Types of archaeological sites in the Australian forest (based on data by Horsfall, 1987b, and Cosgrove, 1996)

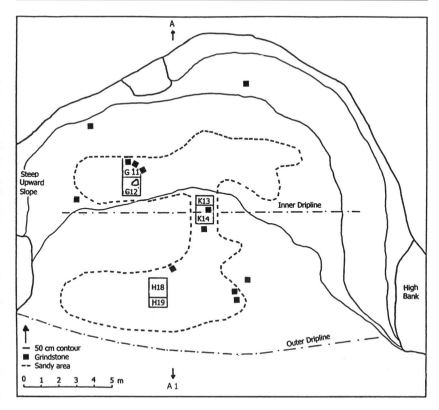

Fig. 6.4. Jiyer Cave, site plan (after Horsfall, 1987b). Reproduced with permission from the author.

remains are less than 0.3 grams from 5130 B.P. to 2650 B.P. (data from Horsfall, 1987b). The oldest piece of nutshell is from approximately 4000 B.P. but is unidentifiable. The oldest identifiable pieces are from the toxic *Elaeocarpus bancroftii* and dated between 3200 B.P. and 3000 B.P.; and fragments of nutshell from the genus *Endiandra* dated between 3500 B.P. and 3200 B.P. (Horsfall, 1996: 179). Shells from mangrove, freshwater, and coastal environments were identified at the site and are largely confined to the top 20 cm of the site. Two shell implements were recovered: one a shell scraper, the other a shell fish hook (Horsfall, 1996: 181).

Faunal remains are from a range of rain forest species, including wallaby, possums, rats, platypuses, fruit bats, snakes, lizards, frogs, and fish, some of which are thought to represent noncultural additions. Bone fragments are represented in small amounts in earlier levels, but increase in the top 20 cm. Minimum numbers of individuals could not be calculated. One bone point from a macropod was recovered at the site in the recent levels (Horsfall,

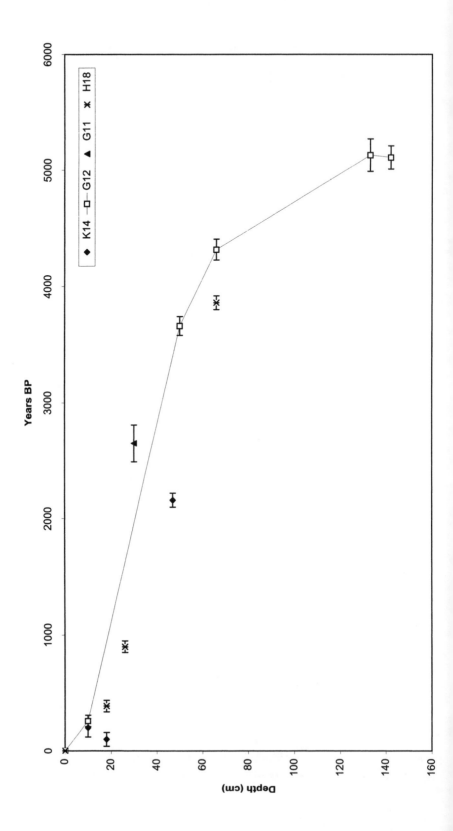

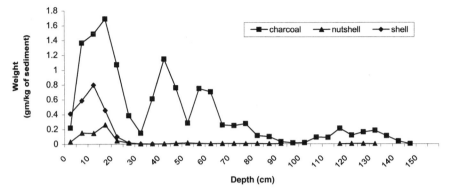

Fig. 6.6. Jiyer Cave, distribution of charcoal, nutshell, and shell in depth

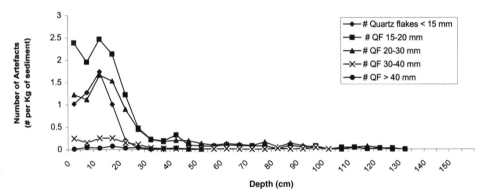

Fig. 6.7. Jiyer Cave, distribution of quartz flakes in depth

1996: 181). Glass and metal, obtained in the colonization period, are also represented in the top 20 cm of the site.

Quartz artifacts comprise a substantial part of the deposit, especially in the top 30 cm. Horsfall divided the number of quartz flakes into size categories and excavation units. Unfortunately, the quantities of quartz flakes less than 15 mm in size are not given for all pits (only 2 of 6 are available), and this must be taken into consideration when interpreting figure 6.7. There is a consistent peak in the discard of all size categories of flaked quartz at a depth of 15 cm at the site, and flakes between 15 and 20 mm in length are the most numerous. Eleven grindstones lay on the surface layer of the site, while four were excavated, one per 10 cm in the top 30 cm and the lowest at 60 cm with adhering red pigment (Horsfall, 1987b).

Mulgrave River 2 is a large, open site covering about 70 m² and located on a raised area on the bank of an unnamed creek near the Mulgrave River

Fig. 6.5. Jiyer Cave, age-depth plots

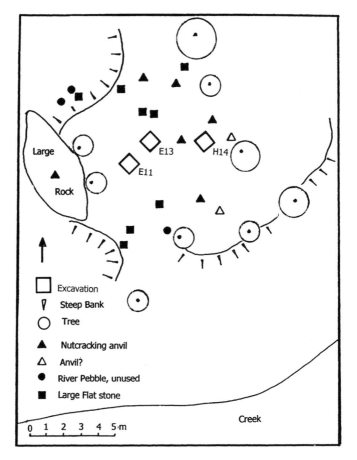

Fig. 6.8. Horizontal plan of Mulgrave River 2 site (after Horsfall, 1987b). Reproduced with permission from the author.

(Horsfall, 1996: 183; see fig. 6.8). The site is believed to be above flood level (Horsfall, 1996). The date obtained for the lower excavated level, between 60 and 65 cm, is 2690 ± 100 B.P. (Horsfall, 1987b, 1996: 184; see fig. 6.9). The site was excavated in 5-cm units, to a total depth of 65 cm (Horsfall, 1996). Three stratigraphic levels encountered are a granular level, 20 cm thick, followed by a thick band of grayish brown soil which cannot be distinguished from the third layer (Horsfall, 1996). Carbon dates obtained from the site are internally consistent.

No faunal remains were found at this site, but otherwise the range of archaeological materials is very similar to Jiyer Cave (fig. 6.10). Increases in most deposited items are confined to the last 300 years, with the exception

Fig. 6.9. Mulgrave River 2, age-depth plots.

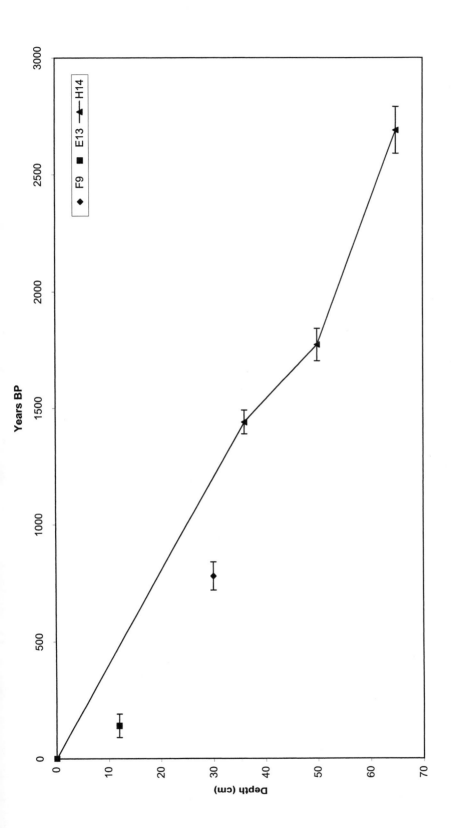

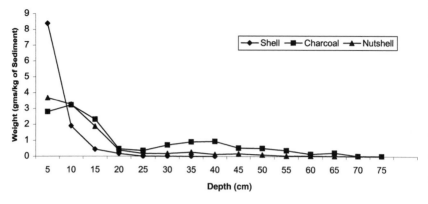

Fig. 6.10. Mulgrave River 2, distribution of shell, nutshell, and charcoal in depth

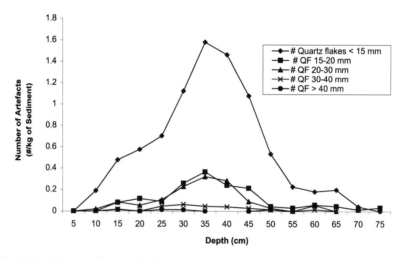

Fig. 6.11. Mulgrave River 2, distribution of quartz flakes in depth

of quartz artifacts and ochre (Horsfall, 1996). Horsfall divided the stone flakes into number per size category per excavation unit (Horsfall, 1987b). The quartz artifacts peak in deposition at approximately 1440 B.P. Quartz flakes greater than 15 mm occur in the greatest quantity throughout all levels of the site (fig. 6.11). There are several peaks in the distribution of ochre at the site, some of which appear to coincide with quartz flake deposition. Shell, nutshell, and charcoal, although present with depth in the deposit, show increased deposition in the upper 20 cm of the site (fig. 6.10). Nutshell remains begin to increase at about 300 B.P., and most are deposited in the top 15 cm. Five different types of nutshell were identified from the deposit, two or three of these are toxic (Horsfall, 1996). Shellfish from mangrove environments were identified, increasing significantly in the top 5 cm.

No grindstones were excavated from the site, although 20 were located on the surface. Artifacts made from European goods are also represented in the top 20 cm of this site (Horsfall, 1996).

Models of Rain Forest Occupation

Two researchers have developed models for northern Queensland rain forest prehistory: N. Horsfall and R. Cosgrove. Both researchers have based their models on ethnographic depictions and reconstructions of rain forest societies at European contact. Both have made strong claims for rain forest–adapted and specialized semi-sedentary hunter-gatherers and for the early occupation of rain forests.

Horsfall's Model

Horsfall has interpreted the data from excavated rain forest sites as matching the patterns of Aboriginal rain forest utilization reported in the early days of European settlement (Horsfall and Furay, 1988: 26). Ethnographic descriptions are seen as suggesting the existence of a sophisticated and unique cultural adaptation to local rain forest ecosystems in which the environment was skillfully managed and utilized. Management of floral resources provided a predictable resource base on which elaborate cultural systems could be built. Horsfall argues that a range of toxic plant species endemic to these rain forests were utilized in great quantities on an everyday basis (Horsfall, 1987a: 57).

Although a lot of time was required to process these toxic nuts, this had several beneficial "flow on effects" (Horsfall, 1987a: 58). Plant resources, specifically nuts, provided the main staple over much of the year and were seen as the main determining factor in rain forest occupation. The dependence on plant resources supported high-density populations in a semi-sedentary lifestyle. A range of plants had seasonal fruiting cycles so that plants were available over several months of the year, and some could also be stored to increase quantities of food available in lean periods (Harris, 1978; Horsfall, 1987a: 58–59 and 1984: 169). A range of material culture documented and collected from 1880 to 1910 (Khan, 1993; McInnes, 1995) has been seen as specialized and uniquely adapted to extracting resources from the rain forest environment and included nutcrackers and incised grindstones, used to crack hard nutshells and grind the nut kernels, and baskets used to leach the toxic nuts.

Horsfall suggests the rain forest "culture" of ethnographic times could only have developed since the expansion of the rain forests from relict areas 9000 years ago (Horsfall and Furay, 1988: 30). She considers that the earliest direct evidence for the occupation of the rain forest is at 5100 B.P. at Jiyer

Cave, with an intensive utilization of toxic plants in the last 1000 years, indicated by increased deposition of archaeological materials (Horsfall, 1996). Horsfall suggests there is great potential for the initial occupation of rain forests before 38,000 B.P. when the tablelands were covered in rain forest. Based on the paleoenvironmental data, Horsfall suggests that a rain forest–based occupation was not possible for 30,000 years, although the resources available in restricted rain forest areas could have provided "useful additions" to the resource base during this time (Horsfall, 1987b).

Horsfall also attempts to address the relevance of rain forest archaeology for debates regarding economic intensification (Horsfall, 1987b). Some archaeologists have argued that a dramatic amplification of social and economic aspects of hunter-gatherer societies occurred during the last 4000 years (Lourandos, 1980, 1983, 1985, 1997; Lourandos and Ross, 1994; David and Lourandos, 1999). Archaeological indices of this intensification include increases in site numbers/site establishment, increases in intensity of site use, broadening of the resource base, the use of marginal environments, and increasing regionalization (see Lourandos, 1997, for details; see also Bird and Frankel, 1991; Frankel, 1993).

Horsfall suggests that there are several features of rain forest archaeology that might be related to economic intensification. These include the remains of toxic food plants, deposited in increasing quantities over time; stone artifacts associated with the processing of toxic plants, also deposited in increasing quantities; more intensive site use, as indicated by the increased deposition of stone artifacts; and a high proportion of relatively young sites (Horsfall, 1987b: 263). The idea of a north-south break in cultural systems during the last 1000 years (Horsfall, 1987b) and regional links in art styles between Kennedy A and Laura (Brayshaw, 1995; Horsfall and Furay, 1988; see also Cole and David, 1992) might also be added.

Cosgrove's Model

Cosgrove also draws upon ethnographic reports describing a "unique" rain forest culture. Cosgrove argues that nut use and a specialized material culture were the primary features of the rain forest occupation. He interprets the archaeological evidence of "abundant charred nutshells" (Cosgrove, 1996: 907) as implying the use of rain forest nut species as early as 3500 B.P. and toxic species at 1000 B.P. Unique material culture is described as a "catalyst" whereby a range of specialized resource options could be exploited. This development is linked to semi-sedentary cultures and a rich and dependable food resource (Cosgrove, 1996). Some stone artifacts were made from local sources, and this has been interpreted to indicate the development of a territorially bounded rain forest culture. He also suggests that clearing areas for campsites and storing large quantities of nuts at these campsites may have resulted in the present-day clumped location of economically important nut trees and that such areas may indicate the locations of past campsites (Cosgrove, 1996).

The main difference, compared to Horsfall's model, is that Cosgrove makes a bolder argument for long-term, continuous rain forest occupation. In his view, the archaeological evidence suggests that human occupation and exploitation of rain forest resources occurred by at least 5000 B.P. In addition, successful exploitation and manipulation of tropical rain forests of Australia may have a very long antiquity, before the drowning of the Torres Strait land bridge, before 9000 B.P. A "fully functioning" rain forest society could have existed during the pre-glacial, glacial, and post-glacial periods, based on a hypothetical environmental model that suggests rain forests may have existed in a fragmented form along the coastline (Cosgrove, 1996). The presence of a linked network of rain forest vegetation types and relict patches of rain forest are seen as supporting an "endemic" Aboriginal culture which "may have formed the core of the Holocene rain forest society documented at contact" (Cosgrove, 1996: 909).

Discussion

At present there is no direct archaeological evidence for a Pleistocene occupation of the rain forests of northeast Queensland, although this is unsurprising given the lack of research in pertinent sites. Hunter-gatherers colonized Australia sometime during the last glacial cycle (Roberts et al., 1990; Smith and Sharp, 1993; Fullagar et al., 1996; O'Connell and Allen, 1998; Thorne et al., 1999; Cosgrove, 1999; Turney et al., 2001) and prior to the onset of severe climatic aridity (Bowman, 2000: 226). There is also evidence that a wide range of niches were occupied around the continent before 30,000 B.P. (Smith and Sharp, 1993). Archaeological research in open woodlands immediately west of the rain forest block suggests that hunter-gatherers had occupied the region by 39,000 B.P. (Rosenfeld et al., 1981; David, 1994; Morwood and Hobbs, 1995). The oldest archaeological evidence for occupation of Southeast Asia is roughly contemporaneous with that of Greater Australia (Bowdler, 1993; Lourandos, 1997: 81). Pleistocene occupation dates of around 40,000 B.P. have been obtained in Thailand, Sarawak, Southern Sulawesi, the Malay Peninsula, island Malaysia and North Vietnam, West New Britain, New Ireland, and New Guinea (Bowdler, 1993; Pavlides and Gosden, 1994: 610; see also chapters 4 and 5 in this book).

Two indirect lines of evidence argue for early occupation of the northeast Queensland rain forests. Twenty-five waisted axes have been collected from the surface in open, undated rain forest–open forest locations around Mackay (Lampert, 1981: 190). Groube has argued that similarities exist between the Mackay and New Guinea axes and that the Mackay axes may have been used in northeast Queensland in forest clearance to promote the growth of many forest fringe plants (Groube, 1989; Cosgrove, 1996; Lourandos, 1997). Stylistic similarities have been used to argue that the Australian tools may have a Pleistocene antiquity (Groube, 1986; Cosgrove, 1996). However, the antiquity and function of these tools are far from resolved. There is little supporting evidence that the Mackay tools have a great antiquity or that they were used in forest clearance. None of the northeastern Queensland axes have been excavated and no residue analysis has been attempted.

Evidence from deep-sea and tableland pollen cores has also been used to support the idea of an early occupation of northeast Queensland rain forests. Kershaw has interpreted dramatic increases in quantities of microscopic charcoal particles in pollen cores as evidence for anthropogenic burning. One increase occurs at approximately 120,000 B.P. in the deep-sea core Ocean Drilling Program 820, 80 km to the northeast of Cairns (Kershaw et al., 1993), and the other at 38,000 B.P. at Lynch's Crater (Kershaw, 1974, 1976, 1978, 1985, 1986). The idea of anthropogenic burning at 120,000 B.P. is seen as problematic by many Australian archaeologists (Anderson, 1994; Hope, 1994; White, 1994), but the idea of anthropogenic burning at 38,000 B.P. appears to coincide with other evidence of regional occupation. However, this data does not provide direct evidence that the rain forests were being occupied. It may represent the burning of other non–rain forest areas (Butler, 1998; Walker, 1991). Hopefully, future palynological research will clarify this (Chen, 1986). Ongoing archaeological work by Cosgrove may also provide direct evidence for occupation of this antiquity.

These models seek to explain changes in the Holocene archaeological record by referring to descriptive ethnographic sources. Yet there is nothing in the archaeological record from excavated sites that could be interpreted to indicate occupation by rain forest specialists. A more critical appraisal of the ethnographic data renders problematic the idea of semi-sedentary, nut-dependent specialists.

Ethnographic descriptions were made during the time of European incursion into the rain forest during the 1860s, when societies had already suffered great dislocation, reduction of territories and foraging ranges, famine, disease, and warfare (Dixon, 1976: 219; Bottoms, 1999). When this is taken into account, it seems likely that sedentism, resource use, territorial boundaries, and occupation strategies over a year had been extensively altered by post-colonial processes and did not represent "traditional" patterns.

Early researchers argued that rain forest populations were physically and culturally different from their neighbors, and they were inclined to emphasize the differences between rain forest and non–rain forest populations (Birdsell, 1967, 1977; Tindale, 1974). However, studies of blood proteins and cranial morphology suggest similarities between forest people and non-forest people from Queensland and New South Wales (Macintosh et al., 1973; Simmons, 1973). At present there is little compelling evidence for long-term isolation and separation between rain forest groups and neighbors, and more sophisticated genetic analysis is required to further test this claim. Early linguistic studies have also created the impression of an extremely distinctive language (Tindale and Birdsell, 1941); however, subsequent studies show that although there are differences in dialects, the degree of distinctiveness and variability is comparable to other linguistic regions throughout the continent, and there are clear linguistic links to neighboring groups (Dixon, 1976).

Material culture documented and collected at contact also cannot be seen as specifically adapted to extracting endemic rain forest resources. Material culture, including waterproof huts, bark cloth, wooden swords and shields, has direct correlates in other non–rain forest environments. Other items argued as being adapted to the "unique resource base" include nut-cracking stones, incised grindstones, and baskets (Harris, 1978: 121; Horsfall, 1984: 165; Horsfall and Furay, 1988; Cosgrove, 1996). The apparent differences of these items seem to be partly due to the collection practices of ethnographers (Khan, 1993; McInnes, 1995) and to superficial stylistic differences rather than functional differences. Again, all of these items have direct functional equivalents in other environments and were used with similar resource items, including toxic and non-toxic nuts (see McCarthy, 1967; Morwood, 1980). Other items, including "ooyurkas," do appear to be "different," but the extent and reason for the differences remain unclear. These artifacts are from undated contexts and their uses are not known.

Given the problems with ethnographic accounts, it is clear that these models of rain forest use need to be tested against archaeological data. This presents us with a range of methodological challenges. We need to assess the impact of taphonomic factors and differentiate between the anthropogenic and natural contributions to sites (Asmussen, in preparation). Residue and use-wear analysis of stone tools can identify those items used for processing animals and plants. Also, inter-site comparisons will be an important part of this type of work, including comparisons with overseas data (Asmussen, in preparation).

Conclusions

Rain forest archaeology has received little attention in Australian literature and has been discussed in relation with the intensification debate (Lourandos, 1997). However, the amount and type of archaeological data from the rain forest cannot address the issue of intensive plant use or how the forest compares to other regions (Asmussen, in preparation). Moreover, to assess whether there was a population increase of the type defended by Lourandos (1997), we would have to be able to identify changes in population densities inside the rain forest; this is not possible at present. When was the rain forest initially occupied? The only radiocarbon dates available to answer this question date back to the mid-Holocene and were collected from the basal strata at one site, but do not represent the initial settlement of the whole Australian forest. Likewise, a late Pleistocene antiquity is indeed possible, but it remains to be shown with direct evidence. Ecological data, for their part, suggest that in the past the Australian forest has been heterogeneous and very productive, undermining the claims of unity and marginality of this type of environment. Nonetheless, new paleoenvironmental data are required to evaluate resource availability, the impact of climate change, and human intervention in these patchy ecosystems.

Acknowledgments

Thanks to Julio Mercader and Peter Hiscock for providing the opportunity to contribute to this volume. The comments of Julio Mercader, Peter Hiscock, Paul McInnes, Cedney Kinney, Chris Clarkson, and anonymous referees improved the quality of this chapter. Many thanks to Chris Clarkson for technical support.

References

Adam, P. (1992): *Australian Rainforests.* Oxford: Clarendon Press.
Allen, J., and P. Kershaw (1996): The Pleistocene-Holocene transition in Greater Australia. In *Humans at the end of the Ice Age: The archaeology of the Pleistocene-Holocene transition,* edited by L. Guy, 175–199. New York: Plenum Press.
Anderson, A. (1994): Comment on J. Peter White's paper "Site 820 and the evidence for early occupation in Australia." *Quaternary Australasia* 12:30–31.
Archer, M., and H. Brayshaw (1978): Recent local faunas from excavations at Hervey's Range, Kennedy, Jourama, and Mount Roundback, northeastern Queensland. *Memoirs of the Queensland Museum* 18(2):165–177.
Ash, J. (1988): The location and stability of rainforest boundaries in north eastern Queensland. *Australian Journal of Biogeography* 15:619–630.
Asmussen, B. (in preparation): Re-analysis of archaeological sites from Carnarvon Gorge, Central Queensland Highlands. Working title for Ph.D. diss., School of Archaeology and Anthropology, Australian National University.
Bell, F. C., W. Winter, L. I. Pahl, and R. G. Atherton (1987): Distribution, area, and tenure of rainforest in northeastern Australia. *Proceedings of the Royal Society of Queensland* 98:27–39.
Bird, C.F.M., and D. Frankel (1991): Problems in constructing a prehistoric regional sequence: Holocene south-east Australia. *World Archaeology* 23(2):179–192.

Birdsell, J. B. (1967): Preliminary data on the tri-hybrid origin of the Australian Aborigines. *Archaeology and Physical Anthropology in Oceania* 2:100–155.

———. (1977): The recalibration of a paradigm for the first peopling of Greater Australia. In *Sunda and Sahul prehistoric studies in Southeast Asia, Melanesia, and Australia*, edited by J. Allen, J. Golson, and R. Jones, 113–167. London: Academic Press.

Bohte, A., and A. P. Kershaw (1999): Taphonomic influences on the interpretation of the palaeoecological record from Lynch's Crater, northeastern Australia. *Quaternary International* 57–58:49–59.

Boland, D. J., M.I.H. Brooker, G. M. Chippendale, N. Hall, B.P.M. Hyland, R. D. Johnston, D. A. Kleining, and J. D. Turner (1984): *Forest trees of Australia.* South Melbourne: Commonwealth Scientific Industrial Research Organisation Press.

Bottoms, T. (1999): *Djabugay Country: An Aboriginal history of North Queensland.* St. Leonards: Allen and Unwin.

Bowdler, S. (1992): *Homo sapiens* in Southeast Asia and the antipodes: Archaeological versus biological interpretations. In *The evolution and dispersal of modern humans in Asia and Japan*, edited by T. Akazawa., K. Aoki, and T. Kimura, 559–589. Tokyo: Hokusen-sha.

Bowdler, S. (1993): Sunda and Sahul: A 30 K yr B.P. culture area? In *Sahul in review: Pleistocene archaeology in Australia, New Guinea, and Island Melanesia*, edited by M. Smith, M. Spriggs, and B. Fankhauser, 60–70. Occasional Papers in Prehistory, no. 24. Canberra: Department of Prehistory, Research School of Pacific Studies Press.

Bowman, D.M.J.S. (2000): *Australian rainforests: Islands of green in a land of fire.* New York: Cambridge University Press.

Brayshaw, H. (1995): *Well-beaten paths: Aborigines of the Herbert Burdekin district, north Queensland, an ethnographic and archaeological study.* Townsville: James Cook University Press.

Butler, D. (1998): Environmental change in the Quaternary. In *Ngarrabulgan: Geographical investigations in Djungan Country, Cape York Peninsula*, edited by B. David, 78–97. Clayton: Monash University Publications in Geography and Environmental Science.

Campbell, J. B. (1982): Automatic sea-food retrieval systems: Evidence from Hinchinbrook Island and its implications. In *Coastal archaeology in eastern Australia*, edited by S. Bowdler, 96–107. Canberra: Australian National University Press.

Catling, P. C., and A. E. Newsome (1981): Responses of the Australian vertebrate fauna to fire: An evolutionary approach. In *Fire and the Australian biota*, edited by A. M. Gill., R. H. Groves, and I. R. Noble, 273–310. Canberra: Australian Academy of Science.

Chen, Y. (1986): Early Holocene vegetation dynamics at Lake Barrine basin, northeast Queensland, Australia. Ph.D. diss., Australian National University.

———. (1988): Early Holocene population expansion of some rainforest trees at Lake Barrine basin, Queensland. *Australian Journal of Ecology* 13:255–233.

Cole, N., and B. David (1992): "Curious drawings" at Cape York Peninsula: An account of the rock art of the Cape York Peninsula region of north-eastern Australia and an overview of some regional characteristics. *Rock Art Research* 9(1): 3–23.

Cosgrove, R. (1996): Origin and development of Australian Aboriginal tropical rainforest culture: A reconsideration. *Antiquity* 70:900–912.

———. (1999): Forty-two degrees south: The archaeology of the late Pleistocene Tasmania. *Journal of World Prehistory* 13:357–402.

Crome, F.H.J. (1990): Vertebrates and successions. In *Australian tropical rainforests: Science, values, meaning,* edited by J. L. Webb and J. Kikkawa, 53–64. Melbourne: Commonwealth Scientific Industrial Research Organisation.

Crowley, G. M., P. Anderson, A. P. Kershaw, and J. Grindrod (1990): Palynology of a Holocene marine transgressive sequence, lower Mulgrave River valley, northeast Queensland. *Australian Journal of Ecology* 15:231–240.

David, B. (1994): A space-time odyssey: Rock art and inter-regional interaction in northeast Australian prehistory. Ph.D. diss., University of Queensland.

David, B., and H. Lourandos (1999): Landscape as mind: Land use, cultural space, and change in north Queensland prehistory. *Quaternary International* 59:107–123.

Dixon, R.M.W. (1976): Tribes, languages, and other boundaries in northeast Queensland. In *Tribes and boundaries in Australia,* edited by N. Peterson, 207–236. Canberra: Australian Institute of Aboriginal Studies Press.

Frankel, D. (1993): Pleistocene chronological structures and explanations: A challenge. In *Sahul in review: Pleistocene archaeology in Australia, New Guinea, and Island Melanesia,* edited by M. Smith, M. Spriggs, and B. Fankhauser, 24–36. Occasional Papers in Prehistory, no. 24. Canberra: Department of Prehistory, Research School of Pacific Studies.

Fullagar, R., D. Price, and L. Head (1996): Early human occupation of northern Australia: Archaeology and thermoluminescence dating of Jimium rock shelter, Northern Territory. *Antiquity* 70:751–773.

Gill, A. M., J.R.L. Hoare, and N. P. Cheney (1990): Fires and their effects in the wet-dry tropics of Australia. *Ecological Studies* 84:159–178.

Groube, L. (1989): The taming of the rain forest: A model for late Pleistocene forest exploitation in New Guinea. In *Foraging and farming: The evolution of plant exploitation,* edited by D. R. Harris and G. C. Hilman, 292–317. London: Unwin Hyman.

Hamilton, A. (1989): African forests. In *Tropical rain forest ecosystems, biogeographical and ecological studies, ecosystems of the world,* edited by H. Lieth and M.J.A. Werger, 155–182. Amsterdam: Elsevier.

Harris, D. R. (1978): Adaptation to a tropical rainforest environment: Aboriginal subsistence in northeastern Queensland. In *Human behaviour and adaptation,* edited by N. Blurton-Jones and V. Reynolds, 113–134. Symposia for the Society for the Study of Human Biology, vol. 18. London: Halstead Press.

———. (1987): Aboriginal subsistence in a tropical rainforest environment: Food procurement, cannibalism, and population regulation in northeastern Australia. In *Food and evolution: Toward a theory of human food habits,* edited by M. Harris and E. B. Ross, 357–385. Philadelphia: Temple University Press.

Harrison, S. P., and J. Dodson (1993): Climates of Australia and New Guinea since 18,000 yr B.P. In *Global climates since the last glacial maximum,* edited by W. E. Wright Jr., J. E. Kutzbach, T. Webb, W. F. Ruddman, F. A Street-Perrott, and P. J. Bartelin, 265–293. Minneapolis: University of Minnesota Press.

Hiddins, L. (1999): *Explore wild Australia with the Bush Tucker Man.* Sydney: Viking Press.

Hope, G. S. (1994): Comment on ODP site 820 and inference of early occupation in Australia. *Quaternary Australasia* 12:32–33.

Hopkins, M. S. (1990): Disturbance—the forest transformer. In *Australian tropical rainforests: Science, values, meaning,* edited by J. L Webb and J. Kikkawa, 40–52. Melbourne: Commonwealth Scientific Industrial Research Organisation Press.

Hopkins, M. S., A. W. Graham, and R. Hewett (1990): Short note: Evidence of late Pleistocene fires and *Eucalypt* forests from a north Queensland humid tropical rainforest site. *Australian Journal of Ecology* 15:345–347.

Hopkins, M. S, Ash, J, Graham, A. W, Head, J. and Hewett, R. K. (1993): Charcoal evidence of the spatial extent of the *Eucalyptus* woodland expansions and rainforest contractions in north Queensland during the late Pleistocene. *Journal of Biogeography* 20:357–372.

Horsfall, N. (1983): Excavations at Jiyer Cave, northeast Queensland: Some results. In *Archaeology at ANZAAS,* edited by M. Smith, 172–178. Perth: Western Australian Museum Publications.

———. (1984): Theorising about north Queensland prehistory. *Queensland Archaeological Research* 1:164–172.

———. (1987a): Aborigines and toxic north-eastern Queensland rainforest plants. In *Toxic plants and animals: A guide for Australia,* edited by J. Covacevich, P. Davie, and J. Pearn, 57–63. Brisbane: Queensland Museum Publications.

———. (1987b): Living in rainforest: The prehistoric occupation of north Queensland's humid tropics. Ph.D. diss., James Cook University of North Queensland.

———. (1996): Holocene occupation of the tropical rainforests of north Queensland. In *The archaeology of northern Australia,* edited by P. Veth and P. Hiscock, 175–190. Tempus, 4. Brisbane: Anthropology Museum, University of Queensland.

Horsfall, N., and M. Fuary (1988): The Aboriginal heritage values of Aboriginal archaeological sites and associated themes in and adjacent to the area nominated for world heritage listing in the wet tropical rainforest region of north east Queensland. A report to the State of Queensland.

Kershaw, A. P. (1970): A pollen diagram from Lake Euramoo, northeast Queensland, Australia. *The New Phytologist* 69:785–805.

———. (1971): A pollen diagram from Quinkan Crater, north-east Queensland, Australia. *The New Phytologist* 70:669–681.

———. (1974): A long continuous pollen sequence from north-eastern Australia. *Nature* 251:222–223.

————. (1975): Late Quaternary vegetation change and climate in northeastern Australia. *Royal Society of New Zealand Bulletin* 13:181–187.

————. (1976): A late Pleistocene and Holocene pollen diagram from Lynch's Crater, northeastern Queensland, Australia. *The New Phytologist* 77:469–498.

————. (1978): Record of the last interglacial-glacial cycle from northeastern Queensland. *Nature* 272:159–160.

————. (1983): A late Holocene pollen diagram from Lynch's Crater, northeastern Queensland, Australia. *The New Phytologist* 94:669–682.

————. (1985): An extended late Quaternary vegetation record from northeastern Queensland and its implications for the seasonal tropics of Australia. *Proceedings of the Ecological Society of Australia* 13:179–189.

————. (1986): Climatic change and Aboriginal burning in north-east Australia during the last two glacial/interglacial cycles. *Nature* 322:47–49.

————. (1995): Environmental change in Greater Australia. *Antiquity* 69:656–675.

Kershaw, A. P., G. M. McKenzie, and A. McMinn (1993): A Quaternary vegetation history of northeastern Queensland from pollen analysis of ODP site 820. *Proceedings of the Ocean Drilling Program, Scientific Results* 133:107–114.

Khan, K. (1993): *Catalogue of the Roth collection of Aboriginal artefacts from north Queensland.* Technical Reports of the Australian Museum, no. 12, vols. 1 and 2. Sydney: Australian Museum Press.

Kikkawa, J. (1990): Specialisation in the tropical rainforest. In *Australian tropical rainforests: Science, values, meaning,* edited by J. L. Webb and J. Kikkawa, 67–73. Melbourne: Commonwealth Scientific International Research Organisation.

Lampert, R. (1981): *The great Kartan mystery: Terra Australis 5.* Canberra: Department of Prehistory, Research School of Asian and Pacific Studies, Australian National University Press.

Latz, P. (1995): *Bushfires and Bushtucker: Aboriginal plant use in central Australia.* Alice Springs: IAD Press.

Lourandos, H. (1980): Change or stability? Hydraulics, hunter-gatherers, and population in temperate Australia. *World Archaeology* 11:245–266.

————. (1983): Intensification: A late Pleistocene-Holocene archaeological sequence from southeastern Victoria. *Archaeology in Oceania* 18:81–94.

————. (1985): Intensification and Australian prehistory. In *Prehistoric hunter-gatherers: The emergence of cultural complexity,* edited by T. D Price and J. Brown, 385–423. Orlando: Academic Press.

Lourandos, H. L. (1997): *Continent of hunter-gatherers: New perspectives in Australian prehistory.* Cambridge: Cambridge University Press.

Lourandos, H., and A. Ross (1994): The great "intensification debate": Its history and place in Australian archaeology. *Australian Archaeology* 39:54–63.

Macintosh, N.W.G., and S. L. Lanarch (1973): A cranial study of the Aborigines of Queensland with a contrast between Australian and New Guinea crania. In *The human biology of Aborigines in Cape York,* edited by R. L. Kirk, 1–12. Australian Aboriginal Studies, no. 44, Human Biology Series, no. 5. Canberra: Australian Institute of Aboriginal Studies.

McCarthy, F. D. (1967): *Australian Aboriginal implements, including stone, shell and teeth implements.* Sydney: Australian Museum Press.

McInnes, H. (1995): Through Roth-coloured glasses. Honors diss., University of Queensland.

Miller, J. B., K. W. James, and P.M.A. Maggiore (1993): *Tables of composition of Australian Aboriginal foods.* Canberra: Aboriginal Studies Press.

Morwood, M. (1980): Art and stone: Towards a prehistory of central western Queensland. Ph.D. diss., Australian National University.

Morwood, M. J., and D. R. Hobbs (1995): *Quinkan prehistory: The archaeology of Aboriginal art in southeast Cape York, Australia.* Tempus, 3. St. Lucia: Anthropology Museum University of Queensland.

O'Connell, F., and J. Allen (1998): When did humans first arrive in Greater Australia and why is it important to know? *Evolutionary Anthropology* 6:132–146.

Pavlides, C., and C. Gosden (1994): 35,000-year-old sites in rainforests of West New Britain, Papua New Guinea. *Antiquity* 68:604–610.

Prance, G. T. (1989): American tropical rainforests. In *Tropical rain forest ecosystems, biogeographical and ecological studies, ecosystems of the world,* edited by H. Lieth and M.J.A. Werger, 99–132. Amsterdam: Elsevier.

Roberts, R., R. Jones, and M. Smith (1990): Thermoluminescence dating of a 50,000-year-old human occupation site in northern Australia. *Nature* 345:153–156.

Rosenfeld, A., D. Horton, and J. Winter (1981): *Early man in north Queensland.* Terra Australis 6. Canberra: Department of Prehistory, Research School of Pacific Studies.

Simmons, R. T. (1973): Blood group and genetic studies in the Cape York area. In *The human biology of Aborigines in Cape York,* edited by R. L. Kirk, 13–24. Australian Aboriginal Studies, no. 44, Human Biology Series, no. 5. Canberra: Australian Institute of Aboriginal Studies Press.

Smith, M. A. (1993): Sahul in review: An introduction. In *Sahul in review: Pleistocene archaeology in Australia, New Guinea, and Island Melanesia,* edited by M. Smith, M. Spriggs, and B. Fankhauser, 1–7. Occasional Papers in Prehistory, no. 24. Canberra: Department of Prehistory, Research School of Pacific and Asian Studies.

Smith, M. A., and N. D. Sharp (1993): Pleistocene sites in Australia, New Guinea, and Island Melanesia: Geographic and temporal structure of the archaeological record. In *Sahul in review: Pleistocene archaeology in Australia, New Guinea, and Island Melanesia,* edited by M. Smith, M. Spriggs, and B. Fankhauser, 37–59. Occasional Papers in Prehistory, no. 24. Canberra: Department of Prehistory, Research School of Pacific and Asian Studies.

Smith, M. A., M. Spriggs, and B. Fankhauser (1993): Sahul in review: An introduction. In *Sahul in review: Pleistocene archaeology in Australia, New Guinea, and Island Melanesia,* edited by M. Smith, M. Spriggs, and B. Fankhauser, 1–7. Occasional Papers in Prehistory, no. 24. Canberra: Department of Prehistory, Research School of Pacific and Asian Studies.

Stocker, G. C. (1981): Regeneration of a north Queensland rain forest following felling and burning. *Biotropica* 13(2):86–92.

Stocker, G. C., and G. L. Unwin (1989): The rainforests of northeastern Australia — their environment, evolutionary history, and dynamics. In *Tropical rain forest ecosystems, biogeographical and ecological studies, ecosystems of the world*, edited by H. Lieth and M.J.A Werger, 241–259. Amsterdam: Elsevier.

Strahn, R. (1983): *The Australian Museum complete book of Australian mammals*. Sydney: Australian Museum Press.

Thorne, A., R. Grüün, G. Mortimer, N. A. Spooner, J. J. Simpson, M. McCulloch, L. Taylor, and D. Curnoe (1999): Australia's oldest human remains: Age of the Lake Mungo 3 skeleton. *Journal of Human Evolution* 36:591–612.

Tindale, N. B. (1974): *Aboriginal tribes of Australia*. Canberra: Australian National University Press.

Tindale, N. B., and J. B. Birdsell (1941): Results of the Harvard Adelaide University anthropological expedition, 1938–1939: Tasmanoid tribes in north Queensland. *Records of the Australian Museum* 7:1–9.

Tracey, J. G. (1982): *The vegetation of the humid tropical region of north Queensland*. Melbourne: Commonwealth Scientific Institute Research Organisation Press.

Turney, C., M. Bird, L. Fifield, M. Smith, C. Dortch, R. Grun, E. Lawson, L. Ayliffe, G. Miller, J. Dortch, and R. Cresswell (2001): Early human occupation at Devil's Lair, southwestern Australia, 50,000 years ago. *Quaternary Research* 55:3–13.

Walker, D. (1991): Fine-resolution pollen analysis and rainforest dynamics. In *The rainforest legacy: Australian national rainforests study*, edited by G. Werren and P. Kershaw, 139–145. Special Australian Heritage Publications Series, no. 7(3). Canberra: Australian Government Printing Service.

Walker, D., and Y. Chen (1987): Palynological light on tropical rainforest dynamics. *Quaternary Science Review* 6:77–92.

Webb, L. J., and J. G. Tracey (1981): Australian rainforests: Patterns and change. In *Ecological biogeography of Australia*, edited by A. Keast, 607–691. Boston: Dr. W. Junk Publishers.

Werren, G. L., and A. P. Kershaw (1991): *The rainforest legacy Australian national rainforest study: History, dynamics, and management*. Vol. 3. Canberra: Australian Heritage Commission.

White, J. P. (1994): Site 820 and the evidence of early occupation in Australia. *Quaternary Australasia* 12:21–23.

White, J. P., and J. F. O'Connell (1982): *A prehistory of Australia, New Guinea, and Sahul*. Sydney: Academic Press.

Whitmore, T. C. (1989): Southeast Asian tropical forests. In *Tropical rain forest ecosystems, biogeographical and ecological studies, ecosystems of the world*, edited by H. Lieth and M.J.A. Werger, 195–220. Amsterdam: Elsevier.

Woolston, F. P. (1965): An excavation at Platform Gallery, near Cooktown, April 1965. Manuscript. Canberra: Australian Institute of Aboriginal Studies.

PART III

THE LAST FRONTIER: NEWCOMERS IN A NEW WORLD

CHAPTER 7

Late Glacial and Early Holocene Occupation of Central American Tropical Forests

Anthony J. Ranere and Richard G. Cooke

The focus of archaeological research in Central America, like other tropical regions of the world, has not been on early hunter-gatherer occupants of tropical forests but rather on farmers and village dwellers living in anthropogenically modified habitats. Nonetheless, a gradual accumulation of data, with some acceleration of late, has made it quite clear that human populations were living in Central American forests at least by the late glacial stage (circa 14,000–10,500 B.P.) of the last glaciation (henceforth LGS) and continuously thereafter, initially as hunter-gatherers and later as agriculturalists.

Central America today is a mosaic of forests, savannas, agricultural fields, and pastures, most resulting from human manipulation to one degree or another. The same can be said of the Central America first described by Spanish chroniclers in the sixteenth century A.D., although the distribution and relative proportions of forest to savannas to agricultural fields differed considerably (there were no pastures) (Cooke and Ranere, 1992a; Sauer, 1966). In the absence of human interference, the current climate of Central America would potentially support several tropical vegetation types ranging from very humid forests through various seasonal forest formations to more open habitats, such as wooded savannas and thorn scrub woodlands (fig. 7.1). Paleoclimatic data indicate that any human population attempting an overland passage between North and South America during the LGS would have had to traverse some large expanses of tropical forests floristically unlike modern formations, but forests nonetheless (fig. 7.2; Piperno et al., 1991a; Piperno and Pearsall, 1998: fig. 2.4). The dates for human occupation at Monte Verde in Chile make it clear that people entered South America well before 12,500 B.P. (Dillehay, 1997, 2000: 160–168). Recent reports of late Pleistocene–early Holocene maritime adaptations along the coast of Peru may provide clues about the nature of that entry (Keefer et al., 1998; Sandweiss et al., 1998). Whether or not a maritime-oriented population moved south along the Pacific coast of Central America before 12,500 B.P.—perhaps avoiding tropical forests—is an intriguing question to

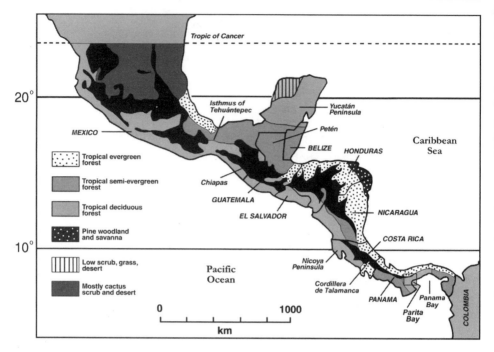

Fig. 7.1. Central America, showing the distribution of potential (non-anthropogenic) modern vegetation types. (Re-drawn from Piperno and Pearsall 1998: figure 2.2). (The rectangular box in Panama highlights the area covered in detail in Figure 7.6).

contemplate, but one whose answer must await the discovery of new evidence. Currently, nearly all the available archaeological evidence for the LGS-Holocene transition in Central America clearly relates to human occupation between about 11,000 and 10,000 B.P. and no earlier.[1] The earliest cultural materials are fluted points, scrapers, and other lithic artifacts technologically and morphologically so similar to those of the Clovis tradition of North America that it is highly likely that they are approximately synchronous with and culturally related to that tradition (Cooke and Ranere, 1992c; Ranere and Cooke, 1991, 1995, 1996).[2] And while there may be some debate about what, exactly, the subsistence economy of Clovis-age populations entailed, no one has yet described them as maritime.

No LGS or early Holocene archaeological assemblages have yet been excavated in Central America with associated faunal or floral remains (Cooke, 1998). Therefore, the case for the first human colonists having occupied tropical forests depends on comparing the distribution of vegetation cover with that of archaeological materials during the appropriate time intervals. Consequently, we will first reconstruct the distribution of forests and other vegetation formations in Central America during the LGS and discuss the

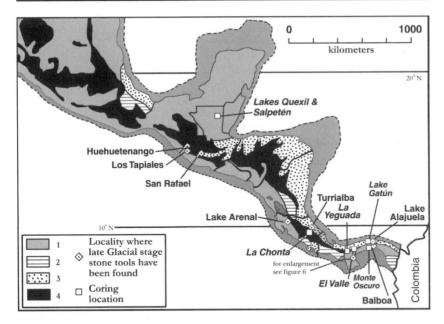

Fig. 7.2. Central America, showing the relationship of pre-10,000 B.P. archaeological sites to coring locations and reconstructed late glacial vegetation and sea levels (redrawn from Piperno and Pearsall, 1998: fig. 2.4).

evidence used in this reconstruction. Next we will talk about those localities that have yielded evidence of human occupation for the 11th and 10th millennia B.P., during which time the LGS-Holocene transition took place, and plot their distribution against hypothetical LGS and early Holocene vegetation. Finally, we will examine the early settlement history of central Pacific Panama, derived from archaeology and lake-core evidence, in order to gain some insights about the lifestyles of Central American hunter-gatherers and the forested and non-forested landscapes they occupied.

The Late Glacial Landscape in Central America

Only a handful of localities in Central America have provided paleoecological records that extend back into the Pleistocene. Nevertheless a general picture of regional climate during the transition from late glacial to early Holocene conditions is emerging from proxy lake-sediment records (Brenner, 1993; Bush and Colinvaux, 1990; Bush et al., 1992; Cooke, 1998; Hooghiemstra et al., 1992; Horn, 1990; Islebe and Hooghiemstra, 1997; Islebe et al., 1995, 1996; Leyden, 1984, 1995, 1997; Leyden et al., 1993; Piperno et al., 1990; Piperno and Pearsall, 1998: 167–182). These provide us with reasonably accurate reconstructions of the vegetation formations that early hunter-gatherer populations in the area would have encountered. Significant shifts away from the vegetation that current climatic conditions could support are indicated for the end of the late glacial maximum (circa 16,000–14,000 B.P.) and for the LGS (circa 14,000–10,500 B.P.), when conditions considerably cooler and drier than today's prevailed. During the LGS, fluctuations in temperatures and precipitation may well have occurred; however, the globally visible Younger Dryas chronozone has only been identified thus far in highland Costa Rica (as the La Chonta stadial), where it was characterized by a temperature depression of 1.5–2.5°C from modern values (Islebe et al., 1995; Islebe and Hooghiemstra, 1997; Leyden, 1995). The beginning of Holocene (10,500 or 9,800 to 8,000 B.P.) appears everywhere to have been somewhat wetter than present.

Cores taken near sea level in the Chagres River basin, Panama, have provided a continuous pollen and phytolith record for the last 11,300 years. The Chagres River was dammed below the coring sites, forming Gatún Lake, during the construction of the Panama Canal. This area today receives approximately 2500 mm of rain annually, seasonally distributed, and has a semi-evergreen forest cover. Its vegetation history is of particular interest because seven whole or broken Paleoindian fluted points have been picked up from the eroded shores of Lake Alajuela (formerly Madden Lake), located 40 km inland along the Chagres River (Bird and Cooke, 1977, 1978; Cooke, 1998; Sander, 1959). A. Bartlett and E. Barghoorn divided the pollen sequence into four periods. Taken together, the pollen record (Bartlett and Barghoorn, 1973) and phytolith record (Piperno 1985, 1988: 204–206) for the earliest paleovegetational horizon (11,300–9000 B.P.) show the presence of a mature forest moist enough to have supported such taxa as *Trichomanes* (Pterophyta: Hymenophyllaceae), palms, *Guatteria* (Annonaceae), and *Tetragastris* or *Protium* (Burseraceae). Other recorded tree taxa are *Phytelephas* (Palmae), *Dialyanthera* and *Virola* (Myristicaceae), *Copaifera* and *Swartzia* (Caesalpinioideae), *Bursera* (Burseraceae), and *Lafoensia* (Lythraceae). Marantaceae (herb understory) and Palmae/Bromeliaceae (epiphytes) are also common, while grass shows low frequencies and sedges (Cyperaceae) are absent. Pollen of Ericaceae and *Iriartea* (Arecaceae), plants which today normally grow at elevations of 1000 and 1200 m, infers that montane forests may have been approximately 700–900 m lower than their modern range would indicate (Ranere, 1980: 43). The canopy might also have been more open than in modern analog forests, reflecting reduced annual precipitation (Piperno, 1988).

A second pollen and phytolith record for Panama, which extends back into the LGS, has been recovered from Laguna de la Yeguada (650 m in elevation) in central Pacific Panama (Bush et al., 1992; Piperno et al., 1990,

1991a,b). This lake, which has no outlet, measures 1.5 × 0.75 km and drains an area of 5.5 km². With a mean annual temperature of 23°C and annual rainfall in excess of 3500 mm (over 90% falling during the May to December wet season), its catchment would support premontane wet forest in the absence of human interference (Estadística Panameña, 1986; Tosi, 1971). Four cores were recovered from the lake sediments, the longest (17.5 m) reaching gravel beneath laminated silty gyttja. A series of 19 carbon-14 dates from the 17.5-m core and a 13.0-m core place the beginning of the lake sequence at 14,300 B.P. La Yeguada lies within the watershed of the Santa María River, where Paleoindian and early preceramic (10,000–7000 B.P.) archaeological deposits have been documented; these will be discussed shortly.

Pollen and phytolith spectra allude to a montane flora in the La Yeguada region for the period 14,300–11,050 B.P.; *Quercus, Magnolia, Ilex, Symplocos,* and cf. *Carex*—genera which today normally grow in lower Central America at elevations between 1500 and 2500 m—are among the taxa present. Piperno et al. (1990) infer a minimum vegetation zone depression of 800 m and a minimum temperature depression of 5°C during this period (using a moist-air lapse rate of 0.6°C per 100 m).

Beginning at 11,050 B.P., some lowland taxa (e.g., Chrysobalanaceae) and invasive forest elements (e.g., *Trema, Cecropia,* and Urticaceae/Moraceae) increase in frequency while others (e.g., *Croton*) make their initial appearance. Additional lowland forms first appear shortly thereafter at circa 10,800 B.P.; e.g., *Bocconia* (a lowland forest tree), *Trichomanes* (a lowland forest fern), and the Podostenaceae (a family of aquatic plants which grow below 1000 m) (Piperno et al., 1990). Over this same time period (circa 11,050–10,800 B.P.), most of the montane taxa drop from the record. Only *Quercus* and *Ilex* continue to contribute pollen to the sequence, disappearing at 8600 B.P.

In addition to documenting a substantial lowering of temperature and vegetation zonation during the LGS, and a rather rapid transition toward present temperatures during the interval from 11,050 to 10,490 B.P., the La Yeguada sequence provides data on effective moisture as well. The banded, silty clays at the base of the sequence appear to have been deposited in a shallow lake from 14,300 to 10,490 B.P. The alternating green and orange laminations are thought to represent algal blooms *(Coelastrum reticulatum, Botryococcus spp.,* and *Pediastrum spp.),* responding to greatly reduced lake levels during the dry season, and oxidized clay washed into the lake at the onset of the wet season from the exposed sediments at the lake's edge. Increases in *Typha* and *Cyperaceae* pollen between 11,050 and 10,490 B.P., coupled with more weakly developed sediment laminations, suggest a rise in lake level during this period (Piperno et al., 1990). The disappearance of the silty clay laminations altogether at 10,490 B.P. and their replacement by weakly banded gyttjas indicate even higher lake levels after this date, presumably reflecting increased rainfall at the beginning of the Holocene.

A third Panamanian paleoenvironmental record comes from a 50-m core raised from a now dry lake basin situated in the volcanic caldera of El Valle, located at an altitude of 500 m, 90 km east of La Yeguada and 75 km southwest of the Lake Gatún coring site (Bush and Colinvaux, 1990; Piperno et al., 1991a). Today the area receives 3000 mm of rain annually and would potentially support a premontane wet forest (Tosi, 1971). The disappearance of planktonic diatoms such as *Melosira* along with the presence of soil diatoms shortly after 16,000 B.P. indicates lowered lake levels. Chenopodiaceae and the marsh genera *Alternanthera* also occur, suggesting that decreased precipitation had reduced the lake to a swamp. At the same time, high levels of *Quercus* pollen and phytoliths from *Magnolia* and *Chusquea* (a bamboo associated with oak in contemporary montane forests) document vegetation which today is found above 1500 m. Thus, El Valle diatom, pollen, and phytolith evidence indicates that between 16,000 and 11,000 B.P. the temperature was circa 5°C cooler and precipitation was reduced by 20 to 30% from modern values (Piperno et al., 1991a).

Moving northward, two proxy climate records are available from bogs in the Cordillera de Talamanca in Costa Rica. Although the core sites are at considerably higher elevations, their results are consistent with the La Yeguada and El Valle sequences in terms of the timing and magnitude of shifts in vegetational zones in the late Pleistocene and early Holocene.

One 13-m core taken at an elevation of 2400 m produced a pollen sequence which extended back through the Holocene well into the late Pleistocene when *páramo* vegetation, found today at elevations 650 m above the core site, contributed the bulk of the pollen (Martin, 1964). A minimum reduction in mean annual temperature of circa 4°C would be required to support *páramo* vegetation at this elevation today. Beginning around 11,000 B.P., montane forest genera, including *Quercus, Alnus,* and *Podocarpus,* began appearing in the pollen record. The montane rain forest, which characterizes the area surrounding the core site today, became established between 9000 and 8000 B.P. At the La Chonta bog (elevation 2310 m) Islebe et al. (1995) found evidence for a cooler, drier interval between circa 11,100 and 10,400 B.P., alluding to global Younger Dryas cooling (see also Horn, 1990; Leyden, 1995). Full Holocene conditions are inferred by 9800 B.P.

We conclude this section with a consideration of lake cores in the karstic Petén region of Guatemala (Brenner, 1993; Leyden, 1984, 1997). Around Lake Salpetén (32 m deep, covering 2.6 km², and 104 m in elevation) the annual rainfall of 1600 mm, seasonally distributed, supports semi-evergreen forest. A 15-m core was raised below 26 m of water; it contained a sediment sequence that extended back into the late Pleistocene. Dating of the lower section of the core was done through palynological correlations with a core from Lake Quexil, 20 km distant, which produced a carbon-14 date on wood of 10,750 ± 460 B.P. at the Pleistocene-Holocene boundary. The dominance of herbs and aquatics induced Leyden to infer the presence of open scrub along with *Alternathera* marshes in the exposed lake basins prior to 11,000 B.P. The presence of *Juniperus*-type pollen at Salpetén in the LGS sequence also suggested to her that the low hills of the surrounding region (maximum elevation 400 m) were covered by a *Juniperus comitana* scrub. Since this species now occurs only above 1200 m (and 200 km distant), a depression of elevational ranges in excess of 800 m is required for it to occupy the uplands near the coring area. The LGS lake levels, 30–40 m lower than today, far exceed any Holocene lake lowering in the Petén and suggest extremely arid conditions (Leyden, 1997). Sometime around 11,000 B.P., lake levels began rising and gyttja containing pollen of *Brosimum (ramón)*, *Chlorophora*-type (both Moraceae types), *Cecropia*, *Trema*, and other semi-evergreen forest taxa was deposited in Lake Salpetén. The percentages of *ramón* pollen from these early Holocene levels exceed modern values, suggesting that these semi-evergreen forests were more mesic than those found there today.

It is not unreasonable to conclude from the data reviewed that rainfall was reduced from modern levels throughout Central America during the LGS. This reduction (Piperno and Pearsall suggest 35% [1998: 96]) is dramatically reflected in lake-level lowering in closed basins even where modern precipitation is relatively high (e.g., 3500 mm per year at La Yeguada, Panama). All of the lake sediment sequences are also consistent with an atmospheric temperature depression in the order of 4 to 6°C during the LGS and, therefore, with a lowering of vegetation zones between 600 and 1000 m. Only in the Talamanca coring sites, where high elevation (2300–2400 m) and lowered temperatures produce an evaporation rate less than a tenth of that found at lowland sites, are lake levels seemingly immune to LGS-reduced precipitation.

Vegetation does not appear to have been as sensitive as lake levels to reductions in precipitation. During the LGS, the La Yeguada and El Valle watersheds remained forested even though water levels at La Yeguada were much lower than modern and early Holocene levels, and the El Valle lake was reduced to a swamp. The Chagres drainage appears to have remained forested as well, although amelioration of the LGS climate was probably underway (but just barely) when this particular paleobotanical record commenced (11,300 B.P.). In contrast, the vegetation in the Petén under LGS-reduced rainfall was open scrub rather than forest. This is not surprising, in view of the fact that the Petén today is much drier than the other localities we have examined in Central America (1600 mm per year versus 2500–3500 mm per year) and that it has a limestone substrate which results in excessive drainage. Unpublished data from Monte Oscuro (Capira), on the central Pacific coast of Panama, where the climate today is strongly seasonal, indicate that a Holocene mesic forest replaced a Pleistocene vegetation with few trees (Piperno, personal communication, 2000).

To sum up, although the sample of paleoecological records remains small in Central America, the relationship between current mean annual rainfall and hypothetical LGS vegetation at six coring localities discussed suggests that areas now receiving in excess of circa 2000 mm of rainfall annually would have been forested during the LGS. Areas where annual rainfall is currently below 2000 mm are likely to have had more open vegetation, e.g., savannas, low scrub, and open woodland. These open formations would have been located at the base of the Yucatán Peninsula, in the Petén and Belize, and along the Pacific coastal plain from Chiapas to northwestern Costa Rica. Another patch of open vegetation would have covered the coastal plains along Panama Bay. These are not insignificant areas; during the LGS when sea levels were at least 50 m lower (Bartlett and Barghoorn, 1973; Golik, 1968), the Panama Bay littoral plains would have been twice the size that they are today. Nonetheless, large areas on the Pacific side of the isthmus receive over 2000 mm of rain annually (Estadística Panameña, 1986; Vivó Escoto, 1964), including most of the terrain stretching from central Costa Rica to Central Panama and the area near the Colombian border. Most of the Atlantic watershed of Central America receives well over 2000 mm of rain annually. Thus, we conclude that while the LGS landscape of Central America contained more extensive open habitats than at any time during the Holocene (that is, until massive forest clearing for cultivation was under way), it was still a largely forested landscape.

The Evidence for Late Pleistocene and Early Holocene Human Occupation of Central America

Evidence for the human occupation of Central America during the LGS is sparse and poorly situated in time. It is also limited to remains that are all linked to Clovis and related technological traditions (with the exception of one surface-collected artifact [fig. 7.5d; note 2]). There are, however, two aspects of that record that are relatively secure and warrant close examination: (1) the geographical and paleoenvironmental contexts from which Paleoindian remains have been recovered, and (2) the nature of the lithic artifacts themselves (Ranere, 1980; Ranere and Cooke, 1991, 1995).

Central American fluted points occur in dated, stratigraphic contexts at only one site: Los Tapiales in highland Guatemala (Gruhn et al., 1977). Los

Tapiales sits above 3150 m in an alpine meadow. A single fluted point base and a channel flake with an attached "ear" were among 100 tools (and nearly 1500 flakes) recovered from an excavated area of nearly 250 m². Bifaces, uniface points, burins, gravers, end-scrapers (several with lateral spurs), blades, and retouched flakes make up the rest of the assemblage, which is bracketed by carbon-14 dates of 11,170 ± 200 B.P. and 8,810 ± 110 B.P. The authors view the occupation as short term and feel that the 10,710 ± 170 B.P. date provides the best estimate of its age. The case for an LGS occupation at Los Tapiales is strengthened by two carbon-14 dates of 10,650 ± 1350 B.P. and 10,020 ± 260 B.P. from the nearby locality of La Piedra del Coyote, where a test excavation yielded only retouched flakes and a scraper fragment (Gruhn et al., 1977: 254). On the strength of the evidence from both Los Tapiales and La Piedra del Coyote, then, we are probably safe in assigning the Paleoindian occupation of this highland Guatemalan region to the 11th millennium B.P. Extrapolation from the paleoenvironmental record from the Cordillera de Talamanca suggests that at this time these two sites were situated in *páramo*.

The largest Paleoindian site yet reported in Central America is Turrialba (Finca Guardiria), which is located at an altitude of about 700 m on terraces of the Reventazón River on the Atlantic watershed of Costa Rica (Castillo et al., 1987; Pearson, 1998a,b; Snarskis, 1979). The Turrialba Valley receives 4000 mm of rain annually and would be cloaked in evergreen forest if not for human intervention. It was surely forested during the LGS, although we are ignorant of the floristic composition of the vegetation. The entire site, which covers an impressive 10 ha, has been disturbed by cane field activities. Test excavations indicated that cultural deposits were shallow (less than 40 cm) and confined to the plow zone. They were not carbon-14 dated. Coarse-grained cherts, which are still abundant as cobbles and sometimes huge boulders in the adjacent stream beds, were used in the production of a large quantity of tools and associated debris. Snarskis (1979) reports recovering in systematic surface collections and shallow excavations 28,000 lithic specimens, including 18 fluted points (fig. 7.3e,g), large numbers of bifacial performs, and tools often found with other fluted point assemblages, i.e., snub-nosed keeled scrapers, end-scrapers with lateral spurs, large blades, burins, bifacial and unifacial knives, and well-made side-scrapers.

The impressive size of the Finca Guardiria artifact sample has been helpful for determining tool reduction processes. Pearson (1998a,b) studied the collection with a view to identifying variables considered diagnostic of Clovis technology in North America. Bifacial reduction was a major activity at the site, and techniques revealed close similarities with Clovis. Decorticated river cobbles were whittled down into large flake blanks. The fact that many bifacial preforms were fluted indicates that detaching flutes using a

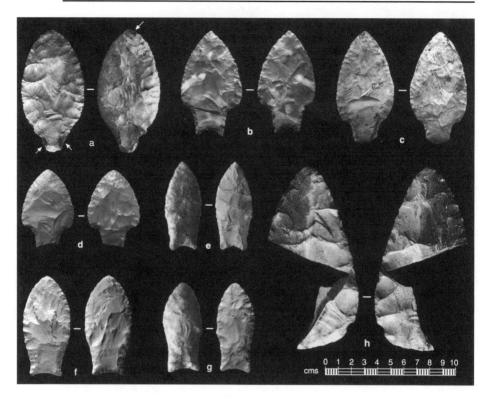

Fig. 7.3. Artifacts: *a–g,* fluted projectile points from Panama and Costa Rica; *a,* from Cañazas, Veraguas, Panama (the arrows point to flake scars made subsequent to the artifact's original patination; the tool was perhaps found fortuitously and reused in pre-Columbian times); *b,* from San Juan, Lake Madden, Panama; *c,* from Isla Macapalé, Lake Madden, Panama; *d,* from Isla Butler, Lake Madden, Panama; *e and g,* from Finca Guardiria, Turrialba, Costa Rica; *f,* from Isla Macapalé, Lake Madden, Panama; *h,* large bifacial artifact from Lake Madden–East End. Photos a and h by R. Cooke, others by Junius B. Bird.

ground nipple platform occurred well before the points were finished. Pearson proposes that they resemble those from the Ready site in Illinois (Morrow, 1996) and that thinning flake scars on large bifaces recall Clovis removal sequences outlined by Bradley (1982). The variety of finished tools, many of which have been reworked and/or show edge damage and polish, identify Turrialba as a habitation site as well as a workshop. Castillo et al. (1987) identified one area of the site as a bone-working locus based on the concentration index of burinated flakes, gravers, and beaked scrapers. Diagnostic tools include snub-nosed and spurred end-scrapers similar to examples reported from Los Tapiales and La Mula-West (Cooke, 1998: fig. 5h,m–n; Cooke and Ranere, 1992c: fig. 5a). They were made from proximal and distal platform segments and flattened on the ventral surface by the detachment of long, straight flakes ending in step fractures (Pearson, 1998b: fig. 8). Very similar examples recovered on the eroded surfaces of Lago Alajuela by Junius Bird, near the fluted-point find sites, are housed at the Smithsonian Tropical Research Institute in Panama. Additional Paleoindian materials have been recovered 2 km upstream from Turrialba at La Florencia-1, but here they occur as a minor component in a predominantly Archaic site (Acuña, 1983).

Lake Alajuela is a reservoir formed by damming of the Chagres River in order to provide water for the Panama Canal. During the December to April dry season, the lake is gradually drained to maintain the level of Lake Gatún (also created by a dam lower down on the Chagres River), on whose surface ships sail while traversing the waterway. Wave action during the annual raising and lowering of the water level has eroded the shorelines of both the lake itself and the numerous islands within it, which were once hilltops. At the end of the dry season, great expanses of barren ground are exposed and cultural remains that date from the Paleoindian period to the present are visible on the surface. Periodic examination of these eroded surfaces has resulted in the recovery of seven isolated fluted points (Bird and Cooke, 1977, 1978). Six are stemmed "fishtail" varieties (fig. 7.3b,c,d), similar to specimens from Fell's Cave and Taguatagua in southern Chile (Bird, 1969; Núñez et al., 1994) and El Inga in Ecuador (but cf. Mayer-Oakes, 1986). The blade of a stemless point from La Gloria in northern Colombia (Cooke, 1998: fig. 5b; Correal Urrego, 1983) is similar. The seventh Lake Alajuela point is a waisted Clovis-like specimen (fig. 7.3f; Cooke, 1998: fig. 4c). A workshop for the production of bifaces is located on a small island in the lake (the Westend site; Ranere and Cooke, 1991). This site covered more than 1 ha (it extended below the lake level at the time of our visits) and contained stone tools and ceramics from later prehistoric periods as well as the Paleoindian workshop debris. We recovered large numbers of bifacial thinning flakes, many with heavily ground and "lipped" platforms (Cooke, 1998: fig. 4i), as well as a large bifacially worked leaf-shaped preform (fig. 7.3h). We did not, unfortunately, recover any completed points. Although Cooke (1998) proposed that the Westend site's flaking debris was typical of a "thin flake" reduction technology more appropriate for making fishtail-stemmed points than Clovis-like ones such as those of Turrialba and La Mula-West, Ranere believes that we cannot determine whether Clovis or stemmed fishtail points were the final product. Nonetheless, we would be most surprised if the workshop and isolated fluted points from Lake Alajuela did not document an 11th millennium B.P. occupation of the forested Chagres River basin.

A workshop where Clovis-like points were definitely manufactured has been identified in Central Panama at the edge of Parita Bay, whose coastline would have been approximately 50 km distant at 11,000 B.P. (Golik, 1968). La Mula-West is located on a small hill overlooking an intermittent stream. The area today receives 1000 mm of rainfall annually, which falls almost entirely in the May to December wet season. Erosion has left Paleoindian stone tools, and both ceramics and lithics from later time periods, on the sloping surface of the hillside as a lag deposit. A test excavation dug in 1988 located Clovis-like tools in a thin (10-cm) stratum under a 1 m deep aeolian deposit; but we believe that it was secondarily deposited in antiquity in a small erosion gully. The combination of abundant biface manufacturing debris and rare finished tools identifies the site as primarily a workshop and, secondarily, a short-term campsite. Biface fragments, spurred end-scrapers, burins, gravers, and blades with small platforms prepared by grinding characterize the site assemblage (fig. 7.4; Cooke and Ranere, 1992b: figs. 4, 5a–f).

Translucent agate (or chalcedony), milky white to brown in color, was used almost exclusively by Clovis knappers (but not by ceramic-period knappers). This material occurs as veins in the eroded bedrock of the area and is probably what attracted Clovis-age people to the location. The site activity best documented is the manufacture of bifacial points. We have recovered over 80 biface fragments in total, of which most were broken in the manufacturing process. Twelve of the 15 basal fragments are either fluted or extensively basally thinned (fig. 7.4k,l, 7.5a–c; Cooke and Ranere, 1992c: fig. 4d–g). As at Turrialba, the reduction sequence represented at La Mula-West closely parallels that documented for North American Clovis workshops (Bradley, 1982; Morrow, 1996: 201–215; Pearson, 1998a,b; Ranere, 2000).

According to vegetation reconstruction models (Piperno et al., 1991a; Piperno and Pearsall, 1998), La Mula-West is situated in an area which would have been covered in an open, thorny woodland-scrub vegetation at 11,000 ± 200 B.P.—in our opinion the likely age of the site's occupation. Los Tapiales was set in *páramo*. Both the Turrialba and Lake Alajuela localities would have been under lowland forest when occupied by Paleoindians. We can expand our knowledge about the environments in which Paleoindians operated in Central America by comparing the location of isolated fluted points with a map of vegetation as it would have appeared at the end of the Pleistocene (fig. 7.2). The fluted-point localities of Belize (Cooke, 1998: figs. 1, 4b; Hester et al., 1980; Hester et al., 1981; MacNeish and Nelken-Terner, 1983) are all found in areas where LGS vegetation would have been some sort of thorn woodland, low scrub, or wooded savanna. Similar open landscapes would have been present in northwestern Costa Rica, where a Clovis-like fluted point was recovered (Bird and Cooke, 1977: fig. 3a; Swauger and Mayer-Oakes, 1952). Closed canopy montane forests are reconstructed for the valleys of Guatemala (Bray, 1978) and Ocozocoautla (Brown,

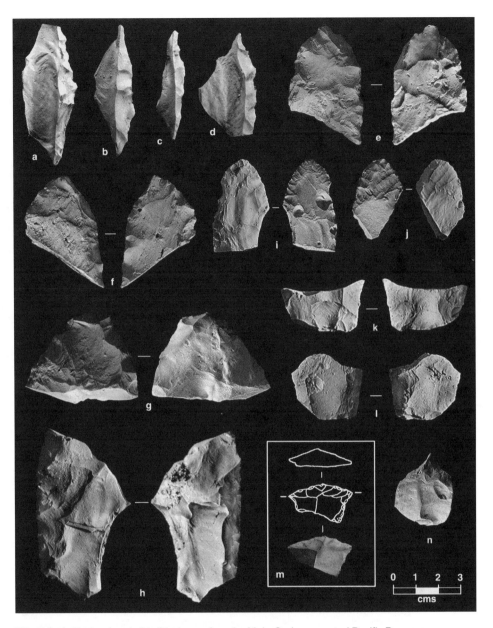

Fig. 7.4. Artifacts of probable Clovis age from La Mula-Sarigua, central Pacific Panama: *a–d,* overshoot flakes from bifacial point manufacture; *e and f,* middle stages of bifacial point manufacture (note edge-to-edge thinning flakes); *g,* large biface that represents the earlier stages of bifacial point reduction; *h,* bifacial point fragment which has already been fluted (prior to the final reduction of the artifact); *i and j,* proximal fragments from bifacial projectile points; k and l, fluted point bases (note lack of "ears"); *m,* double-spurred end-scraper; *n,* borer or graver made on a bifacial thinning flake. All artifacts are covered with ammonium chloride; all are made of translucent agate except i and j (brown jasper).

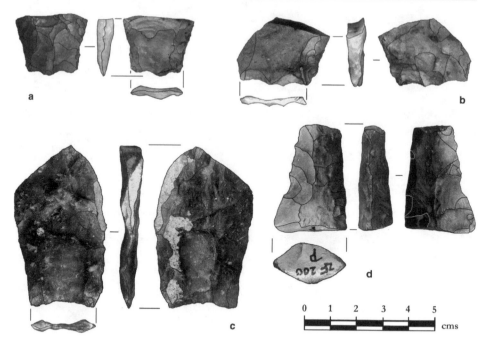

Fig. 7.5. Probable late glacial artifacts recently found in Panama: *a–c,* fluted-point bases, made of translucent agate, from La Mula-Sarigua, central Pacific Panama; *d,* medial fragment of a bi-lenticular "Joboid" projectile point, made of purple jasper, from the surface of Lake Madden, Panama. The black blotches on *a* and *b* represent an orange-colored mineral. Photos by R. Cooke

1980) in the Guatemalan Highlands, as well as for the Arenal region in north-central Costa Rica (Sheets and McKee, 1994)—areas where Clovis-like fluted points have also been reported. The central part of Panama Bay (Balboa), where a blade of a Clovis-like point was found in dredge sediments (Bird and Cooke, 1977: fig. 4d), would have been located, according to Piperno's model and the new Monte Oscuro data, in open vegetation. Another almost complete fishtail point is from Cañazas, Veraguas Province, Panama (fig. 7.3a). This point—from a private collection and purportedly found in a rock shelter—was probably used in later pre-Columbian times (the arrows in figure 7.4a point to flake scars that were made after the surface was patinated). Vegetation in this zone of central Pacific Panama is hypothesized to have been dry forest or thorn scrub during the LGS.

We sum up. Although not many sites in Central America have reported Paleoindian artifacts, they are found in a diversity of geographic regions and environments: Caribbean and Pacific coastal lowlands; highlands on both sides of the continental divide; relatively open thorn scrub–savanna vegetation; lowland and montane closed canopy forests; and *páramo.* In other words, Paleoindians roamed through most of the range of environmental and geographic settings present in Central America during the LGS. One infers from this evidence that tropical forests were as hospitable for late Pleistocene hunter-gatherers as were other Central American environments.

Unfortunately, reconstructions of potential *animal prey* cannot be as precise as those of potential *vegetation cover* in Central America (Webb, 1997). All we can safely say on the basis of the paleontological record is that Central America during the Pleistocene was home to at least 15 genera of large herbivores (Janzen and Martin, 1982) which are now extinct, including the mastodont-like gomphotheres *(Cuvieronius tropicus* and *Haplomastodon)*, mammoth *(Mammuthus)*, megalonychid and megatheriid giant ground sloths, including the huge *Eremotherium* (Webb, 1997: fig. 4.2), glyptodonts (Glyptodontidae), giant armadillos *(Pampatherium* and *Chlamytherium)*, a horse (Equus *[Amerhippus])*, and, perhaps, the rhinoceros-like *Toxodon* (Webb, 1997: fig. 4.5a), the camel-sized *Macrauchenia, Paleolama,* giant capybara *(Neochoerus)* (Webb, 1997: fig. 4.3c), flat-headed peccary *(Platygonus)*, and bear *(Arctodus).* Included in this list are browsers, grazers, and omnivores adapted to a range of forest, forest edge, and open thorn scrub–savanna environments (Cooke and Ranere, 1992a; Janzen and Martin, 1982: table 1). In northern South America some of these taxa have been found associated with Paleoindian and perhaps earlier stone tool complexes. For example, extinct *Cuvieronius, Haplomastodon,* and *Equus (Amerhippus)*, as well as extant fox *(Cerdocyon)* and white-tailed deer *(Odocoileus)*, were associated with human activities at Tibitó (Colombia), dated by a single carbon-14 sample to 11,740 ± 110 B.P. (Correal Urrego, 1981). *Haplomastodon* was reported in association with a possibly pre-Clovis lanceolate point at Taima-Taima (Bryan et al., 1978; Oschenius and Gruhn, 1979). A. Jaines (1999) has reported lanceolate Jobo points (Cooke, 1998: fig. 5a) in direct association with the giant ground sloth *Eremotherium rusconii.* If human hunters are in any way implicated in the extinctions of these herbivores (and we believe they are), then the disappearance of both forest-adapted species and others that would have lived in thorn scrub or savannas would confirm inferences from the paleovegetational record, namely, that all Central American environments—including forests—were utilized by hunter-gatherer populations in the late Pleistocene (Cooke, 1998; Ranere, 2000; Ranere and Cooke, 1991).

Human Interactions with Tropical Forest Environments: A Panama Example

One of the most intensively studied regions of Central America is the Santa María River watershed in central Pacific Panama and adjacent basins (fig. 7.6;

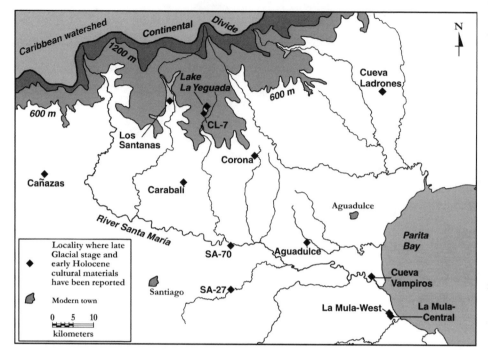

Fig. 7.6. Central Pacific Panama, showing localities where late glacial and early Holocene cultural remains have been located. See figure 7.1 for location within Panama.

Cooke and Ranere, 1992b). We have already summarized paleoecological data obtained at the Laguna de La Yeguada, located within the Santa María drainage, data which indicate that the foothill and mountain areas of central Pacific Panama and those of El Valle, 90 km to the east, have been under forest cover for at least the last 14,000 years and which demonstrate that these zones were forested for the whole of the last glaciation. We indicated that the coastal plain (elevations below 200 m) before 11,000 B.P. was likely to have held thorn woodland and scrub associations near the coast and perhaps some areas of savanna near the foothills (Piperno et al., 1991a; Piperno and Pearsall, 1998: fig. 2.4). However, shortly after 11,000 B.P. the coastal plain would have been invaded by dry tropical forest associations as the increased rainfall that characterized the onset of the Holocene began to affect plant community structure. Plant microfossil records from the Laguna de La Yeguada core and from several stratified and dated rock shelters in the region document the presence of swidden agriculture in the region by 7000 B.P. (Cooke and Ranere, 1992b; Piperno et al., 1991b, 1992; Piperno and Pearsall, 1998; Ranere, 1992). Moreover, it appears that some crops were being cultivated in small house garden plots in Central Panama by the ninth millennium B.P. (Piperno and Holst, 1998; Piperno et al., 2000). The following discussion will, therefore, focus on the evidence for human occupation in Central Panama for the period from 11,000 to 9000 B.P., at the time when subsistence was entirely based on hunting and gathering, and from 9000 to 7000 B.P., when subsistence was primarily based on hunting and gathering but did include some plant cultivation.

The earliest contextualized evidence for human occupation synchronous with the Paleoindian horizon in Panama is provided by the analysis of pollen, phytoliths, and carbon flux at La Yeguada (Piperno et al., 1990). A sharp rise in the amount of charcoal deposited in the lake sediments at 11,050 B.P. and an increase in phytoliths from weedy plants (e.g., *Heliconia*) are strongly suggestive of human interference with the natural vegetation. Moreover, 90% of the weed phytoliths have been burnt. This record of disturbance continues and intensifies throughout the early Holocene in spite of the increased moisture indicated by the rebirth of the lake. Conversely, there is no evidence for disturbance of vegetation in the watershed from 14,300 to 11,050 B.P. Thus, Piperno et al. (1990) conclude that the disturbance recorded at 11,050 B.P. probably represents activities of the earliest human inhabitants of the La Yeguada watershed.

The initial archaeological survey of the lakeshore recorded three sites with bifacial thinning flakes (Weiland, 1984, 1985); one of these (CL-7) produced the blade of a stemmed bifacial point (Cooke and Ranere, 1992b: fig. 6e), which must be either Paleoindian or early preceramic in age (i.e., 11,000–7,000 B.P.) since bifacial thinning as a lithic reduction strategy for cryptocrystalline stone was not practiced in Panama after 7000 B.P. (Ranere and Cooke, 1995, 1996). A more intensive survey of terrain adjacent to the lake was undertaken by G. Pearson (1999a), who located several large quarries/workshops for stone tools. Several tools—including a stemmed, fluted, or basally thinned point similar to Elvira points from Colombia (fig. 7.7, inset; Pearson, 1999b: 8; cf. Gnecco Valencia and Mohammed, 1994), broken bifaces, bifacial thinning flakes, spurred end-scrapers, and keeled scrapers ("limaces")—vouch for pre-7000 B.P. occupations. While none of these La Yeguada preceramic sites has yet been dated, their numbers are consistent with the pollen, phytolith, and carbon flux records, which indicate increased clearing and burning in the 5.5 km^2 watershed after such activities were initiated in 11,050 B.P.

The Corona shelter in Central Panama is one of only two sites on the Isthmus of Panama that has radiocarbon-dated cultural deposits in excess of 10,000 years B.P. (Cooke and Ranere, 1984, 1992b; Valerio, 1985, 1987). It is located in the foothills (elevation 270 m) within 15 km of La Yeguada in an area which today would support tropical moist forest (Tosi, 1971). Three

Fig. 7.7. Lake La Yeguada, situated at 650 m above sea level in the central cordillera of Panama. Paleoecological data from cores made here have indicated that human occupation of the local seasonal premontane forests began at the end of the late glacial stage and that deforestation was continuous thereafter until Spanish conquest, in the early 16th century, which drastically reduced Native American populations. *Inset:* fluted point found on the north shore of the lake (Pearson, 1999b). It is similar to Elvira points reported from western Colombia.

radiocarbon dates were reported from a small test pit: 10,440 ± 650 B.P. near the base of the deposits (130–145 cm below modern surface), 7440 ± 280 B.P. midway through the deposits (80–85 cm below modern surface), and 5980 ± 100 B.P. near the top (20–25 cm below modern surface). Unfortunately, the earliest deposits were restricted to an area of less than 1 m² of the excavated trench due to the bedrock configuration of the shelter floor. Only flakes, including several bifacial thinning flakes with ground platforms, were associated with the Paleoindian-age occupation of the site. The succeeding early preceramic (circa 10,000–7000 B.P.) levels produced a number of bifacial thinning flakes, a series of small flake knives and scrapers, gravers, and the tip of a broad, thin bifacial point which had seen heavy use as a knife.

A second Panamanian site, the Aguadulce shelter, has yielded a sequence of carbon-14 dates that range from 10,725 ± 80 B.P. and 10,529 ± 184 B.P. at the base of the cultural deposits (zone D) to 2570 ± 95 B.P. near the top

Fig. 7.8. The Aguadulce shelter, 1998 excavations. This small overhang in the central Pacific plains of Panama was occupied during the late glacial stage and early and middle Holocene.

(Piperno et al., 2000; Ranere and Hansell, 1978, 2000). This shelter is formed by a huge boulder atop a 25 m knoll in the middle of the coastal plain. Very little cultural material was found in zone D, the oldest occupation, but among the lithic debris were bifacial thinning flakes with crushed and ground platforms. Additional bifacial thinning flakes as well as a tip from a broad, thin bifacial point were recovered from the early preceramic deposits. The area surrounding the Aguadulce shelter may have been thorn scrub/savanna for the initial LGS occupation, but by 10,000 B.P., if not earlier, it would have been under dry tropical forest cover, the vegetation formation that would potentially be supported by today's annual rainfall of 1600 mm (Tosi, 1971).

There are seven localities in Central Panama with clear evidence of occupation during the early preceramic (circa 10,000–7000 B.P.); these include sites in the previously discussed La Yeguada watershed and also the Corona and Aguadulce shelters. The Carabal' and Los Santanas rock shelters, located in the Pacific foothills, provided single charcoal radiocarbon dates of 8040 ± 390 B.P. and 7100 ± 230 B.P. near the base of their deposits. Cueva de los Vampiros, where a charcoal date of 8560 ± 160 B.P. was recovered from basal (but not necessarily bottom-most) deposits, is located near the present Parita Bay coastline. All these deposits contained bifacial chalcedony thinning flakes, while at Los Santanas a limace-type scraper was recovered (Cooke and Ranere, 1992c: fig. 6f). Similar scraping tools were surface-collected at the SA-27 and SA-70 sites lower down in the Veraguas foothills (Cooke and Ranere, 1992c: fig. 5g–k). Another locality near the Parita Bay coastline is the quarry/workshop La Mula-Central, where stemmed bifacial points were manufactured (Cooke and Ranere, 1992c: fig. 6a–d; Hansell, 1988). Thirteen additional sites from Central Panama can be assigned to either the early preceramic or Paleoindian periods on the basis of the bifacial thinning flakes included in their surface collections. We have already emphasized that bifacial thinning as a lithic reduction technique for *cryptocrystalline stone* disappears from the Central Panama sequence after 7000 B.P. (Ranere and Cooke, 1996). Since the ratio of sites with radiocarbon-dated early preceramic occupations to sites with dated Paleoindian-age occupations is 5:2, it is likely that most of these surface-collected sites are early preceramic in age.

The carbon flux and microbotanical records from La Yeguada indicate a significant increase in forest burning early in the ninth millennium B.P. (Piperno et al., 1991b). It appears likely, then, that Central Panama populations at this point in time had begun to augment their food supplies by planting preferred plant species in newly burned forest patches. Paleobotanical data from archaeological sites support this hypothesis. Piperno identified phytoliths from *Calathea allouia* (lerén, a domesticated tuber), *Cucurbita spp.* (domesticated squash) and *Lagenaria secaria* (bottle gourd) in pre-7000 B.P. deposits at the Aguadulce shelter, and phytoliths from *Maranta arundinacea* (arrowroot) in pre-8600 B.P. layers at the Cueva de los Vampiros (Piperno and Pearsall, 1998). Arrowroot and manioc *(Manihot es-*

culenta) starch grains, embedded in one-hand grinders known in local par-
lance as "edge-ground cobbles," which were found in stratified deposits at
the Aguadulce shelter, are certainly preceramic and may be as old as 7000–
6000 B.P. (Piperno et al., 2000). These specialized grinding tools appeared
for the first time during the ninth millennium B.P. at the Carabal' shelter
(Valerio, 1987). Along the coast in the Cueva de los Vampiros, estuarine resources—sea catfish *(Ariidae),* mullet *(Mugil),* and shellfish—are present in
the 8600 B.P. deposits and were presumably being used for food.

We summarize: the archaeological and paleobotanical record from Central Panama is consistent with the following three-stage diachronic model:
(1) a small initial Clovis population that was highly mobile and focused on
hunting as the dominant subsistence activity, (2) a slightly larger population of hunter-gatherers who used fire as a major tool in managing forest
resources, and (3) a still larger population who were primarily hunter-
gatherers, albeit ones who now grew some of their food. The population of
Central Panama grew and flourished even as the climate became hotter and
wetter and tropical forests expanded to cover the entire region.

Concluding Remarks

The expansion of *Homo sapiens* into nearly all corners of the globe and nearly
all environments by the end of the Pleistocene now seems self-evident. The
fact that it was only recently that archaeologists began to document the Pleistocene occupation of tropical forested regions says more about the nature
of archaeological research than it does about early human adaptations to
tropical forests. Central America shares this archaeological legacy with other
tropical forested regions of the world and, like these other regions, the long
history of human habitation in its forests is gradually coming to light. There
may well have been a pre-Clovis episode of exploration and exploitation
of Central American tropical forests that has yet to be documented. Similar
observations can be made about most regions in the Americas (Dillehay,
2000). Whether Central America was unoccupied or sparsely or coastally occupied before 11,500 B.P., however, makes little difference to our model
of the demic expansion into this region—and as far as central Venezuela
(Jaines, 1999)—of peoples who made tools very similar to the Clovis tradition of North America. Not only do we currently lack identifiable residential populations, which could have adopted Clovis technology and Clovis
subsistence strategies, but also the very detailed similarities between Central
American and North American Clovis technology strongly suggest that all
populations associated with Clovis lithic assemblages are historically related
(Morrow and Morrow, 1999; Pearson, 1998a,b; Ranere, 2000).

Clovis in Central America is best explained as a cultural-behavioral pattern established elsewhere and carried into the region. Theoretical considerations and empirical evidence support a focus on animal procurement for Clovis populations in the open environments of North America. We have previously argued that hunting would remain the subsistence pursuit of choice by Paleoindians as they moved from open habitats into forested ones (Ranere and Cooke, 1991, Ranere, 2000). Here it is particularly important to distinguish between initial colonizing populations and the subsequent descendant populations in a region (cf. Kelly and Todd, 1988, Dincauze, 1993). The subsistence strategies of the initial colonizers should be very similar to the strategies pursued in their previous homeland. Subsequent strategies will reflect changes in those initial strategies which more efficiently utilize the resources of the newly occupied territories.

One can argue that Clovis populations were able to move so rapidly through the range of environments found in the Americas precisely because of the nature of hunting and meat consumption (Kelley and Todd, 1988; Morrow and Morrow, 1999; Ranere and Cooke, 1991). Clovis colonizers were not so much adapted to particular environments as they were to a particular mode of food procurement, i.e., hunting. Hunting techniques can be transferred, with minor modifications, from environment to environment; preparation and consumption of animals can certainly be transferred without difficulty. Clovis populations may have also fished and used other aquatic resources. They undoubtedly ate some plant foods as well, particularly fruits and perhaps young shoots which needed no special preparation to make them edible for humans. However, tropical plant parts (e.g., leaves, underground storage organs, seeds) are often chemically or mechanically defended and frequently need special methods of processing in order to render their consumption harmless. The absence of any specialized plant-processing tools in Clovis assemblages is consistent with the proposition that plants were dietary supplements rather than dietary staples.

Our reply to the contrary idea, i.e., that hunting is not now—nor was it ever—a viable subsistence option in forested areas, particularly in the tropics, is that recent inventories of animal biomass in the Neotropics reveal more resources than previously suspected. This is not to say that animals are today abundant; they are not. However, they seem sufficient for supporting widely dispersed and highly mobile hunting bands. D. Piperno and D. Pearsall (1998) have calculated that modern seasonal forests can sustain human population densities of nearly two persons per 1 km^2, with animals supplying most of the calories. This is not entirely unexpected, since terrestrial game provide the Aché foragers of Paraguay with over half of their dietary calories (Hawkes et al., 1982; Piperno and Pearsall, 1998).

One can only speculate about the situation of prey species during the LGS and earliest Holocene. But clearly, large forest fauna were not immune to the mass extinctions which occurred at the end of the Pleistocene (Martin and Guilday, 1967). Moreover, if one takes into account the capacity of contemporary large herbivores (like elephants) to affect the openness of the environment (Owen-Smith, 1987), it is likely that terminal Pleistocene forests in Central and South America were much more attractive to hunters and their prey species than is currently the case (Ranere and Cooke, 1991).

Another point that should be emphasized is that Paleoindians likely began altering the nature of tropical forests as soon as they entered them. This is evident at La Yeguada in Panama, where there are no indications of fire around the lakeshore from 14,000 to 11,050 B.P., the driest part of the sequence; burning began at approximately 11,050 B.P. and continued thereafter with ever increasing intensity (Piperno et al., 1991a, 1991b). This suggests that Paleoindian populations were purposefully burning patches of moist tropical forest in the vicinity of the lake, perhaps to replicate the pattern found in drier forests where natural fires were common. The use of fire in prehistory for opening up forested areas and initiating successional changes (among other uses) has been well documented (Lewis, 1972; Sauer, 1966). It is, of course, the early successional stages which provide the most abundant and most nutritious plants for consumption by humans and by the animals upon which humans prey. Núñez et al. (1994) have provided convincing evidence from Chile on the efficiency of Paleoindian gomphothere hunting near lacustrine environments during times of climatic stress.

What needs to be clearly understood is the fact that Clovis populations occupied and exploited tropical forest resources throughout Central America just as they occupied and exploited savanna, thorn scrub, and alpine tundra resources in the region. Moreover, Clovis descendant populations continued to live in these forests and became more closely adapted to forest resources over time. Documentation still leaves much to be desired, but the eco-manipulation found during the early Holocene in the isthmus parallels similar evidence found in other tropical regions and variously described as "firestick farming" by R. Jones (1969) or "environmental domestication" by D. Yen (1989) for Australia, "taming of the rain forests" by L. Groube (1989) for New Guinea, and agroecology by D. Rindos (1984). By the ninth millennium B.P. in some parts of Central America, but most clearly documented in Central Panama, populations began to grow some of their food. However, it was only after 7000 B.P. (and in many areas considerably after) that farming became a major subsistence activity. Until that time and beginning at least 11,000 years ago, hunting and gathering sustained growing numbers of people in the tropical forests of Central America.

Notes

1. Throughout this presentation dates will be given in radiocarbon years B.P. (before present), uncorrected for the fluctuations in concentration of carbon 14 in the biosphere, fluctuations which cause radiocarbon years to differ from sidereal years. Both archaeological and paleoecological chronologies for the late Pleistocene and early Holocene of the Americas have been established using radiocarbon years, and, in the interests of clarity, this convention is followed here. It is, however, useful to note that the time period most critical for the initial appearance of Paleoindians in Central America, circa 11,500–10,500 B.P., is more or less equivalent to 1000 sidereal years (11,510–10,520 B.C. [Stuiver et al., 1998]). In contrast, the radiocarbon plateaus at circa 10,000, 9500, and 8900 radiocarbon years B.P. result in 2000 radiocarbon years (from 10,500 to 8,500 B.P.) being the equivalent of approximately 3000 sidereal years (10,520–7,560 B.C. [Bartlein et al., 1995, Goslar et al., 1995; Stuiver et al., 1998]). This means that the proposed transition from hunting to generalized foraging to incipient horticulture was not quite as rapid as the carbon-14 chronology might suggest.

2. In the early 1970s Junius Bird picked up a bifacial fragment off the eroded shores of Lago Alajuela (fig. 7.5d); this fragment appears to be the midsection of a lanceolate projectile point like those in the El Jobo series from Venezuela (Dillehay, 2000: photo 5.1; Jaines, 1999).

References

Acuña, V. (1983): Florencia-1, un sitio precerámico en la vertiente atlántica de Costa Rica. *Vínculos* 9(1–2):1–14.

Bartlein, P., M. Edwards, S. Shafer, and E. Barker Jr. (1995): Calibration of radiocarbon ages and the interpretation of paleoenvironmental records. *Quaternary Research* 44:417–424.

Bartlett, A., and E. Barghoorn (1973): Phytogeographic history of the Isthmus of Panama during the past 12,000 years (a history of vegetation, climate, and sea level change). In *Vegetation and vegetational history of northern Latin America*, edited by A. Graham, 203–299. New York: Elsevier.

Bird, J. (1969): A comparison of South Chilean and Ecuadorian "fishtail" projectile points. *The Kroeber Anthropological Society Papers* 40:52–71.

Bird, J. B., and R. G. Cooke (1977): Los artefactos más antiguos de Panamá. *Revista Nacional de Cultura* 6:7–31.

Bird, J., and R. Cooke (1978): The occurrence in Panama of two types of Paleo-Indian projectile points. In *Early man in America from a circum-Pacific perspective*, edited by A. Bryan, 263–272. Occasional Papers no. 1. Edmonton: Department of Anthropology, University of Alberta.

Bradley, B. (1982): Flaked stone technology and typology. In *The agate basin site*, edited by G. Frison and D. Stanford, 181–208. New York: Academic Press.

Bray, W. (1978): An eighteenth-century reference to a fluted point in Guatemala. *American Antiquity* 43:457–460.

Brenner, M. (1993): Lakes Salpetén and Quexil, Guatemala, Central America. In *Global geographical records of lake basins,* vol. 1, edited by E. Gierlowski and H. Kelts, 377–380. Cambridge: Cambridge University Press.

Brown, K. (1980): A brief report on Paleoindian-Archaic occupation in the Quiche Basin, Guatemala. *American Antiquity* 45(2):313–324.

Bryan, A., R. Casimiquela, J. Cruxent, R. Gruhn, and C. Oschenius (1978): An El Jobo mastodon kill at Taima-Taima, Venezuela. *Science* 200:1275–1277.

Bush, M., and P. Colinvaux (1990): A pollen record of a complete glacial cycle from lowland Panama. *Journal of Vegetation Science* 1:105–118.

Bush, M., D. Piperno, P. Colinvaux, P. de Oliveira, L. Krissek, M. Miller, and W. Rowe (1992): A 14,300-year paleoecological profile of a lowland tropical lake in Panama. *Ecological Monographs* 62(2):251–275.

Castillo, D., E. Castillo, M. Rojas, and C. Valldeperas (1987): Análisis de la Lítica Lasqueada del Sitio 9-FG-T. en Turrialba. "Tesis de grado," Escuela de Antropología y Sociología, Universidad de Costa Rica.

Cooke, R. (1998): Human settlement of Central America and northern South America, 14,000–8,000 B.P. *Quaternary International* 49–50:177–190.

Cooke, R., and A. Ranere (1984): The "Proyecto Santa Maria": A multidisciplinary analysis of prehistoric adaptations to a tropical watershed in Panama. In *Recent developments in isthmian archaeology: Advances in the prehistory of lower Central America,* edited by F. Lange, 31–53. Oxford: BAR International Series 212.

———. (1992a): Human influences on the zoography of Panama: An update based on archaeofaunal and documentary data. In *Biogeography of Mesoamerica,* edited by S. Darwin and A. Welden, 21–58. Supplementary Publication no. 1. New Orleans: Tulane Studies in Zoology and Botany.

———. (1992b): Prehistoric human adaptations to the seasonally dry forests of Panama. *World Archaeology* 24(1):114–133.

———. (1992c): The origins of wealth and rank in prehistoric central Panama: Some new data and some old problems. In *Wealth and hierarchy in the intermediate area,* edited by F. Lange, 243–316. Washington: Dumbarton Oaks.

Correal Urrego, G. (1981): *Evidencias culturales y megafauna Pleistocénica en Colombia.* Bogotá: Fundación de Investigaciones Arqueológicas Nacionales, Banco de la República.

Correal Urrego, G. (1983): Evidencia de cazadores especializados en el sitio de La Gloria, Golfo de Urabá. *Revista de la Academia Colombiana de Ciencias Exactas, Físicas y Naturales* 15(58):77–82.

Dillehay, T. (1997): *Monte Verde: A late Pleistocene settlement in Chile.* Vol. 2, *The archaeological context and interpretation.* Washington, D.C.: Smithsonian Institution Press.

———. (2000): *The settlement of the Americas: A new prehistory.* New York: Basic Books.

Dincauze, D. (1993): Pioneering in the Pleistocene: Large Paleoindian sites in the northeast. In *Archaeology of eastern North America: Papers in honor of Stephen Williams,* edited by J. Stoltman, 43–60. Archaeological Report no. 25. Jackson: Mississippi Department of Archives and History.

Estadística Panameña (1986): Estadística Panameña. Situación Física, Metereología: Año 1984. Panama.

Gnecco Valencia, C., and A. Mohammed (1994): Tecnología de cazadores-recolectores subandinos: Análisis functional y organización tecnológica. *Revista Colombiana de Antropología* 31:7–31.

Golik, A. (1968): History of Holocene transgression in the Gulf of Panama. *The Journal of Geology* 76(5):497–507.

Goslar, T., M. Arnold, E. Bard, T. Kus, M. Pazdur, M. Ralska-Jasiewiczowa, K. Rozanski, N. Tisnerat, A. Walanus, B. Wicik, and K. Wieckowski (1995): High concentration of atmospheric ^{14}C during the Younger Dryas cold episode. *Nature* 377:414–417.

Groube, L. (1989): The taming of the rain forests: A model for late Pleistocene forest exploitation in New Guinea. In *Foraging and farming: The evolution of plant exploitation*, edited by D. Harris and G. Hillman, 292–304. London: Unwin Hyman.

Gruhn, R., A. Bryan, and J. Nance (1977): Los Tapiales: A Paleo-Indian campsite in the Guatemalan Highlands. *Proceedings of the Philosophical Society* 121(3):235–273.

Hansell, P. (1988): The rise and fall of an early formative community: La Mula-Sarigua, central Pacific Panama. Ph.D. diss., Temple University, Philadelphia.

Hawkes, K., K. Hill, and J. O'Connell (1982): Why hunters gather: Optimal foraging in the Ache of eastern Paraguay. *American Ethnologist* 9:379–398.

Hester, T., H. Shafer, and T. Kelley (1980): Lithics from a preceramic site in Belize. *Lithic Technology* 9:9–10.

Hester, T., T. Kelley, and G. Ligabue (1981): A fluted Paleo-Indian projectile point from Belize, Central America. Working papers no. 1. Center for Archaeological Research, University of Texas, San Antonio.

Hooghiemstra, H., A. Cleff, G. Noldus, and M. Kappelle (1992): Upper Quaternary vegetation dynamics and paleoclimatology of the La Chonta bog area (Cordillera de Talamanca, Costa Rica). *Journal of Quaternary Science* 7:205–225.

Horn, S. (1990): Timing of deglaciation in the Cordillera de Talamanca. *Climate Research* 1:81–83.

Islebe, G., and H. Hooghiemstra (1997): Vegetation and climate of montane Costa Rica since the last glacial. *Quaternary Science Reviews* 16:589–604.

Islebe, G., H. Hooghiemstra, and K. van der Borg (1995): A cooling event in the Younger Dryas Chron in Costa Rica. *Palaeogeography, Palaeoclimatology, Palaeoecology* 117:73–80.

Islebe, G., H. Hooghiemstra, M. Brenner, J. Curtis, and D. Hodell (1996): A Holocene vegetation history from lowland Guatemala. *Holocene* 6:265–271.

Jaines, A. (1999): Nuevas evidencias de cazadores-recolectores y aproximación al entendimiento del uso del espacio geográfico en el noroccidente de Venezuela: Sus implicaciones en el contexto suramericano. *Arqueología del Area Intermedia* 1:83–120.

Janzen, D., and P. Martin (1982): Neotropical anachronisms: Fruits the gomphotheres ate. *Science* 215:19–27.

Jones, R. (1969): Firestick farming. *Australian Natural History* 16:224–228.

Keefer, D., S. deFrance, M. Moseley, J. Richardson III, D. Satterlee, and A. Day-Lewis (1998): Early maritime economy and El Niño events at Quebrada Tacahuay, Peru. *Science* 281:1833–1835.

Kelly, R., and L. Todd (1988): Coming into the country: Early Paleoindian hunting and mobility. *American Antiquity* 53:231–244.

Lewis, H. (1972): The role of fire in the domestication of plants and animals in Southwest Asia: A hypothesis. *Man* 7:(2):195–222.

Leyden, B. (1984): Guatemalan forest synthesis after Pleistocene aridity. *Proceedings of the National Academy of Sciences* (USA) 81:4856–4859.

———. (1985): Quaternary aridity and Holocene moisture fluctuations in the Lake Valencia basin, Venezuela. *Ecology* 66(4):1279–1295.

———. (1995): Evidence of the Younger Dryas in Central America. *Quaternary Science Reviews* 14:833–839.

———. (1997): Man and climate in the Maya lowlands. *Quaternary Research* 28:407–414.

Leyden, B., M. Brenner, D. Hodell, and J. Curtis (1993): Late Pleistocene climate in the Central American lowlands. In *Climate Change in Continental Isotopic Records,* edited by P. Swart, K. Lohmann, J. McKenzie, and S. Savin, 78:165–178. Geophysical Mongraph, no. 78.

MacNeish, R., and A. Nelken-Terner (1983): *Final annual report of the Belize Archaic Archaeological Reconnaissance.* Boston: Center for Archaeological Studies, Boston University.

Martin, P. (1964): *Paleoclimatology and a tropical pollen profile.* Vol. 2, *Palaeo-climatological section,* 319–323. Report of the 6th International Congress on Quaternary (Warsaw, 1961). Lodz, Poland.

Martin, P., and J. Guilday (1967): A bestiary for Pleistocene biologists. In *Pleistocene extinctions: The search for a cause,* edited by P. Martin and H. Wright, 1–62. New Haven: Yale University Press.

Mayer-Oakes, W. (1986): Early man projectile and lithic technology in the Ecuadorian Sierra. In *New evidence for the Pleistocene peopling of the Americas,* edited by A. Bryan, 133–156. Orono: Center for the Study of Early Man, University of Maine.

Morrow, J. (1996): The organization of early Paleoindian lithic technology in the confluence region of the Mississippi, Illinois, and Missouri Rivers. Ph.D. diss., Washington University, St. Louis.

Morrow, J., and T. Morrow (1999): Geographic variation in fluted projectile points: A hemispheric perspective. *American Antiquity* 64(2):215–230.

Núñez, L., J. Varela, R. Casimiquela, V. Schiappacasse, H. Niemeyer, and C. Villagrán (1994): Cuenca de Taguatagua en Chile: El ambiente del Pleistoceno superior y ocupaciones humanas. *Revista Chilena de Historia Natural* 67:503–519.

Oschenius, C., and R. Gruhn (1979): Taima-Taima: A late Pleistocene kill-site in northernmost South America. Final report on the 1976 excavations. Programa CIPICS, Monografias Cientificas, Universidad Francisco de Miranda, Coro, Venezuela.

Owen-Smith, N. (1987): Pleistocene extinctions: The pivotal role of megaherbivores. *Paleobiology* 13:351–362.

Pearson, G. (1998a): Pan-American Paleoindian dispersals as seen through the lithic reduction strategies and tool manufacturing techniques at the Guardiria site, Turrialba Valley, Costa Rica. Paper presented at the 63rd Annual Meeting of the Society for American Archaeology, Seattle.

―――. (1998b): Reduction strategy for secondary source lithic raw materials at Guardiria (Turrialba, FG-T-9), Costa Rica. *Current Research in the Pleistocene* 15: 84–85.

―――. (1999a): Isthmus be here somewhere. *Anthropology News* 40(6):22.

―――. (1999b): Where north meets south: Seeking a "unified theory" in Panama. *Mammoth Trumpet* 14:8–11.

Piperno, D. (1985): Phytolithic analysis of geological sediments from Panama. *Antiquity* 59:13–19.

―――. (1988): *Phytolith analysis: An archaeological and geological perspective.* San Diego: Academic Press.

Piperno, D., A. Ranere, I. Holst, and P. Hansell (2000): Starch grains reveal early root crop horticulture in the Panamanian tropical forest. *Nature* 407 (19 October):894–897.

Piperno, D., and D. Pearsall (1998): *The origins and development of agriculture in the lowland Neotropics of Latin America.* San Diego: Academic Press.

Piperno, D., and I. Holst (1998): The presence of starch grains on prehistoric stone tools from the humid Neotropics: Indications of early tuber use and agriculture in Panama. *Journal of Archaeological Science* 25:765–776.

Piperno, D., M. Bush, and P. Colinvaux (1990): Paleoenvironments and human settlement in late-glacial Panama. *Quaternary Review* 33:108–116.

―――. (1991a): Paleoecological perspectives on human adaptation in Panama. I. The Pleistocene. *Geoarchaeology* 6:201–226.

―――. (1991b): Paleoecological perspectives on human adaptation in Panama. II. The Holocene. *Geoarchaeology* 6:227–250.

―――. (1992): Patterns of articulation of culture and the plant world in prehistoric Panama: 11,500 B.P.–3000 B.P. In *Archaeology and environment in Latin America*, edited by O. Ortiz-Troncoso and T. van der Hammen, 109–128. Amsterdam: I.P.P., Universiteit van Amsterdam.

Ranere, A. (1980): Human movement into tropical America at the end of the Pleistocene. In *Anthropological papers in honor of Earl H. Swanson, Jr.*, edited by L. Harten, C. Warren, and D. Tuohy, 41–47. Pocatello: Idaho State University Press.

―――. (1992): Implements of change in the Holocene environments of Panama. In *Archaeology and environment in Latin America*, edited by O. Ortiz-Troncoso and T. van der Hammen, 25–44. Amsterdam: I.P.P., Universiteit van Amsterdam.

―――. (2000) Paleoindian expansion into Central America: The view from Panama. In *Archaeological passages: A volume in the honor of Claude N Warren*, edited by J. Schneider, R. Yohe III, and J. Gardner, 110–122. Publications in Archaeology, no. 1. Western Center for Archaeology and Paleontology.

Ranere, A., and P. Hansell (1978): Early subsistence patterns along the Pacific coast of Panama. In *Prehistoric coastal adaptations,* edited by B. Stark and B. Voorhies, 43–59. New York: Academic Press.

————. (2000): The Aguadulce shelter revisited: Early and middle Holocene occupation in the Llanos of Gran Cocle. Paper presented at the 65th annual meeting of the Society for American Archaeology, Philadelphia.

Ranere, A., and R. Cooke (1991): Paleoindian occupation in the Central American tropics. In *Clovis: Origins and human adaptation,* edited by R. Bonnichsen and K. Fladmark, 237–253. Corvallis: Center for the Study of the First Americans, Department of Anthropology, Oregon State University.

————. (1995): Evidencias de occupación humana en Panamá a postrimerías del pleistoceno y a comienzos del holoceno. In *Ambito y Ocupaciones Tempranas de la América Tropical,* edited by I. Cavelier and S. Mora, 5–26. Bogotá: Fundación Erigiae y Instituto Colombiano de Antropología.

————. (1996): Stone tools and cultural boundaries in prehistoric Panamá: An initial assessment. In *Paths to Central American prehistory,* edited by F. Lange, 49–78. Niwot: University Press of Colorado.

Rindos, D. (1984): *The origins of agriculture.* New York: Academic Press.

Sander, D. (1959): Fluted points from Madden Lake. *Panama Archaeologist* 2:39–51.

Sandweiss, D., H. McInnis, R. Burger, A. Cano, B. Ojeda, R. Paredes, M. del Carmen Sandweiss, and M. Glascock (1998): Quebrada Jaguay: Early South American maritime adaptations. *Science* 281:1830–1832.

Sauer, C. (1966): *The early Spanish main.* Berkeley: University of California Press.

Sheets, P., and B. McKee, eds., (1994): *Archaeology, volcanism, and remote sensing in the Arenal region, Costa Rica.* Austin: University of Texas Press.

Snarskis, M. (1979) Turrialba: A Paleo-Indian quarry and workshop site in eastern Costa Rica. *American Antiquity* 44:125–138.

Stuiver, M., P. Reimer, E. Bard, J. Beck, G. Burr, K. Hughen, B. Kromer, F. McCormac, J. v. d. Plicht, and M. Spurk (1998): INTCAL98 radiocarbon age calibration 24,000–0 cal B.P. *Radiocarbon* 40:1041–1083.

Swauger, J., and W. Mayer-Oakes (1952): A fluted point from Costa Rica. *American Antiquity* 17:264–265.

Tosi, J. (1971): Inventariación y Demostraciones Forestales, Panamá. Zonas de Vida. FO:SF Pan 6, technical report. Food and Agricultural Organization, Rome.

Valerio, L. W. (1985): Investigaciones preliminares en dos abrigos rocosos en la Región Central de Panamá. *Vínculos* 11(1–2):17–29.

————. (1987): Análisis Funcional y Estratigráfico de Sf-9 (Carabalí), un Abrigo Rocoso en la Región Central de Panamá. "Tesis de Grado," Escuela de Antropología y Sociología, University of Costa Rica, San José.

Vivó Escoto, J. (1964): Weather and climate of Mexico and Central America. In *Natural environment and early cultures,* edited by R. West, vol. 1, *Handbook of middle American Indians,* 87–215. Austin: University of Texas Press.

Webb, S. (1997): The great American fauna interchange. In *Central America: A natural and cultural history,* edited by A. Coates, 97–122. Princeton: Princeton University Press.

Weiland, D. (1984): Prehistoric settlement patterns in the Santa Maria drainage of central Pacific Panama: A preliminary analysis. In *Recent developments in isthmian archaeology: Advances in the prehistory of lower Central America,* edited by F. Lange, 31–53. Oxford: BAR International Series 212, British Archaeological Reports.

———. (1985): Preceramic settlement patterns in the Santa Maria basin, central Pacific Panama. Paper presented at the 45th International Congress of Americanists, Bogota.

Yen, D. (1989): The domestication of the environment. In *Foraging and farming: The evolution of plant exploitation,* edited by D. Harris and G. Hillman, 55–75. London: Unwin Hyman.

CHAPTER 8

Holocene Climate and Human Occupation in the Orinoco

William P. Barse

This chapter provides a review of the evidence for Holocene climatic change in the northern lowlands of South America and the results of archaeological and geomorphological investigations conducted in the Orinoco River Valley, Amazonas State, Venezuela. Three related topics will be addressed: (1) What was the nature of climate change during the Holocene, taken here to mean events occurring after about 10,000 B.P.? (2) How are such climatic changes represented in the depositional record of the Orinoco River Valley as marked by buried paleosols in archeological sites? (3) What was the nature of the associated human occupation within these paleosols? This data will be used to address a more general question, to wit, What implications do these climatic changes have on the general trajectory of preceramic (or Archaic) human adaptation to the tropical lowlands of the Orinoco River Valley? This chapter will include a brief review of the available pollen record from the lowlands of northern South America and the results of archaeological testing that I conducted during the last twelve years along the Orinoco River in Amazonas State, Venezuela.

A word on the nature of the available information for the above noted topics is in order. Paleoclimatic research in the tropical lowlands of northern South America is still in its early stages. Reconstructions of past events are broad in scope and in need of much basic fieldwork on a regional basis. My goal is to cite the more pertinent material (such as Van der Hammen's work in Colombia and Guyana and Behling and associates' recent work in Colombia) that may pertain to climatic events that occurred in the Orinoco drainage, rather than to provide an exhaustive review of research done throughout the lowlands. Work conducted in Brazil (cf. works in Prance and Lovejoy, 1985) has general implications for the lowlands as a whole and supports data that point to major shifts in vegetation throughout the late Pleistocene and Holocene, shifts that also occurred in the Orinoco drainage.

Some of the paleoenvironmental work conducted in the lowlands does not have direct implications to the study of climatic shifts in the Orinoco, given the geographical distance and types of ecosystems involved. For example, the research by Colinvaux and others (Colinvaux, 1989; Bush and

Fig. 8.1. Location of study area on the Colombian-Venezuelan border

Colinvaux, 1988; Bush et al., 1990) on the Ecuadorian Amazon close to the foot of the Andes bears little relevance for Orinocan climatic events. Thus, the argument offered by Bush et al. (1990) that vegetation changes in the Ecuadorian lowlands during the late Pleistocene were not related to periods of aridity does not have direct implications for the region under study.

The archeological record for the early to mid Holocene of the Orinoco is still being worked out (cf. Barse, 1989, 1990, 1995). Most of the data that I offer come from strata cuts excavated at a number of archeological sites in the Puerto Ayacucho vicinity of Venezuela. However, these data (dated depositional episodes with associated archeological remains) provide an excellent snapshot of major climatic events that took place during the last 10,000 years. And, while obtained during a program of research directed mostly toward archeological goals, they provide support for climatic models being developed for the northern lowlands. To date, few investigators have attempted to correlate archeological data with climatic events in the lowlands. A notable exception is B. J. Meggers, who reviewed gaps in radiocarbon date sequences and argued that they correlated with periods of Holocene aridity (Meggers, 1987; Meggers and Danon, 1988). It was suggested that they reflected an adaptive response marked by settlement dispersal. This chapter is the first attempt for the northern lowlands (in particular, the Orinoco drainage) to dovetail climatic events as measured by pollen sequences and depositional markers for these events as indicated by paleosol data and to discuss how such events may have impacted Holocene archeological populations.

Evidence for Climatic Shifts

Several authors (Van der Hammen, 1972, 1974, and 1982; Wijmstra and Van der Hammen, 1966; Behling et al., 1999; Behling and Hooghiemstra, 1998, 1999, 2000) reported evidence for late Pleistocene and Holocene climatic shifts for the northern tropical lowlands of South America. The evidence they presented, based on a limited number of pollen cores taken during the late 1960s and 1970s, and more recently in the 1990s, suggested that changes in climate (and prevailing vegetation) were characterized by shifts between wet and dry environmental regimes, triggered by changes in temperature (cf. Wijmstra and Van der Hammen, 1966; Van der Hammen, 1972; Behling and Hooghiemstra, 1998). Van der Hammen has constructed a long and reliable sequence of climatic oscillations for the Sabana de Bogotá, Colombia, indicating that climatic events could be linked to Andean glacial advances and retreats (Van der Hammen, 1974). An interest in the kinds of impact that these changes might have had on the lowlands east of the Andean region prompted Van der Hammen to seek data from the *llanos* to see if comparable shifts in climate could be documented (Van der Hammen, 1982). Here, too, he found evidence for marked climatic fluctuations in several pollen sequences. These research efforts were linked with broader questions concerning the investigation of Pleistocene and Holocene refugia in the Amazon basin. And the results of these investigations can be used to evaluate depositional events in the Orinoco River Valley of Venezuela.

Pollen cores are available from the Colombian Llanos close to the headwaters of the Meta River and from the Ireng River drainage in the western part of Guyana. Evidence of climatic changes in these pollen cores is supported by additional data from Brazil and the Lake Valencia area. Given that similar kinds of shifts are seen in pollen sequences from Guyana and the western llanos of Colombia, it can be assumed that the climatic events they mark pertain to the lowlands of the intervening Orinoco region reviewed herein as well. Two of these pollen sequences are of particular interest for understanding climatic changes in the Orinoco Valley and northern lowlands of South America. One is the sequence from Lake Moriru in Guyana and the second from the Laguna de Agua Sucio in Colombia; both have been radiocarbon dated. The more important of the two sequences is that from Lake Moriru. According to T. A. Wijmstra and T. Van der Hammen (1966), this column spans the entire Holocene and latter part of the Pleistocene. Of interest in the column reported for this lake sequence is the marked increase in Byrsonima pollen and other tree pollen between circa 11,000 and 8700 B.P. at the expense of grass pollen.

The above noted authors suggest that at that time there was probably no open savanna in the lake's vicinity, a region that is currently open savanna grassland. During the Holocene, Wijmstra and Van der Hammen (1966) suggest that the time of closed dry forest or closed savanna woodland with Byrsonima as the principal pollen-producing tree was between circa 10,000 to 9000 B.P. After about 8700 B.P. the record shows increasing dryness as indicated by notable increases in grass and herb pollen. This is especially clear from about 5000 B.P. and later; though as the paleosol data indicate, this pattern may have been interrupted by brief periods of slightly wetter conditions that may not be well represented in currently available pollen sequences. Whether they may be indicated by paleosol data is a question still under investigation.

The pollen sequence obtained from the Laguna de Agua Sucio in the Colombian Llanos did not extend as far back in time as the one documented from Lake Moriru. However, it clearly shows a strong shift to drier conditions, again marked by a great increase in grass pollen from about 6000 B.P. and later. That sequence shows minor shifts between savanna woodland and open savanna during the Holocene, though from about 3000 B.P. open savannas predominated. Some of these fluctuations between savanna and savanna-woodland or dry forest could be preserved as paleosols. In fact, as recent archeology has shown (Barse, 1989, 1995), many are preserved.

Van der Hammen's work in the Colombian Llanos was followed by that of

Behling and associates, who provided additional data bearing on late Pleistocene to Holocene vegetation changes. Pollen cores reported by Behling
and Hooghiemstra from Laguna Sardinas and Angel reflect more humid
conditions at the Pleistocene-Holocene transition up to about 10,600 radiocarbon years B.P. Significant expansion of savanna elements at the expense
of tropical gallery forest is evident in the two lake cores up until a period between about 5200 to 3850 B.P., when climatic conditions returned to more
humid conditions (Behling and Hooghiemstra, 1998). Similar climatic reconstructions were obtained from Laguna El Pinal and Laguna Carimagua
(Behling and Hooghiemstra, 1999) and from Laguna Loma Linda (Behling
and Hooghiemstra, 2000). It is significant that the pollen cores from all the
lakes sampled by Behling showed clear evidence of more humid conditions
at the Pleistocene-Holocene transition, followed by a subsequent period of
extensive savanna expansion.

Climatic data from Lake Valencia, at the northern edge of the Venezuelan Llanos and out of the lowlands proper, also shed light on vegetation
conditions at the onset of the Holocene. M. L. Salgado-Labouriau (1982:
77) noted that an increase in forest pollen elements in this lake's sequence
could suggest that a more humid environment characterized the Pleistocene-Holocene transition. Thus, her data are in concordance with the Lake
Moriru sequence from the Rupununi Savanna area in Guyana. Recent work
conducted by Van der Hammen and others (Van der Hammen et al., 1990);
Behling, Berrio, and Hooghiemstra (1999); and Urrego (1991) in the Caquetá River Valley of eastern Colombia has shown that the onset of the Holocene also was characterized by a more humid climate (see also Cavelier
et al., 1995, for a review of the macro-botanical species from the early Holocene site of Peña Roja on the Caquetá). Data from the middle Caquetá
area, now characterized largely by dense tropical forest, show a more continuous humid climate throughout the Holocene than that in the savanna
regions found north of the Guaviare River and in the Orinoco region under
consideration herein.

The Caquetá evidence suggests (as alluded to in the introduction) that
climatic events will have to be reconstructed on a local scale (by river drainage or portion thereof) to understand processes that took place on a more
geographically widespread scale. However, in view of the general concordance of climatic events between the Colombian Llanos and the Ireng River
of Guyana, similar conditions can be assumed to have been present in the
Orinoco region throughout the Holocene, at least in the areas north of
where the Guaviare River enters at San Fernando de Atabapo. The following section on depositional events in Orinocan archeological sites explores
this link.

Depositional Evidence for Climatic Shifts

Data that show changes in climate at the Pleistocene-Holocene boundary are available from other areas, though not in abundance. However, it is clear that the Holocene was subjected to marked changes in the nature of the dominant vegetation in the northern South American lowlands. These shifts, from wetter to drier conditions, or at times from drier to wetter conditions, undoubtedly left their mark on the landscape. According to Van der Hammen (1982), these shifts between wet and dry climates may be nothing more than lengthening and/or shortening of the rainy season. Given shifts in prevailing vegetation, what can be inferred about depositional processes or events in the Orinoco Valley region?

The Orinoco Valley region north of Puerto Ayacucho in Amazonas State and extending around the great bend of the river into the state of Bolivar is characterized by massive sand sheet formations that extend outwards from the base of the Guiana Shield westward toward the alluvial deposits of the river. Currently, the typical vegetation on the sand sheet formations is characterized mostly by savanna, though tropical forest is present along all tributary streams that drain into the Orinoco. Deposition in these sandy formations may occur at variable rates depending on the nature of the prevailing vegetation cover. Processes of erosion are also mediated by the nature of the surface vegetation, either enhanced by denudation of the landscape, or hindered by the cover of dense foliage.

For instance, closed-canopy forested ecosystems versus open-canopy vegetation regimes (either in dry tropical forest or woody savanna as opposed to open, bunch-grass savanna) present markedly different implications for processes of erosion and consequent depositional processes. Landscapes characterized by a closed canopy (e.g., forested or woody savanna ecosystems) would have allowed stable weathering conditions to occur in the soil column, thus resulting in clear expression of relatively distinct soil horizons. On the other hand, open savanna conditions (i.e., open-canopy ecosystems) would foster considerable sediment relocation due to seasonal wind and rain action. Rapid changes in prevailing vegetation to more open conditions could lead to mass wasting of soil. Such events undoubtedly would have resulted in higher sediment loads in rivers, which then would have led to (1) faster rates of alluvial deposition in riverine settings during seasonal flood episodes and (2) greater rates of colluviation, with the latter simply effecting considerable horizontal movement of sediment downslope.

Colluviation in the Orinocan region characterized by the above noted extensive sand sheet formations has resulted in the buildup of thick deposits on benches and toe-slopes overlooking streams tributary to the river. This pattern of mass wasting (or deflation) of higher elevations and the buildup of colluvial materials on the lower bench formations paralleling the Orinocan tributaries has been well documented by the author in the course of numerous bucket and auger surveys. Nearly all the west-draining tributaries of the Orinoco north of Puerto Ayacucho have variably thick packages of unweathered colluvium capping archeological deposits. Simply put, denuded landscapes have led to cycles of erosion-deposition while vegetated surfaces fostered weathering processes leading to clearly expressed zonation in soil profiles.

Considering the above, depositional evidence for climatic change during the Holocene is abundant in the Orinoco Valley. Such evidence can be seen in alluvial terrace formations along the river, as well as in the sand sheet formations that stretch between the shield and the alluvial formations bordering the Orinoco. These climatic events are marked by well-preserved paleosols and by thick, undifferentiated sediment packages sealing paleosols. Some of the paleosols documented in archeological sites in the Puerto Ayacucho vicinity date to the early Holocene, while others clearly date to the later Holocene or Formative stage when small horticultural settlements prevailed in the region. Paleosols are important markers for climatic events. As noted by G. Rapp and C. L. Hill (1998: 34), they are indicative of stable landscape environments. M. R. Waters (1992: 59–60) emphasized the importance of paleosols as markers of landscape stability where little (or minimal) deposition and erosion took place. L. B. Leopold, M. G. Wolman, and J. P. Miller (1995: 471) emphasis the importance of widespread paleosols as reflections of general climatic events as opposed to localized phenomena.

In sum, buried paleosols in the Orinoco Valley area, represented either by fossil B-horizons or buried A-horizons, can be interpreted as marking periods of landform stability. Such periods of stability are seen here as representing more humid climatic conditions when closed- or nearly closed–canopy tropical forest or, minimally, woody savanna situations prevailed. These surfaces received minimal deposition and were not subject to extensive erosion given the inferred presence of closed-canopy tropical forest environments. On the other hand, thick and undifferentiated packages of sediment, either in sand sheet formations or alluvial terraces, point to rapid deposition and/or lack of stable vegetation cover. These episodes of rapid sediment buildup capping buried paleosols are indirect evidence of erosion that occurred in more open (i.e., savanna) environments that developed subsequent to onset of the Holocene. Preceramic (or Archaic stage) archaeological sites that contain such depositional situations preserving paleosol surfaces dating to the early to mid Holocene are briefly reviewed below.

Archeological survey and test excavations conducted during the last 15 years in Amazonas State, Venezuela, have documented several sites that contain preceramic occupations associated with paleosols dating to the early

Holocene. Of particular note is the site of Provincial, located about 15 km north of Puerto Ayacucho on a sand sheet formation (refer to fig. 8.1). The site is situated close to the edge of the formation, overlooking a relict channel of the river. Excavations at Provincial conducted in 1987, 1990, and 1991 revealed a well-defined fossil B-horizon radiocarbon dated to circa 9200 B.P. (Barse, 1989, 1990, and 1995). Several hearths and associated lithic debris belonging to the Atures tradition, the local manifestation of the Archaic in the Orinoco (cf. Barse, 1989, 1990), were contained within the upper portion of the sandy loam B-horizon (zone 4) at the site. Capping the B-horizon was approximately 80 cm of an undifferentiated loamy sand C-horizon (undifferentiated in that it lacked any visible soil stratigraphy marking former land surfaces). A low-density Atures II component was present in the lower portion of this horizon. Stratified well above the Archaic deposits was a ceramic component dating to about 2000 B.P. These components and their stratigraphic contexts are illustrated in figure 8.2a. The ceramics from this component, found within the zone IIT A/C transition horizon, belong to the Pozo Azul phase, a local member of the Barrancoid tradition in the Puerto Ayacucho area.

A comparable situation was documented at the Culebra site, located on an alluvial terrace close to the Atures Rapids (refer to fig. 8.1). Excavations here revealed a preceramic living floor that was found resting on a fossil B-horizon approximately 1 m below the current ground surface (refer to fig. 8.2b). This paleosol (zone 5) was, in turn, capped by approximately 35 to 40 cm of sterile sediment, a sandy loam C-horizon, zone 4, indicating an extended (or continuous) period of alluvial deposition. Although no radiocarbon dates were obtained from this living floor, it is thought to be early Holocene in age. This chronological assessment is based on its stratigraphic correlation with Provincial and the presence of dated occupations in the upper part of the profile documented at the Culebra site. In the profile illustrated by figure 8.2b, a ceramic occupation radiocarbon dated to approximately 1700 B.P. was documented at the base of zone 2 (called II-b). This occupation was identified as a Saladoid-related component. This was beneath a late Barrancoid tradition ceramic occupation designated the Culebra phase and radiocarbon dated to about 1200 B.P. (Barse, 1989). Zone 3 in the profile is a more recently buried A-horizon paleosol stratified above the zone 4 C-horizon. This layer contained a later Archaic component with an assemblage composed of quartz flake scrapers and two chert projectile points. These two paleosols were the basis for the two-part subdivision of the Archaic tradition in the region, designated Atures I and Atures II of the Atures tradition noted above.

Two additional sites containing paleosols have been documented in the study area, each of which supported a preceramic living floor. One example

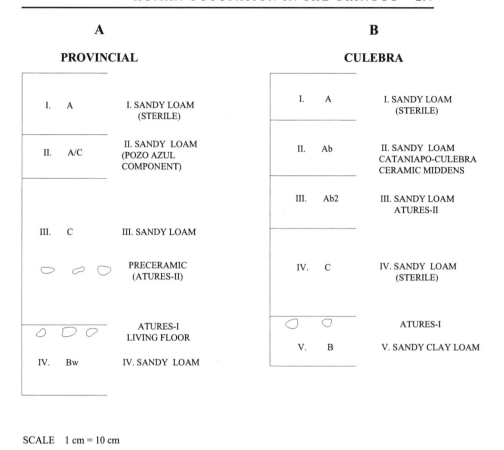

Fig. 8.2. Stratigraphic sections from the sites of Provincial and Culebra. Reprinted with permission from *Science*.

was encountered during test excavations at the site of Pozo Azul Sur-2, and a second at Lucero-3 (or Lucero). Both sites are north of Puerto Ayacucho and located on massive sand sheet formations bordering streams draining west to the Orinoco River (refer to fig. 8.1). Test excavations at the site of Pozo Azul Sur-2 contained a weakly expressed fossil C/B- to Bw-horizon sequence with a preceramic living floor radiocarbon dated to circa 7500 B.P. (see fig. 8.3a). As a paleo-surface, the fossil B-horizon (zones 4 and 5) is likely contemporaneous with the one at the Provincial site, though the occupation on it is about 1500 years younger. It was not as well expressed in terms of texture and color as the B-horizon at Provincial. But like the profile

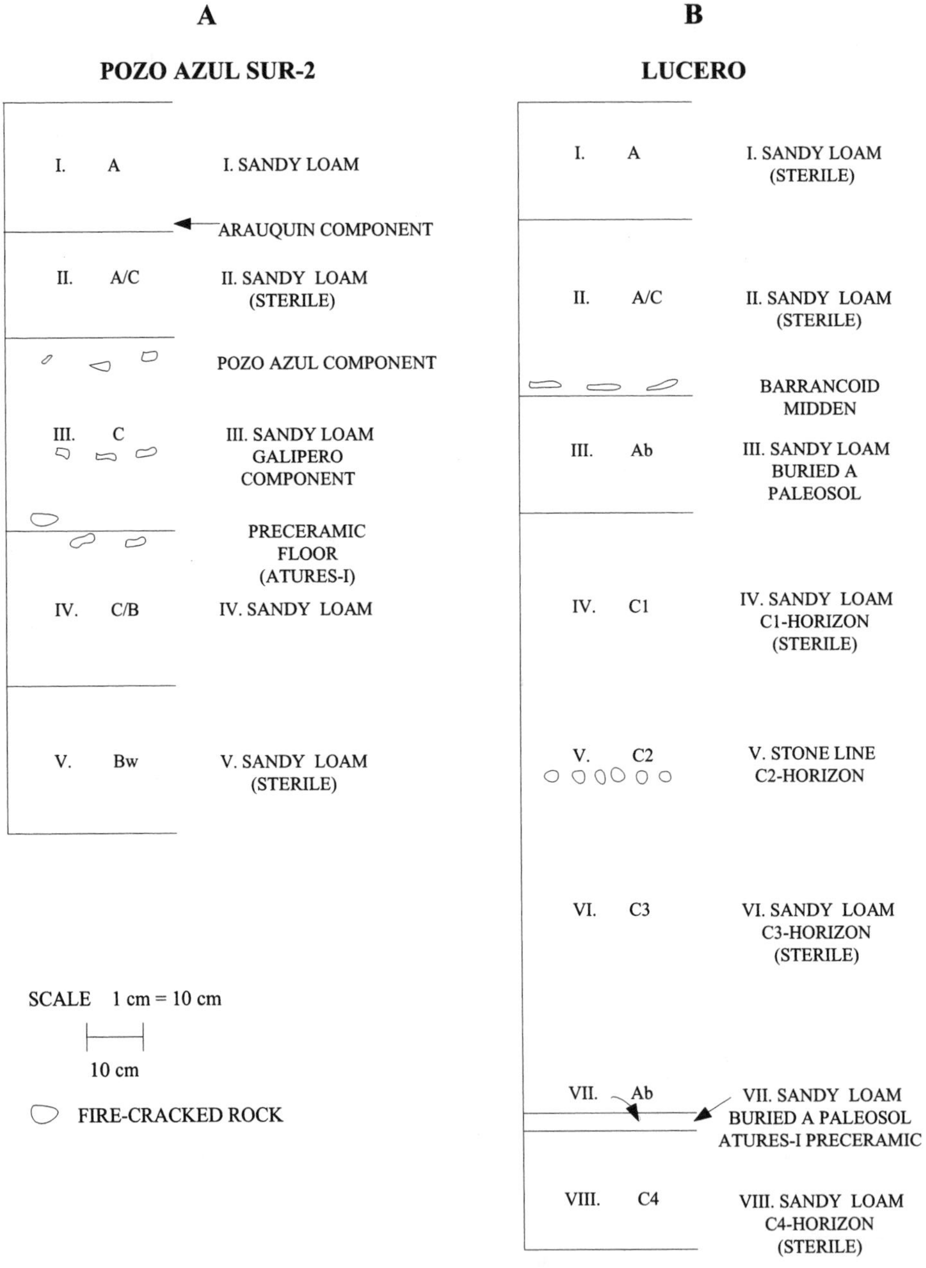

Fig. 8.3. Stratigraphic sections from the sites of Pozo Azul and Lucero. Reprinted with permission from *Science*.

at Provincial, it was capped by a thick and undifferentiated zone of loamy sand, the zone 3 C-horizon in figure 8.3a, that contained an early Formative Galipero complex ceramic occupation that has been radiocarbon dated to between 3500 and 2800 B.P. (Barse, 1996). These ceramics are tempered with clay pellets and are distinctly different than those belonging to the Saladoid and Barrancoid wares.

An older paleosol may be present at the site of Lucero-3. This is of particular interest because it contains a deeply buried paleosol A-horizon extending from about 1.75 m to nearly 2.00 m below surface (refer to fig. 8.3b). The Lucero paleosol, zone 7 in figure 8.3b, was exposed in a large borrow pit and dipped slightly to the west. It clearly marks the A-horizon of a gently sloping former surface that dropped in elevation toward Caño Parhueña, an Orinocan tributary stream. Three quartzite flakes and a fire-cracked rock fragment were recovered from this deeply buried A-horizon. Although undated radiometrically (at the time this chapter was written), the Lucero paleosol is assumed to be early Holocene in age. It was found more than 1 m below a younger paleosol containing a ceramic occupation belonging to the Barrancoid tradition and thought to date to about 2000 B.P., based on comparative ceramic typology. The latter paleosol is zone 3 in figure 8.3b. The ceramic midden was found resting on its upper surface immediately below the zone 2 A/C-horizon.

Assuming depositional processes to be more or less comparable in the massive Orinocan sand sheet deposits (holding other variables equal), the depth below surface of the Lucero paleosol, greater than the preceramic paleosols at both Provincial and Pozo Azul Sur-2, would argue for a very early Holocene placement. It is notable that between the deeply buried paleosol A-horizon and the ceramic paleosol are three well-defined albeit thin stone lines marking periods of deflation. These thin stone lines are relatively continuous across an expanse of nearly 20 m in the profile of the borrow pit and likely mark widespread landscape denudation and deflation of formerly exposed surfaces. Zone 5 marks the best expressed of the three; of the other two, one was in zone 4 and the other midway through zone 6.

The paleosols (either fossil B-horizons or buried A-horizons) documented at the above noted sites are interpreted as marking stable land surfaces that could only have existed under stable weathering conditions that would have been provided by a more densely vegetated landscape. Extensive woody savanna or relatively closed canopy tropical forest may have characterized such a landscape. The radiocarbon dates from Provincial and Pozo Azul Sur-2 clearly place these paleosols at the onset of the Holocene, when climatic conditions were more humid. The well-developed B-horizon at Culebra at the base of the stratigraphic column at the site is unarguably early Holocene in age. The paleosol at Lucero and its associated preceramic component may likely date to this period, though this will have to await radiocarbon dating for verification. The critical point is that these paleosols provide corroboration of more humid conditions at the Pleistocene-Holocene boundary as indicated by the limited palynological data available from the northern tropical lowlands of South America. Of interest as well are the moderately thick packages of undifferentiated sediment capping the paleosols at these sites. Such thick horizons may pertain to periods of more rapid sediment buildup that occurred in the mid-Holocene as climatic conditions became drier (or shifted to periods of shorter rainy seasons).

Preceramic Assemblages, Settlement, and Regional Comparisons

The preceramic components in the above noted sites (Provincial, Culebra, and Pozo Azul Sur-2) all have assemblages characterized by flake tools and few formal tools or other artifact categories. Most common in the Orinocan Archaic sites documented to date (south of the Meta River) are lithic assemblages dominated by crystalline quartz. Other raw materials such as quartzite and chert are rare. Flakes were generated mostly through bipolar reduction of small cobble cores, many of which can be obtained from cracks in the granite bedrock forming the Atures Rapids. Quartz also occurs as veins running through the local bedrock of the shield. The following tables provide summary counts of the assemblages from the Archaic paleosol contexts at Provincial, Culebra, and Pozo Azul Sur-2. Table 8.1 presents debitage and related reduction debris from Provincial. This material, and that in table 8.2, were recovered from a total of 16 m^2 of living floor sampled over three separate field seasons. Figures 8.4 and 8.5 present a sample of these artifacts.

Table 8.1. Provincial: Debitage Categories from 9000 B.P. Paleosol Contexts

Quartz flakes	Quartz shatter	Quartz chunks	Quartzite flakes	Chert flakes	Total debitage
64	7	6	5	4	86

Table 8.2. Provincial: Tool Categories from 9000 B.P. Paleosol Contexts

Quartz cores	Quartz scrapers	Quartz choppers	Ground stone	Faceted hematite
9	3	4	3	2

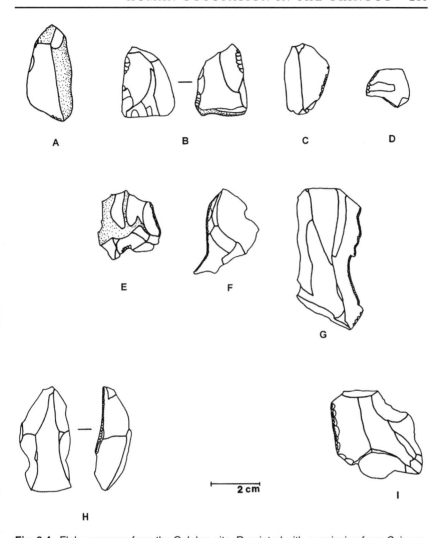

Fig. 8.4. Flake scrapers from the Culebra site. Reprinted with permission from *Science*.

The quartz chunks are angular fragments of vein quartz that probably represent large shatter fragments left over from the reduction of material taken from vein sources. The quartzite and chert in the sample are all flake fragments. Table 8.2 summarizes the various tool categories obtained from the excavations at Provincial. One of the quartz cores was modified from a small river cobble; the rest of the items were worked from quartz fragments recovered from vein sources in the local granite bedrock. The chopping implements were vein nodules with one end worked by percussion flaking into a steep-angled cutting edge. The ground stone tools included a fragment of

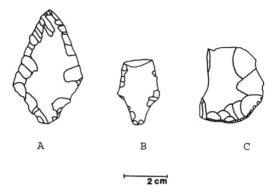

A   B   C

2 cm

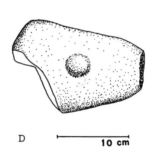

D

10 cm

Fig. 8.5. Artifacts: *a and b,* projectile points from Culebra; *c,* scraper from Provincial; and *d,* pitted nutting stone from Provincial. Reprinted with permission from *Science.*

a small celt, a trapezoidal-shaped and well-ground pitted nutting stone with hammer stone use at one end, and a spheroid-shaped bola stone. The pitted nutting stone is illustrated in figure 8.5d. The hematite nodule exhibited rubbed facets, indicating its use as a source of red pigment.

The assemblage from the Culebra site at the mouth of the Cataniapo River produced a much larger flake assemblage from paleosol contexts. The available assemblage was obtained from a total of 6 m² of excavated deposit that sampled a well-stratified sequence of preceramic to ceramic components (see above). Debitage from Atures I and Atures II contexts are summarized in table 8.3.

As an assemblage, the flakes are all derived from bipolar reduction of small cobbles. Unlike the small assemblage from Provincial, most of the flakes from the Culebra site were derived from riverine cobbles, many examples of which were (and still are) readily available during low water, lodged in the cracks of the bedrock forming the Atures Rapids. Addition-

Table 8.3. Culebra: Debitage Categories from Atures I and II Contexts

Periods	Quartz flakes	Quartz shatter	Quartzite flakes	Chert flakes
Atures I	134	5	10	1
Atures II	270	44	0	2

Table 8.4. Culebra-Tool Categories from Atures I and II Contexts

Periods	Quartz cores	Quartz scrapers	Projectile points
Atures I	1	2	0
Atures II	1	9	2

Table 8.5. Pozo Azul Sur-2: Artifact Categories from Atures I Paleosol

Quartz flakes	Quartz scrapers	Faceted hematite
12	1	1

ally, the very low quantities of quartzite and chert are of interest, a pattern also evident at Provincial. Tool categories from the two preceramic components at the Culebra site are presented in table 8.4.

Both cores are riverine cobble fragments. The scrapers, all evidently removed from cores through bipolar techniques, exhibit scars from use or slight retouching. The two points are notable in that they are both manufactured from chert, a material not local to the Puerto Ayacucho area. Both are contracting, stemmed specimens and are illustrated in figure 8.5a and b.

A smaller preceramic assemblage was obtained from Pozo Azul Sur-2, located to the north of the Provincial site. Although 15 m of closely spaced excavation sampled the paleosol at the site, only a few artifacts, clustered near the remains of a small hearth, were recovered. Charcoal from this hearth produced a radiocarbon date of 7010 B.P. It is assigned to the Atures I period of the Atures tradition. These few items are summarized in table 8.5. The remaining Archaic site discussed in this chapter, Lucero, only produced three small quartzite flakes and a fire-cracked rock from the buried paleosol.

In most cases, the flake scrapers recovered from Culebra, Provincial, and the Pozo Azul Sur sites were used expediently, resulting in minimal edge wear. In some cases, flake scrapers exhibit lateral margins that were slightly retouched by pressure flaking. A number of the flake tools in Archaic assemblages from the above noted sites have concave blade margins. Such spokeshave-like working edges would have been suitable for processing wood or bone shafts. As discussed in earlier works, the flake scraper assemblages dating to the early Holocene formed the basis for the Atures tradition of the Orinocan Archaic (cf. Barse, 1989, 1990). This tradition is comparable to the Abriense in Colombia (cf. Correal, 1979).

Bifaces and other formal flaked stone tools have not yet been documented from the earliest Holocene preceramic components in the Orinoco Valley. Despite the area sampled at Provincial (the area excavated to date measures 16 m²), no points, other bifaces, or fragments thereof were recovered. In addition, the debitage is clearly not the result of biface thinning activities. Although the area in the deeper Archaic component at Culebra was small, the debitage clearly was not derived from biface reduction, nor is the debitage from the 7010 B.P. contexts at Pozo Azul Sur-2. In sum, debitage from all early Holocene contexts that have been dated from about 9000 to 7000 B.P. lack flaking debris indicative of the production of a bifacial tool industry. Nearly all the flakes in the assemblages appear to have been produced for use as expedient tools. In terms of morphology, most specimens have a flake axis perpendicular to the platform.

Stemmed projectile points, however, have been found in later preceramic occupations that date to the mid-Holocene. Projectile points that exhibit long, tapered stem elements have been recovered from surface contexts in the Puerto Ayacucho region by the author and by Cruxent (Rouse and Cruxent, 1963). These items supplement the two examples with shorter tapered stem elements that were recovered from Atures II preceramic contexts at the Culebra site noted above. Such points have been documented across a wide geographic expanse in northern South America. Although few in number, when reliable contexts are available, they appear to fall into the later Holocene, and not from contexts that date to its onset (Barse, 1997). Ground stone tools are present in the early Holocene assemblages of the Orinoco. Fragments of a ground stone celt and a pitted nutting stone were recovered from the fossil B-horizon at Provincial. Other common items in the early Holocene assemblages are hematite nodules exhibiting rubbed facets. In view of the rubbed facets, these items were likely used for the production of red pigment, perhaps for ornamental-ritual purposes. Of interest is the widespread occurrence of these items in preceramic contexts from sites on the Venezuelan coast and Trinidad (cf. Boomert, 2000: 72–73).

Both Provincial and Pozo Azul Sur-2 produced moderate quantities of charcoal and carbonized wood fragments in hearth contexts. Many palm fragments are tentatively identified as *Milpeso* or *Seje,* a fruit that has important nutritional value and is widely exploited by current Native Americans in the Orinoco Valley. The sheer quantity recovered from living floors at Provincial and Pozo Azul Sur-2 clearly points to its harvesting as a subsistence item. Preceramic sites in Trinidad also have produced high quantities of carbonized palm nut fragments (Boomert, 2000: 63–64). The wood charcoal from the preceramic assemblages has yet to be identified. Unfortunately, faunal remains have not been recovered from the Orinocan preceramic assemblages.

Early Holocene Archaic settlement in the Orinoco region, as indicated by the available site distribution, is characterized by small occupations containing one or more hearth features (marked by fire-cracked rock clusters) and associated lithic debris. To date, evidence has not been found for large, semi-permanent settlements, nor are they expected to be present. Most Archaic components documented in the Puerto Ayacucho region appear to mark the locations of temporary, short-term camps. Drawing on ethnographic analogy, settlements may have been similar to the small, temporary occupations that Politis (1997) has described for the Maku (or Nukak) groups currently inhabiting the Vaupes region of eastern Colombia. These settlements are characterized by quickly built structures, ephemeral in nature, housing small nuclear to minimally extended family groups. The archaeological signature of these settlements is limited to living floors with hearth features and low quantities of domestic debris. Currently, additional surveying and testing is being carried out to enhance the settlement data available for the Orinoco region. Comparable types of occupation that may serve as an analog for the Archaic sites are the overnight Guahibo hunting-fishing camps that are established along paths leading to Orinoco or one of its cutoff channels. One to three hearths and a simple lean-to kind of structure characterize these small occupations. Discarded materials are few (no lithics, given that it is an extinct industry) and dominated by fire-cracked rock.

Comparisons between the Orinocan preceramic components and other regions in the northern lowland tropics are limited, given the paucity of known sites. Correal et al. (1990) reported on the excavation of a small rock shelter in the middle Guaviare River of eastern Colombia; it contained a flake tool assemblage dated to about 7000 B.P. The authors classified the assemblage as Abriense, after the El Abra rock shelter in the Sabana de Bogotá. A similar assemblage was documented from the middle Caqueta River of eastern Colombia at the Peña Roja site (Cavelier et al., 1995). This assemblage was obtained from buried contexts dated between 9250 and 9100 B.P., thus corresponding with the radiocarbon dates from Provincial. Comparisons also can be extended to early Holocene sites at the southern edge of the Brazilian Shield (Fogoça and Lima, 1990).

In earlier publications (Barse, 1990, 1995), I compared the Orinoco preceramic assemblages and other lowland Archaic complexes to late Pleistocene and early Holocene Archaic sites situated in the Sabana de Bogotá. Sites such as Nemocon and Sueva (Correal, 1979) contained flake scraper assemblages dated between about 10,000 to 8000 B.P. Both sites, though lacking projectile points, contained abundant faunal remains. The extensive scraper assemblage and total lack of flaked stone points suggest that projectiles had either wooden or bone tips. The Native American adaptation marked by these assemblages, dating to the late Pleistocene to early Holocene of the inter-montane valley of Bogotá, is assumed to have expanded into the eastern lowlands of Colombia and Venezuela, thus serving as ancestral complexes for the Orinocan Archaic (preceramic) complexes.

Conclusions

If the above noted interpretations from the pollen record and its correlation with early Holocene paleosols are accurate, then the initial Archaic occupants of the Orinoco Valley were adapted to tropical forest conditions, and not the open savanna–gallery forest mosaic that characterizes the region today. The impact of dry-wet-dry cycles throughout the Holocene (with an increasing focus on drier conditions or shorter rainy seasons) on human populations adapted to the lowland environment needs to be considered to enhance an understanding of the development of tropical forest culture. Such climatic events and related adaptations by early Holocene groups may be widespread across the northern tropical lowlands, extrapolating between Orinocan depositional contexts reported herein and western Colombian palynological data reviewed earlier. Increasing aridity (or periods characterized by more extended dry seasons and shorter rainy seasons) may have been one of the selective factors, if not *the* selective factor, leading to the eventual domestication of manioc and, perhaps, other root crops. Long ago, C. Sauer identified the Venezuelan llanos as the probable origin for manioc's initial cultivation (Sauer, 1969: 46). Holocene climatic changes and their impact on the resident preceramic groups of the northern tropical lowlands need to be explored in greater depth from this point of view. Other issues dealing with early Holocene climate and human occupation need to be explored as well.

For instance, to date, there is no evidence for a Paleoindian (or perhaps more appropriately, a late Pleistocene) occupation in the Orinoco Valley. This apparent gap may be nothing more than lack of an adequate survey of the region. The archeological signature of small, transient camps dating to this time is undoubtedly small and may take extensive surveying and testing to detect. Whether or not early Holocene Archaic–stage occupations are the earliest in the Orinoco region is still an issue that needs to be considered. Roosevelt et al.'s (1996) contention of a separate Paleoindian occupation of Amazonia separate from a Clovis-derived base has been disputed (cf. Barse,

1997; Haynes, 1997; Reanier, 1997). The early Holocene occupations that she reported on from Pedra Pintada are more in line with the early Holocene Archaic components discussed above for the Orinoco.

One characteristic of Archaic assemblages in the Orinoco and other northern lowland sites is the generalized nature of the lithic assemblages, dominated mostly by uniface flake scrapers. Such a non-diagnostic tool kit, composed of nonformal flake tools and an occasional ground stone tool, makes comparisons between regions difficult at best. However, they are widespread, occurring in late Pleistocene to early Holocene sites in the Sabana de Bogotá east to the Orinoco and south in the Brazilian Shield area. One question is whether or not such nondescript lithic assemblages mark adaptations to forested environments. Were such assemblages nothing more than expedient tools used to fabricate an assemblage from perishable materials? More detailed analysis of edge wear, difficult on quartz, is needed to provide data to answer such a question. These flake assemblages also continue into ceramic periods as well, though the latter generally witness a decline in lithic use. Some of the Barrancas-tradition ceramic components have a nearly complete lack of flaking debris in excavated contexts sampled to date.

As noted, such flake tool assemblages are clearly designed for the production of other implements out of perishable materials. And as such, they may be a key to signifying adaptations to more forested tropical conditions. Related to this question is the relative scarcity of stone projectile points in Holocene Archaic assemblages and, when found, their geographically widespread typological similarity. Is the lack of points simply a reflection of a robust industry of projectiles manufactured from perishable materials? And is the widespread typological similarity a historical issue or a result of comparable hafting technologies? More dated assemblages that contain these points need to be described to answer such questions.

Clearly, additional paleoclimatic data are needed from the northern lowlands of South America. Future investigations planned for the Orinoco include detailed stratigraphic tests on a number of sites that contain both preceramic and ceramic components in paleosol contexts. Depositional information, coupled with palynological studies and conducted in a sound geomorphological framework, should provide a greater environmental context in which to explore early to middle Holocene settlement in the Orinoco region and general northern lowlands of South America.

Acknowledgments

I would like to thank Dr. Betty Meggers for commenting on this chapter, as well as to the anonymous reviewers and Julio Mercader for their comments. I alone am responsible for the final conclusions. Fieldwork in Venezuela has been supported by the National Science Foundation, the National Geographic Society, and the Heinz Family Foundation. Their support is graciously acknowledged.

References

Barse, W. P. (1989): A preliminary archeological sequence in the upper Orinoco Valley, Territorio Federal Amazonas, Venezuela. Ph.D. diss., Catholic University of America, Washington, D.C.

———. (1990): Preceramic occupations in the Orinoco River Valley. *Science* 250: 1388–1390.

———. (1995): El periodo Arcaico en el Orinoco y su contexto en el norte de Sud America. In *Ambito y Ocupaciones Tempranas de la América Tropical,* edited by I. Cavelier and S. Mora, 99–113. Bogotá: Fundacion Erigaie and Instituto Colombiano de Antropología.

———. (1996): Test excavations at Pozo Azul Sur-2. Report submitted to the Heinz Family Foundation.

———. (1997): Technical comment: Dating a Paleoindian site in comparison with Clovis culture. *Science* 275:1949–1950.

———. (1999): Environmental history of the Colombian savannas of the Llanos Orientales since the last glacial maximum from lake records El Pinal and Carimagua. *Journal of Paleolimnology* 21:461–476.

———. (2000): Holocene Amazon rainforest-savanna dynamics and climatic implications: High-resolution pollen record from Laguna Loma Linda in eastern Colombia. *Journal of Quaternary Science* 15(7):687–695.

Behling, H., and H. Hooghiemstra (1998): Late Quaternary palaeoecology and palaeoclimatology from pollen records of the savannas of the Llanos Orientales in Colombia. *Palaeogeography, Palaeoclimatology, Palaeoecology* 139:251–267.

———. (1999): Environmental history of the Colombian savannas of the Llanos Orientales since the last glacial maximin from lake records El Pinal and Carimagua. *Journal of Paleolimnology* 21:461–476.

———. (2000): Holocene Amazon rainforest-savanna dynamics and climatic implications: High-resolution pollen record from Laguna Loma Linda in eastern Colombia. *Journal of Quaternary Science* 15(7):687–695.

Behling, H., J. C. Berrio, and H. Hooghiemstra (1999): Late Quaternary pollen records from the middle Caquetá River basin in central Colombian Amazon. *Palaeogeography, Palaeoclimatology, Palaeoecology* 145:193–213.

Boomert, A. (2000): *Trinidad, Tobago, and the lower Orinoco interaction sphere.* Alkmaar: Cairi Publications.

Bush, M. B., and P. A. Colinvaux (1988): A 7000-year pollen record from the Amazon lowlands, Ecuador. *Vegetation* 76:141–154.

Bush, M. B., P. A. Colinvaux, M. C. Wiemann, D. R. Piperno, and K.-B. Liu (1990): Late Pleistocene temperature depression and vegetation change in Ecuadorian Amazonia. *Quaternary Research* 34:330–345.

Cavelier, I., C. Rodríguez, L. F. Herrera, G. Morcote, and S. Mora (1995): No solo de caza vive el hombre: Ocupación del bosque amazónico, holoceno temprano. In *Ambito y Ocupaciones Tempranas de la América Tropical,* edited by I. Cavelier and S. Mora. Bogotá: Fundación Erigaie and Instituto Colombiano de Antropología.

Colinvaux, P. A. (1989): The past and future Amazon. *Scientific American* 260(5): 102–108.

Correal, G. (1979): *Investigaciones Arqueológicas en Abrigos Rocosos de Nemecón y Sueva.* Bogotá: Fundación de Investigaciones Arqueológicas Nacionales, Banco de Republica.

Correal, G., F. Piñeros, and T. Van Der Hammen (1990): Guayabero I: Un sitio preceramico de la localidad Angostura II, San Jose del Guaviare. *Caldasia* 16:245–254.

Fogoça, E., and M. A. Lima (1990): Lábri du Boquete (Bresil): Les premieres industries lithiques de L'Holocene. *Journal Societe Amercanistas* 77:111–123.

Haynes, C. V., Jr. (1997): Technical comment: Dating a Paleoindian site in the Amazon in comparison with Clovis culture. *Science* 275:1948.

Leopold, L. B., M. G. Wolman, and J. P. Miller (1995): *Fluvial processes in geomorphology.* New York: Dover Publications. (Originally published in 1964, San Francisco: W. H. Freeman and Company.)

Meggers, B. J. (1987): Oscilación climática y cronología cultural en el Caribe. In *Actas del Tercer Simposio de la Fundación de Arqueología del Caribe: Relaciones entre la Sociedad y el Ambiente,* edited by M. Sanoja. Washington, D.C.

Meggers, B. J., and J. Danon (1988): Identification and implications of a hiatus in the archeological sequence on Marajó Island, Brazil. *Journal of the Washington Academy of Sciences* 78:245–253.

Politis, G. (1997): Moving to produce: Nukak mobility and settlement patterns in Amazonia. *World Archaeology* 27:492–511.

Prance, G. T., and T. E. Lovejoy (1985): *Amazonia: Key environments.* Oxford, New York, Toronto: Pergamon Press.

Rapp, G., Jr., and C. L. Hill (1998): *Geoarchaeology: The earth science approach to archaeological interpretation.* New Haven and London: Yale University Press.

Reanier, R. E. (1997): Technical comment: Dating a Paleoindian site in comparison with Clovis culture. *Science* 275:1948–1949.

Roosevelt, A. C., C. M. Lima da Costa, C. Lopes Machado, et al. (1996): Paleoindian cave dwellers in the Amazon: The peopling of the Americas. *Science* 272: 373–384.

Rouse, I., and J. M. Cruxent (1963): *Venezuelan archaeology.* New Haven: Yale University Press.

Salgado-Labouriau, M. L. (1982): Climate change at the Pleistocene-Holocene boundary. In *Biological diversification in the tropics,* edited by G. T. Prance, 74–77. New York: Columbia University Press.

Sauer, C. (1969): *Seeds, spades, hearths, and herds.* 2nd ed. Cambridge: Massachusetts Institute of Technology.

Urrego, L. E. (1991): Sucesión holocénica de un bosque de Maurita flexuosa L.F. en el Valle del rio Caqueta. *Colombia Amazónica* 5:99–118.

Van der Hammen, T. (1972): Changes in vegetation and climate in the Amazon

Basin and surrounding areas during the Pleistocene. *Geologische en Mijnbouw* 51:641–643.

———. (1974): The Pleistocene changes of vegetation and climate in tropical South America. *Journal of Biogeography* 1:3–26.

———. (1982): Paleoecology of tropical South America. In *Biological diversification in the tropics,* edited by G. T. Prance, 60–66. New York: Columbia University Press.

Van der Hammen, T., L. E. Urrego, N. Espejo, J. F. Duivenvorden, and J. M. Lips (1990): Fluctuaciones del nivel del agua del río y de la velocidad de sedimentación durante los últimos 13,000 añós en el área del medio Caquetá. *Colombia Amazónica* 5:91–118.

Waters, M. R. (1992): *Principles of geoarchaeology: A North American perspective.* Tucson: University of Arizona Press.

Wijmstra, T. A., and T. Van Der Hammen (1966): Palynological data on the history of tropical savannas in northern South America. *Liedse Geologische Mededelingen* 38:71–90.

CHAPTER 9

Archaeological Hunter-Gatherers in Tropical Forests

A View from Colombia

Santiago Mora and Cristóbal Gnecco

Until recently, the archaeology of tropical hunter-gatherers has relied on ethnographic depictions of present-day foragers. Thus, archaeological research in the tropical forests has often followed ecological and ethnographic variables to explain the past (e.g., resource distribution and availability, soil characteristics). Consequently, current forager features such as low population density, the use of relatively simple agricultural techniques, and forager-farmer interaction and interdependence have all been projected to our knowledge of the past (Mora, 1993). This frozen perspective of the past denies any dynamism and originality to past cultures and lifeways. Indeed, what we call "essentialism" has entirely dominated accepted views of archaeological hunter-gatherers in the tropics. Essentialism takes the notion of "hunter-gatherers" as a discrete, incontingent class of economic organization.[1] From this point of view, foraging societies are perceived as passive entities in the exploitation of the environment. Foragers would be resource users dependent on what nature can offer. This approach has been indirectly fueled by the idea that early hunter-gatherers of the Americas were large-game hunters. Foragers are not conceived as selective manipulators and modifiers of wild resources. As a result, the term "hunter-gatherers" is imprecise to refer to people who not only gathered and hunted but also altered nature to their benefit and increased the natural productivity of ecosystems.[2]

The occupation of tropical forests by hunter-gatherers has been considered the consequence of cultural degradation (Lathrap, 1968; Myers, 1988), the result of ethnic division of labor (Headland and Reid, 1989; Jolly, 1996; Junker, 1996) and strong interaction between farmers and hunter-gatherers in marginal ecosystems (Headland and Reid, 1989; Jolly, 1996; Junker, 1996). Forager-farmer interaction has been approached from different angles, whether archaeological (Junker, 1996), historical (Morey and Morey, 1973; Gordon, 1984; Solway and Lee, 1990), economical (Peterson, 1978a; Milton, 1984), or ideological (Jolly, 1996). Some scholars have also argued that hunting-gathering as a way of life was impossible in tropical forests until after interaction with farmers (Milton, 1984; Bailey et al., 1989; Headland and Reid, 1989). Seasonal scarcity of carbohydrates and animal proteins have been supposed to prevent human occupation of tropical forests prior to agriculture (Lathrap, 1968; Myers, 1988; Bailey et al., 1989, 1991; Sponsel, 1989; Bailey and Headland, 1991).

Another proposition in the archaeology of tropical forest hunter-gatherers is economic specialization. Drawing on Paleoindian data from the North American grasslands, it was traditionally believed that the early hunter-gatherers of tropical America must have been cooperative, specialized, large-game hunters who lived in open environments. Tropical forests would not be attractive regions to these hunters (see Willey, 1971) because their animal biomass would be low to sustain their focal and specialized economies. Yet, new data depict a different scenario in which early hunter-gatherers colonized rain forests at very early ages and practiced generalist economic strategies and low residential mobility.

Paleoecological Insights and Hunter-Gatherer Archaeology

The idea of an early hunter-gatherer occupation of the Neotropical forest was considered an empirical and theoretical oddity, until recently, on the grounds that Paleoindian large-game hunters were not suited for rain forest settlement. The geographical barrier today posed by the Isthmus of Panama separating the South American forest from Central America led archaeologists to propose different colonization models for the early settlement of the southern cone. Thus, C. O. Sauer (1944) and S. K. Lothrop (1961) suggested the existence of an open corridor along the Pacific coast through which southbound hunter-gatherers entered South America (see also Lynch, 1967, 1978). Recent paleobotanical work has indicated, however, that this open corridor may have never existed (Bartlett and Barghoorn, 1973; Piperno, 1985; Colinvaux and Bush, 1991; Piperno et al., 1991a; Behling et al., 1998). Noting that the actual isthmus may have been forested in the past, A. J. Ranere (1980) suggested that early colonists relied on hunting the large mammals that inhabited the highlands; and, therefore, the occupation of tropical biomes south of Mesoamerica required little change in hunting strategies and tools. R. L. Kelly and L. C. Todd (1988) termed this strategy "technology-oriented" instead of "place-oriented." T. F. Lynch (1978: 473) opposed Ranere, realizing that there was not enough archaeological evidence to support his model and that the low biomass typical of the rain forest could not support specialized hunters. Lynch also believed that foraging in the South American forest was possible only after the onset of extensive forager-farmer interaction: "life in Ranere's forest would have been difficult without substantial reliance on fishing or agriculture, both unknown in the Americas at 10,000 B.P." (Lynch, 1978: 473).

Ten thousand years ago, climate change transformed the weather patterns of Amazonia and the Andes. In the Ecuadorian Andes, pollen analysis shows that the local vegetation was affected by cooling (Colinvaux et al., 1997). Climatic change also brought dryer conditions to the lowlands and thus altered the nature and geographical distribution of the rain forest. Forest fragmentation was common during glacial times. Various climatic models suggest that the northern regions of South America supported botanical formations similar to today's tropical savannas with grassy and open landscapes and gallery forests. Large lakes such as Lake Valencia in Venezuela dried up at glacial times (Salgado-Labouriau, 1980). South of the equator, similar changes have been documented (Ledru et al., 1998). Yet, mounting archaeological evidence from Panama (Ranere and Cooke, 1991), Venezuela (Barse, 1990, 1995), Colombia (Cavelier et al., 1995; Gnecco and Mora, 1997; Gnecco, 2000), and Brazil (Roosevelt et al., 1996) indicates an early settlement of various types of rain forest environments, both open and dense, since the late Pleistocene. The two archaeological sites we discuss here were occupied in what at the time was tropical forest, although we are not saying that its composition was in any way identical to that of modern counterparts. Indeed, the species composition of late Pleistocene and early Holocene rain forests was unlike present-day formations (cf. Gnecco, 1995; see Dillehay, 2000).

Late Pleistocene to Holocene Sites in the Colombian Amazon

Two sites have been excavated recently by the authors: San Isidro and Peña Roja (figs. 9.1 and 9.2).

Peña Roja

This site is located on the Caquetá River, 50 km downstream from Araracuara, in the Colombian Amazon region. Peña Roja lies at 170 m above sea level. Current average temperatures range from 24°C to 28°C. Annual rainfall averages 3500 mm. Present-day vegetation is tropical forest (Cavelier et al., 1995; C. Urrego et al., 1995; Gnecco and Mora, 1997). Sedimentological, geomorphological, and paleobotanical data from the region have been published elsewhere (L. E. Urrego, 1991, 1997; T. Van der Hammen et al., 1991a,b; M. C. Van der Hammen, 1992; Duivenvoorden and Lips, 1993; Duivenvoorden and Cleef, 1994; Behling et al., 1999).

The earliest settlers of Peña Roja arrived in the region prior to 9000 B.P., as indicated by carbon 14 dates in association with archaeological remains:

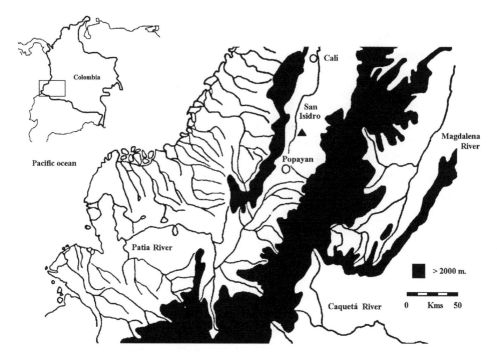

Fig. 9.1. Location of Peña Roja and San Isidro in southwestern Colombia

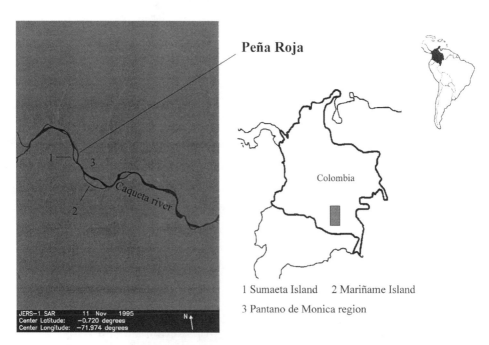

Fig. 9.2. Satellite image showing the Middle Caquetá River and the location of Peña Roja, Mariñame Island, and Pantano de Mónica

9125 ± 250 B.P.; 9160 ± 90 B.P.; 9250 ± 140 B.P. (Cavelier et al., 1995; Gnecco and Mora, 1997); 8090 ± 60 B.P. (Piperno and Pearsall, 1998: 204); 8510 ± 110 B.P.; 8710 ± 110 B.P. (Llanos, 1997: 14). Paleoclimatic and sedimentological data are indicative of vegetational change and major transformations in the Middle Caquetá River drainage system. Such geological and environmental changes triggered terrace formation and landscape remodeling as well as vegetational reassortment.

Pollen analysis at nearby lakes, such as Mariñame, 12 km southwest of Peña Roja, and Pantano de Mónica, 5 km to the south (fig. 9.2), provides reliable late Pleistocene to Holocene environmental data on which to base our reconstructions of the environment at the time of human occupation. The pollen data from the Mariñame core (L. E. Urrego, 1991; T. Van der Hammen et al., 1991a,b) show an increase in forest vegetation and a dramatic decrease in aquatic elements since 11,150 B.P.; *Mauritia flexuosa*, a palm tree that usually grows in swampy areas, was rare at this time. Trees from *Melastomataceae, Cecropia, Alchornea*, and *Euterpe* were also documented (L. E. Urrego, 1991). The Pantano de Mónica M1 pollen diagram (Behling et al., 1999) shows a continuous increase in palms (Arecaceae) in pollen zone 2. This zone corresponds with the initial human occupation of Peña Roja. Here, typical tropical forest elements such as the Anacardiaceae, Sapotaceae, and Malpighiaceae are all present (Behling et al., 1999). Palynological evidence, therefore, leaves no doubt that the initial occupation of Peña Roja took place in a rain forest context.

The site of Peña Roja has two occupation horizons (fig. 9.3). The first horizon corresponds to the agricultural and ceramic occupation of the site, comprising layers 2 and 3. Layer, or stratum, 4 has both ceramic and preceramic materials. Strata 5, 6, 7, and 8 yield preceramic materials alone, and some bioturbation is present. The initial occupation of the site is located in layers 8a to 7b. The stone industry produced by the early inhabitants of the site consists of unifacial industries with little or no retouch (fig. 9.4). Chert is the main raw material. Tools include concave scrapers on thick flakes, wedges, notched flakes, and perforators. But unretouched flakes form the bulk of the artifactual evidence. Some of the unretouched materials bear use-wear. Ground stone technology is also present, with artifacts such as milling stones, flat mortars, and grinding stones, suggestive of seed and root consumption (Llanos, 1997). Overall, the tool kit from Peña Roja indicates a non-specialized extractive technology typical of broad-spectrum economies.

Several features within layers 6a to 5b suggest changing site function over time. Prior to this time, foragers displayed a predilection for chert as the main raw material to make their tools. From this time on, chert was less frequently used, relative to the frequency documented in lower layers. We also

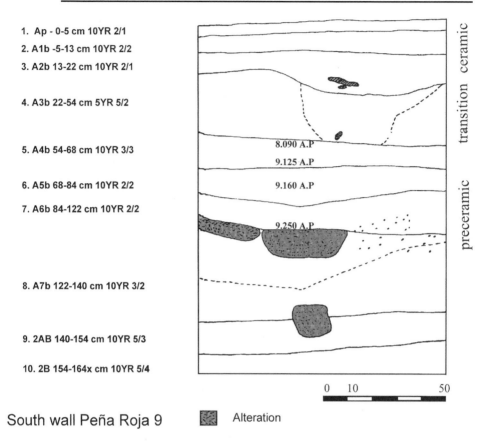

1. Ap - 0-5 cm 10YR 2/1

2. A1b -5-13 cm 10YR 2/2

3. A2b 13-22 cm 10YR 2/1

4. A3b 22-54 cm 5YR 5/2

5. A4b 54-68 cm 10YR 3/3

6. A5b 68-84 cm 10YR 2/2

7. A6b 84-122 cm 10YR 2/2

8. A7b 122-140 cm 10YR 3/2

9. 2AB 140-154 cm 10YR 5/3

10. 2B 154-164x cm 10YR 5/4

8.090 A.P
9.125 A.P
9.160 A.P
9.250 A.P

ceramic

transition

preceramic

0   10   50

South wall Peña Roja 9   Alteration

Fig. 9.3. Stratigraphic section at Peña Roja showing strata, carbon-14 dates, and matrix color

noticed a reduction in charcoal representation, which may suggest fewer burning episodes. This reduction seems concomitant with the introduction of squashes *(Cucurbita spp.)*. D. Piperno (1999), based on the size of the *Cucurbita* phytoliths from this site, has suggested that these squashes were cultivated plants. At a later time, gourd, *Lagenaria siceraria,* lerén, and *Calathea allouia* were also exploited (Piperno and Pearsall, 1998; Piperno, 1999). Lerén has an edible rhizome which is still eaten cooked or roasted in the middle Caquetá (Sánchez, 1997: 255). However, the introduction of these cultivars did not imply the creation of open landscapes or grasslands, since the phytolith evidence points to a forested environment (Piperno and Pearsall, 1998; Piperno, 1999).

Layer 4 shows evidence of stratigraphic discontinuity between the ceramic and preceramic levels at the site, as shown by soil texture (silt increases and clay decreases) and organic matter decrease. The stratigraphic discon-

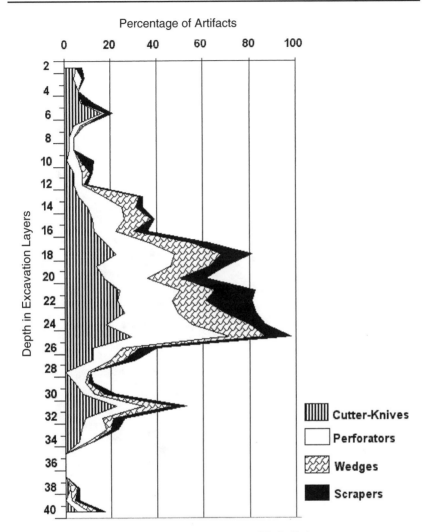

Percentage of Artifacts

Fig. 9.4. Artifact percentages per depth at the site of Peña Roja

tinuity is also evident in the percentages of aluminum, with low values in the preceramic layers. Another sign of discontinuity is the presence of high phosphorus values above 350 ppm (Cavelier et al., 1995) only in the ceramic layers.

Macrobotanical remains from the site are dominated by eight species of palms from four genera (fig. 9.5, table 9.1): *Astrocaryum* (three species: *A. javari, A. aculeatum,* and *A. sciophilum*); *Oenocarpus* (three species: *O. bataua, O. bacaba,* and *O. mapora*); *Mauritia* (one species: *M. flexuosa*); and *Maximiliana* (one species: *M. maripa*). *Astrocaryum, Oenocarpus,* and *Mauritia* are restricted to preceramic levels. *Maximiliana maripa* has similar frequencies in

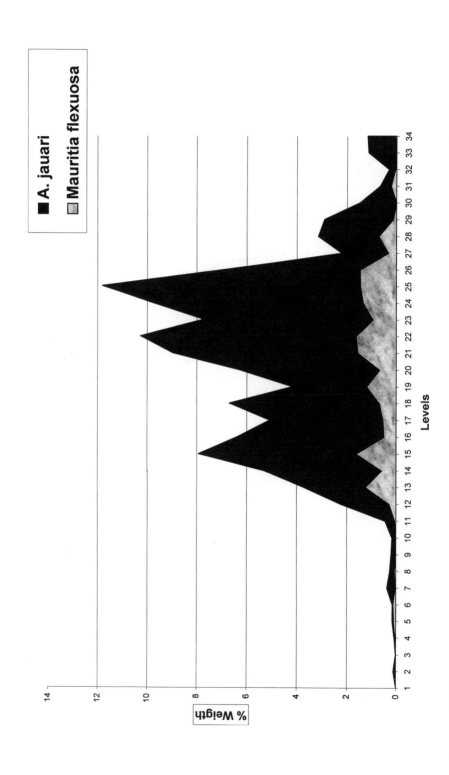

Table 9.1. Taxonomic Groups Identified in Peña Roja's Carbonized Macro Remains

Family	Species
Arecaceae	*Astrocaryum aculeatum*
	Astrocaryum javari
	Astrocaryum sciophilum
	Attalea spp
	Attalea maripa
	Attalea insignis
	Attalea racemosa
	Mauritia flexuosa
	Oenocarpus bataua
	Oenocarpus mapora
	Oenocarpus bacaba
Caryocaraceae	*Caryocar aff. Glabrum*
Humiriaceae	*Vantanea peruviana*
	Humiriastrum sp.
	Sacoglottis sp
Apocynaceae	*Macoubea guianensis*
Lauraceae	*Beilschmiedia brasiliensis*
Annonaceae	*Oxandra euneura*
Chrysobalanaceae	*Licania pyrifolia*
Leguminosae	*Parkia multijuga*
	Inga sp.

SOURCE: Based on Cavelier et al., 1995; Cavelier et al., 1999; Morcote 1994; Morcote et al., 1998.

both ceramic and preceramic levels (Morcote, 1994; Cavelier et al., 1995: 34; C. Urrego et al., 1995; Morcote et al., 1998). These palms are still widely used by local groups. For example, *A. javari* is presently exploited as fish bait. It is thus likely that palms were of economic value to local prehistoric groups (Morcote et al., 1998; Cabrera et al., 1999). In general, the macrobotanical evidence from Peña Roja suggests exploitation strategies very similar to those documented in eastern Brazil among late Pleistocene to Holocene tropical forest foragers (Prous, 1991; Magalhaes, 1994; Roosevelt et al., 1996).

Site location in an alluvial setting would indicate some emphasis on fish exploitation (Gragson, 1992; Rodríguez, 1992), while the surrounding hills would provide a good source of forest plants and animals (M. C. Van der

Fig. 9.5. Distribution of *M. flexuosa* versus *A. jauari* at Peña Roja

Hammen, 1992; Duivenvoorden and Lips, 1993). Nearby swamps with *Mauritia flexuosa* would contribute fruits, fibers, fish bait (Goulding, 1980), hunting prey (Walschburger and Von Hildebrand, 1988), and Coleoptera larvae (Dufour, 1987).

San Isidro

This site is within an inter-Andean valley, 50 km northwest of Popayán, in southwestern Colombia. San Isidro is at 1690 m above sea level. Present rainfall is 1800 mm. Natural forest vegetation has been preserved in parts of the region, and it can be classified as sub-Andean forest (*sensu* Cuatrecasas, 1958). Pollen and macrobotanical data indicate that by 10,000 B.P. the area supported a tropical forest that has no modern analogue (Gnecco, 2000). This site has one archaeological horizon only. We excavated 30% of this site (20 m²) with 1-by-1-m excavation units. Excavation spits were 5 cm thick, and the matrix was screened with a 5-mm mesh. Six kilograms of sediments from each level were floated.

The archaeological deposit stretches through 40 cm of matrix (from 20 to 60 cm below the surface), but cultural materials are scattered between 5 and 85 cm of depth. The highest concentration takes place from 5 to 40 cm below the present-day surface. This type of shallow preceramic deposit is not common in the Popayán Valley, where there is a slow rate of sediment buildup. The sediments at the site have a low percentage of organic matter and lack stratigraphic boundaries. Nonetheless, soil formation provides clear pedological horizons that may have masked the original stratigraphic boundaries at the site. The A horizon is composed of sandy loam without cultural materials and comprises layers 1 and 2. The AB horizon includes layers 3 and 4 and has a sandy texture. The preceramic occupation is restricted to layer 3. Lastly, the B horizon comprises strata 5 to 8, all of which have a clayey texture.

Our analysis indicates that the human occupation of this site took place in soils formed by intense weathering with biochemical activity that affected the pyroclastic materials in the matrix and with a slow sedimentary influx. Organic matter is scarce (10.0%), so is the amount of phosphorus (1.9%). Organic decomposition and/or low input of organic compounds by local foragers could be responsible for this pattern. It is important to note that this site lacks a midden and that human occupation is only evident through the existence of cultural material in what otherwise looks like a natural soil profile. Three radiocarbon dates indicate an early Holocene chronology: charcoal samples from the occupation horizon, dated at 9530 ± 100 B.P. and 10,050 ± 100 B.P., and charred seed, dated at 10,030 ± 60 B.P.

San Isidro yielded more than 65,000 lithic specimens, out of which 752 have clear artifactual features. The majority of the lithic assemblage (98.0%)

Fig. 9.6. Stone tools from San Isidro: *a*, beveled and polished tool; *b*, edge-ground cobble. Macrobotanical remains: *c*, from *Persea americana*, and *d*, from *Erythrina edulis*.

is microdebitage smaller than 1 cm in maximum length. Most of the artifacts are unretouched (38.6%) and 0.7% are cobble stone tools. Bifaces make up 22.0% of the artifacts. Eventually, local foragers produced projectile points. Use-wear analysis of the lithic assemblage (Gnecco, 2000) indicates that tools were used for game butchering and hide preparation and for sawing and grooving. Our lithic assemblage also includes hammer stones; mortars; beveled, polished tools; and edge-ground cobbles (fig. 9.6). Phytoliths and starch grains from economically useful plants, such as legumes, grasses, *Marantha*, and perhaps the edible *M. arundinacea*, were recovered from the surfaces and microfissures in two grinding tools (Dolores Piperno, personal communication), while *Podocarpus* wood was recovered from three small unifacial tools (Nieuwenhuis, 1996).

Almost 4000 carbonized macrobotanical remains were retrieved from this open-air site. A large majority of pieces (92%) are cortical fragments from spherical fruits. Their diameter ranges from 30 to 60 mm. We have reported these fruits as belonging to the Palmae (Gnecco and Mora, 1997: 687), but their final identification awaits in-depth comparative analysis: we are almost certain that this macrobotanical assemblage contains two tropical forest taxa: *Erythrina edulis* and *Persea americana* (fig. 9.6).

"Essentialism" under a New Light

Archaeological data from Peña Roja and San Isidro indicate that human impact and modification of forest ecosystems started before 10,000 B.P. Pollen data from San Isidro (table 9.2) includes pioneer taxa, grasses, and weeds, among a majority of primary forest trees. This suggests that these sites were

Table 9.2. Fossil pollen from San Isidro

Family/genus	N	%	Family/genus	N	%
Leguminosae	11	0.3	Solanaceae	13	3.8
Leguminosae	21	0.3	Cyatheaceae	8	2.3
Leguminosae	31	0.3	Fern	37	10.5
Proteaceae	12	3.5	Moss	2	0.6
Tiliaceae	1	0.3	Acrostichum	1	0.3
Myrtaceae	4	1.1	Alnus sp.	1	0.3
Palmae	14	1.1	Weinmannia sp.	9	2.6
Palmae	23	0.8	Trema sp.	1	0.3
Palmae	31	0.3	Hedyosmum sp.	1	0.3
Gramineae	15	1.4	Vallea sp.	1	0.3
Gramineae	215	4.3	Rapanea sp.	3	0.8
Bignoniaceae	31	9.0	Alchornea sp.	18	5.2
Compositae	97	28.2	Virola sp.	6	1.7
Labiatae	1	0.3	Piper sp.	2	0.6
Melastomataceae	79	23.0	Miconia sp.	4	1.1
Urticaceae/Plantago	1	0.3			
Moraceae	13	3.8	Spigelia type	2	0.6
Caryophyllaceae	4	1.1	Duroia type	6	1.7
Cyperaceae	1	0.3			

surrounded by a mosaic of forest and open spaces. However, the prevalence of mature forest species indicates that there never was a complete forest retreat. At present, we cannot determine whether these open domains were anthropogenic. Disturbance sometimes occurs unintentionally as a result of redundant human use of natural environments (Laden, 1992; Maloney, 1998). But it is also possible that humans contributed to opening the forest by maintaining gaps, perhaps through the use of fire (Piperno, 1999). It is possible that some of the beveled and polished tools found in Peña Roja and San Isidro (figure 9.6a) could be interpreted as tools created to open up the forest.

We also believe that, prior to farming, foragers promoted the artificial concentration of useful plants across their territory. This farming-like behavior focused on species that required little planting or tending (Cabrera et al., 1999) and resembles present-day practices among forager and incipient Amazonian farmer groups (Gutiérrez, 1996; Politis, 1996a; Posey, 1984). Posey indeed calls this type of exploitation "nomadic agriculture," in which humans allow the colonization of forest plots by useful plants such as medicinals and perennials with large edible roots and stems (Posey, 1983: 877–894, 1984: 114–117; see also Piperno, 1989: 541). This type of exploitation also favors artificial concentrations of useful animals, as noted by

O. F. Linares (1976) and others (see Posey, 1983, 1984; Cooke and Piperno, 1993: 30).

The *Persea* seeds (maximum length 6 cm) from San Isidro are likely to be from a cultivar (fig. 9.6c), as they are larger than the average size for specimens from a wild *Persea* population (see Smith, 1966, 1969). The same observation applies to *Erythrina edulis* specimens (fig. 9.6d). *Marantha* phytoliths found in grinding tools from San Isidro could also belong to a cultivar, perhaps *M. arundinacea*. Piperno believes that this root was domesticated 8600 years ago in Panama, as shown by phytoliths from this species retrieved from Cueva de los Vampiros (1995: 139–141; Piperno et al., 1991b: 238; Cooke 1992: 44). Gourd and *lerén* from Peña Roja could also represent early cultivars.

Discussion

C. Levi-Strauss (1950) noted that farming societies of South America complemented their farming economies with wild resources (Sponsel, 1989). If this was the case among ethnographic farmers, it is reasonable to assume that it was more the case with prehistoric hunter-gatherers with mixed forager-farming economies. We believe that late Pleistocene and early Holocene hunter-gatherers were efficient managers of tropical forests and enhanced the natural productivity of the ecosystems in which they lived. Tropical hunter-gatherers did not have to become sedentary farmers to increase the productivity of their resource base. South American tropical forest foragers hunted, gathered, and, somehow, produced their resources. Binford (1980) suggested that residential mobility among hunter-gatherers was almost exclusively contingent upon resource distribution. Yet, mobility was determined by many factors. If residential mobility among Peña Roja and San Isidro foragers influenced local resource distribution through intentional manipulation, then access to resources may not have been free for all individuals from all groups. Late Pleistocene to Holocene Amazonian foragers could have regulated access to resources and exercised some kind of territoriality (cf. Kelly, 1995: 14–15). It is possible that various forms of low residential mobility and territorial societies could have evolved in neighboring Andean societies since the late Pleistocene (Keefer et al., 1998; Sandweiss et al., 1998). R. C. Bailey and T. N. Headland (1991: 268) predicted that if foragers were living in tropical rain forests before the introduction of agriculture, they would have had to be more mobile than Pygmies, Agta, Batek, and Punan groups are today. We do not believe this holds true for the Colombian instance 10,000 years ago. Mobility is determined by the type of exploitation and control over local resources.

The authors of this chapter think that we need to reexamine currently accepted models of prehistoric forager societies in tropical environments. Decades ago Lathrap (1968) challenged the idea of pristine hunter-gatherer groups living in tropical forests based on the ecological information he possessed; he portrayed them as decultured farmers. At present, his hypothesis is increasingly unsustainable, especially in the light of the obvious diversity among prehistoric tropical forest groups and the questionable idea of "primitiveness" (see Headland and Reid, 1989: 43, 49–51; Sponsel, 1985: 96–97). Certainly, present-day forest dwellers interact with farmers, and it is likely that some prehistoric hunter-gatherers also did (Headland and Reid, 1989; Junker, 1996). This inter-ethnic division of labor, as A. Testart (1988: 7) calls it, cannot be denied (see Peterson, 1978a,b; Milton, 1984). Yet, we cannot accept the transhistorical nature of this division. Thus, the question is not whether hunter-gatherers ever existed in tropical forests without farmers, but how they lived in these ecosystems.

Conclusion

Colombian prehistoric hunter-gatherers used, managed, and enhanced the productivity of tropical forests. San Isidro and Peña Roja foragers changed the ecosystem through forest clearing, selective planting, and quasi-domestication (Rindos, 1984). This type of economic behavior has ecological and social repercussions and is linked to territoriality and low residential mobility. This type of forest society dates back to the late Pleistocene. Early foragers of the Colombian Amazon were forest managers who maximized the natural productivity of rain forests in ways that departed from earlier foraging strategies to enter a new cultural domain, that of complex hunter-gatherers.

Notes

1. The intentional definition of the type, as used by archaeologists, is exclusively economical, to the exclusion of alternative characteristics—social, political, ideological—explored by ethnographers.

2. Therefore, the use we make of this concept is purely conventional. See T. Ingold (1991) and R. Kelly (1995) for a critical evaluation of this concept.

References

Bailey, R. C., and T. N. Headland (1991): The tropical rain forest: Is it a productive environment for human foragers? *Human Ecology* 19:261–285.

Bailey, R. C., G. Head, M. Jenike, B. Owen, R. Rechtman, and E. Zechentes (1989):

Hunting and gathering in tropical rain forest: Is it possible? *American Anthropologist* 91:59–82.

Bailey, R. C., M. Jenike, and R. Rechtman (1991): Reply to Colinvaux and Bush. *American Anthropologist* 93:160–162.

Bamforth, D. B. (1986): Technological efficiency and tool curation. *American Antiquity* 51:38–50.

Barse, W. P. (1990): Preceramic occupations in the Orinoco River valley. *Science* 250:1388–1390.

———. (1995): El período arcaico en el Orinoco y su contexto en el norte de Sud América. In *Ambito y Ocupaciones Tempranas de la América Tropical,* edited by I. Cavelier and S. Mora, 99–113. Bogotá: Instituto Colombiano de Antropología-Fundación Erigaie.

Bartlett, A. S., and E. S. Barghoorn (1973): Phytogeographic history of the Isthmus of Panama during the past 12,000 years (a history of vegetation, climate, and sea-level change). In *Vegetation and vegetational history of northern Latin America,* edited by A. Graham, 203–299. New York: Elsevier.

Behling, H., H. Hooghiemstra, and A. J. Negret (1998): Holocene history of the Chocó rain forest from Laguna Piusbi, southern Pacific lowlands of Colombia. *Quaternary Research* 50:300–308.

Behling, H., J. C. Berrio, and H. Hooghiemstra (1999): Late quaternary pollen records from the middle Caquetá River basin in central Colombian Amazon. *Palaeogeography, Palaeoclimataology, Palaeoceology* 145:193–213.

Binford, L. R. (1973): Interassemblage variability—the Mousterian and the functional argument. In *The explanation of culture change,* edited by C. Renfrew, 227–254. London: Duckworth.

———. (1977): Forty-seven trips. In *Stone tools as cultural markers,* edited by R. V. Wright, 24–36. Canberra: Australian Institute of Aboriginal Studies.

———. (1980): Willow smoke and dog's tails: Hunter-gatherer settlement systems and archaeological site formation. *American Antiquity* 45:1–17.

Cabrera, G., C. E. Calvo, and D. Mahecha (1999): *Los Nukak: Nómadas de la Amazonia Colombiana.* Bogotá: Universidad Nacional.

Cavelier, I., C. Rodriguez, S. Mora, L. F. Herrera, and G. Morcote (1995): No sólo de caza vive el hombre: Ocupación del bosque amazónico, Holoceno temprano. In *Ambito y ocupaciones Tempranas de la América Tropical,* edited by I. Cavelier and S. Mora, 27–44. Bogotá: Instituto Colombiano de Antropología, Fundación Erigaie.

Cavelier, I., L. F. Herrera, S. Rojas, and F. Montejo (1999): Las palmas como mediadoras en el origen de las plantas cultivadas en el Caquetá, noroeste amazónico. Paper presented to the I Congreso de Arqueología en Colombia, Manizales.

Colinvaux, P. A., M. B. Bush, M. Steinitz-Kannan, and M. C. Miller (1997): Glacial and postglacial pollen records from the Ecuadorian Andes and Amazon. *Quaternary Research* 48:69–78.

Colinvaux, P., and M. B. Bush (1991): The rain-forest ecosystem as a resource for hunting and gathering. *American Anthropologist* 93:153–162.

Cooke, R. (1992): Etapas tempranas de la producción de alimentos vegetales en la baja Centroamérica y partes de Colombia (región histórica Chibcha-Chocó). *Revista de Arqueología Americana* 6:35–70.

Cooke, R., and D. R. Piperno (1993): Native American adaptations to the tropical forests of Central and South America before the European colonization. In *Tropical forest people and food: Biocultural interactions and applications to development*, edited by C. M. Hladik, A. Hladik, O. Linares, H. Pagezy, A. Semple, and M. Hadley, 25–36. Paris: Parthenon.

Cuatrecasas, J. (1958): Aspectos de la vegetación natural de Colombia. *Revista de la Academia Colombiana de Ciencias Exactas, Físicas y Naturales* 10(4).

Dillehay, T. (2000): *The settlement of the Americas*. New York: Basic Books.

Dufour, D. (1987): Insects as food: A case study from the northwest Amazon. *American Anthropologist* 89:383–397.

Duivenvoorden, J., and A. M. Cleef (1994): Amazonian savanna vegetation on the sandstone plateau near Araracuara, Colombia. *Phytocoenologia* 24:197–232.

Duivenvoorden, J. F., and J. M. Lips (1993): *Ecología del Paisaje del Medio Caquetá: Estudios en la Amazonia Colombiana*. Bogotá: TropenBos.

Gnecco, C. (1995): Paleoambientes, modelos individualistas y modelos colectivos en el norte de los Andes. *Gaceta Arqueológica Andina* 24:5–11.

———. (2000): *Ocupación Temprana de Selvas Tropicales de Montaña*. Popayán: Universidad del Cauca.

Gnecco, C., and S. Mora (1997): Early occupations of the tropical forest of northern South America by hunter-gatherers. *Antiquity* 71:683–690.

Gordon, R. J. (1984): The Kung in the Kalahari exchange: An ethnohistorical perspective. In *Past and present in hunter-gatherers studies*, edited by C. Schrire, 175–193. Orlando: Academic Press.

Goulding, M. (1980): *The fish and the forest*. Los Angeles: University of California Press.

Gragson, T. L. (1992): Fishing the water of Amazonia: Native subsistence economies in a tropical rain forest. *American Anthropologist* 94:428–440.

Gutiérrez, R. (1996): Los Nukak y el uso de los recursos. *Diversa* 2:4–7.

Headland, T. N., and L. A. Reid (1989): Hunter-gatherers and their neighbors from prehistory to the present. *Current Anthropology* 30:43–66.

Ingold, T. (1991): Notes on the foraging mode of production. In *Hunters and gatherers: History, evolution, and social change*, vol. 1, edited by T. Ingold, D. Riches, and J. Woodburn, 269–285. Oxford: Berg.

Jolly, P. (1996): Symbiotic interaction between black farmers and south-eastern San. *Current Anthropology* 37:277–306.

Junker, L. L. (1996): Hunter-gatherer landscapes and lowland trade in the prehispanic Philippines. *World Archaeology* 27:389–410.

Keefer, D. K., S. D. deFrance, M. E. Moseley, J. B. Richardson, D. R. Satterlee, and A. Day-Lewis (1998): Early maritime economy and El Niño events at quebrada Tacahuay. *Science* 281:1833–1835.

Kelly, R. (1983): Hunter-gatherer mobility strategies. *Journal of Anthropological Research* 39:277–306.

————. (1995): *The foraging spectrum: Diversity in hunter-gatherer lifeways.* Washington, D.C.: Smithsonian.

Kelly, R. L., and L. C. Todd (1988): Coming into the country: Early Paleoindian hunting and mobility. *American Antiquity* 53:231–244.

Laden, G. (1992): Ethnoarchaeology and land-use ecology of the Efe (Pygmies) of the Ituri rain forest, Zaire: A behavioral ecological study of land-use patterns and foraging behavior. Ph.D. diss., Harvard University.

Lathrap, D. W. (1968): The "hunting" economies of the tropical forest zone of South America: An attempt at historical perspective. In *Man the hunter,* edited by R. B. Lee and I. DeVore, 23–29. Chicago: Aldine.

Ledru, M. P., M. L. Salgado-Labouriau, and M. L. Lorscheitter (1998): Vegetation dynamics in southern and central Brazil during the last 10,000 yr B.P. *Review of Palaeobotany and Palynology* 99:131–142.

Levi-Strauss, C. (1950): The use of wild plants in tropical South America. In *Handbook of South American Indians,* vol. 6, edited by J. Steward, 465–486. Washington, D.C.: Bureau of American Ethnology, Smithsonian Institution.

Linares, O. F. (1976): Garden hunting in the American tropics. *Human Ecology* 4:331–349.

Llanos, J. M. (1997): Artefactos de molienda en la región del Medio río Caquetá (Amazonia Colombiana). *Boletín de Arqueología* 12(2):3–95.

Lothrop, S. K. (1961): Early migrations to Central and South America: An anthropological problem in the light of other sciences. *Journal of the Royal Anthropological Institute* 91:97–123.

Lynch, T. F. (1967): *The nature of the central Andean preceramic.* Pocatello: Occasional Papers of the Idaho State University Museum 21.

————. (1978): The South American Paleoindians. In *Ancient Native Americans,* edited by J. Jennings, 455–489. San Francisco: W. H. Freeman and Co.

Magalhaes, M. (1994): *Archaeology of Carajas: The prehistoric presence of man in Amazonia.* Rio de Janeiro: Companhia Vale do Rio Doce.

Maloney, B. (1998): The long-term history of human activity and rainforest development. In *Human activities and the tropical rainforest,* edited by B. Maloney, 65–85. The Hague: Kluwer Academic Publishers.

Milton, K. (1984): Protein and carbohydrate resources of the Maku Indians of northwestern Amazonia. *American Anthropologist* 86:7–27.

Mora, S. (1993): Cold and hot, green and yellow, dry and wet: Direct access to resources in Amazonia and the Andes. *Florida Journal of Anthropology* 18:51–60.

Morcote, G. (1994): Estudio paleoetnobotánico en un yacimiento precerámico del medio río Caquetá, Amazonia colombiana. B.A. thesis, Universidad Nacional, Bogotá.

Morcote, G., G. Cabrera, D. Mahecha, and I. Cavelier (1998): Las palmas entre los grupos cazadores-recolectores de la amazonia colombiana. *Caldasia* 20:57–74.

Morey, N., and R. Morey (1973): Foragers and farmers: Differential consequences of Spanish contact. *Ethohistory* 20:229–246.

Myers, T. (1988): Una visión de la prehistoria de la Amazonia superior. In *I seminario de investigaciones sociales en Amazonia,* 37–87. Iquitos: Editorial CETA.

Nieuwenhuis, C. J. (1996): Microwear analysis of a sample of stone artefacts from San Isidro. Manuscript, Departamento de Antropología, Universidad del Cauca, Popayán.

Peterson, J. (1978a): *The ecology of social boundaries: Agta foragers of the Philippines.* Urbana: University of Illinois Press.

———. (1978b): Hunter-gather/farmer exchange. *American Anthropologist* 80: 335–351.

Piperno, D. (1985): Phytolithic analysis of geological sediments from Panamá. *Antiquity* 59:13–19.

———. (1989): Non-affluent foragers: Resource availability, seasonal shortages, and the emergence of agriculture in Panamanian tropical forests. In *Foraging and farming: The evolution of plant exploitation,* edited by D. R. Harris and G. C. Hillman, 538–554. London: Unwin Hyman.

———. (1990): Paleoenvironments and human occupation in late-glacial Panamá. *Quaternary Research* 33:108–116.

———. (1995): Plant microfossils and their application in the New World tropics. In *Archaeology in the lowland American tropics,* edited by P. W. Stahl, 130–153. Cambridge: Cambridge University Press.

———. (1999): Report on phytoliths from the site of Peña Roja, western Amazon basin. Manuscript, Fundación Eriagie, Bogotá.

Piperno, D., and D. Pearsall (1998): *The origins of agriculture in the lowland Neotropics.* Orlando: Academic Press.

Piperno, D. R., M. B. Bush, and P. A. Colinvaux (1991a): Paleoecological perspectives on human adaptation in Panama 1. The Pleistocene. *Geoarchaeology* 6: 201–226.

———. (1991b): Paleoecological perspectives on human adaptation in central Panama I. The Holocene. *Geoarchaeology* 6:227–250.

Politis, G. (1996a): Moving to produce: Nukak mobility and settlement patterns in Amazonia. *World Archaeology* 27:492–511.

———. (1996b): *Nukak.* Bogotá: SINCHI.

Posey, D. A. (1983): Indigenous knowledge and development: An ideological bridge to the future. *Ciencia e Cultura* 35:877–894.

———. (1984): A preliminary report on diversified management of tropical forests by the Kayapo Indians of the Brazilian Amazon. *Advances in Economic Botany* 1:112–126.

———. (1993): The importance of semi-domesticated species in post-contact Amazonia: Effects of the Kayapo Indians on the dispersal of flora and fauna. In *Tropical forest people and food: Biocultural interactions and applications to development,* edited by C. M. Hladik, A. Hladik, O. Linares, H. Pagezy, A. Semple, and M. Hadley, 63–71. Paris: Parthenon.

Prous, A. (1991): Fouilles de L'abri du Boquete, Minas Gerais, Brasil. *Journal de la Societe des Americanistes* 77:77–109.

Ranere, A. J. (1980): Stone tools and their interpretation. In *Adaptive radiations in prehistoric Panama,* edited by O. Linares and A. J. Ranere, 118–138. Cambridge: Peabody Museum Monographs 5, Harvard University Press.

Ranere, A. J., and R. Cook (1991): Paleoindian occupation in the Central American tropics. In *Clovis: Origins and adaptations*, edited by R. Bonnichsen and K. L. Turnmire, 237–253. Corvallis: Center for the Study of the First Americans, Oregon State University.

Rindos, D. (1984): *The origins of agriculture: An evolutionary perspective.* San Diego: Academic Press.

Rodríguez, C. A. (1992): *Bagres, Malleros y Cuerderos en el Bajo Río Caquetá.* Bogotá: TropenBos.

Roosevelt, A. C., M. Lima, C. Lopes, M. Michab, N. Mercier, H. Valladas, J. Feathers, W. Barnett, M. Imazio, A. Henderson, J. Sliva, B. Chernoff, D. S. Reese, J. A. Holman, N. Toth, and K. Schick (1996): Paleoindian cave dwellers in the Amazon: The peopling of the Americas. *Science* 272:373–384.

Salgado-Labouriau, M. (1980): A pollen diagram of the Pleistocene-Holocene boundary of Lake Valencia, Venezuela. *Review of Palaeobotany and Palynology* 30: 297–312.

Sánchez, M. (1997): *Catálogo Preliminar Comentado de la Flora del Medio Caquetá.* Bogotá: TropenBos.

Sandweiss, D. H., H. McInnis, R. L. Burger, A. Cano, B. Ojeda, R. Paredes, M. C. Sandweiss, and M. D. Glascock (1998): Quebrada Jaguay: Early South American maritime adaptations. *Science* 281:1830–1832.

Sauer, C. O. (1944): A geographic sketch of early man in America. *The Geographical Review* 34:529–573.

Shott, M. J. (1996): An exegesis of the curation concept. *Journal of Anthropological Research* 52:259–280.

Smith, C. E. (1966): Archaeological evidence for selection in avocado. *Economic Botany* 20:169–175.

———. (1969): Additional notes on pre-conquest avocados in Mexico. *Economic Botany* 23:135–140.

Solway, J. S., and R. B. Lee (1990): Foragers, genuine or spurious? *Current Anthropology* 33:187–224.

Sponsel, L. E. (1985): Ecology, anthropology, and values in Amazonia. In *Cultural values and human ecology in Southeast Asia,* edited by K. L. Hutterer, A. T. Rambo, and G. Lovelace, 77–121. Ann Arbor: Michigan Papers on South and Southeast Asia 27, University of Michigan Press.

———. (1989): Farming and foraging: A necessary complementarity in Amazonia? In *Farmers as hunters: The implications of sedentism,* edited by S. Kent, 37–45. Cambridge: Cambridge University Press.

Testart, A. (1988): Some major problems in the social anthropology of hunter-gatherers. *Current Anthropology* 29:1–31.

Torrence, R. (1983): Time budgeting and hunter-gatherer technology. In *Pleistocene hunters and gatherers in Europe,* edited by G. Bailey, 11–22. Cambridge: Cambridge University Press.

Urrego, C., L. F. Herrera, S. Mora, and I. Cavelier (1995): *Informática y arqueología: Un modelo para el Manejo de Datos Básicos.* Bogotá: Colcultura.

Urrego, L. E. (1991): Sucesión holocénica de un bosque de Maurita flexuosa L. F.

en el valle del río Caquetá (Amazonia Colombiana). *Colombia Amazonica* 5(2): 99–118.

———. (1997): Los Bosques Inundables del Medio Caquetá: Caracterización y sucesión. Bogotá: TropenBos.

Van der Hammen, M. C. (1992): *El Manejo del Mundo: Naturaleza y sociedad entre los Yukuna de la Amazonia Colombiana*. Bogotá: TropenBos.

Van der Hammen, T., L. E. Urrego, N. Espejo, J. F. Duivenvorden, and J. M. Lips (1991a): El cuaternario tardio del rea del medio Caquetá (Amazonia Colombiana). *Colombia Amazónica* 5(1):91–118.

———. (1991b): Fluctuaciones del nivel del agua del río y de la velocidad de sedimentación durante los últimos 13,000 años en el área de medio Caquetá (Amazonia Colombiana). *Colombia Amazónica* 5(1):119–130

Walschburger, T., and P. Von Hildebrand (1988): Observaciones sobre la utilización estacional del bosque húmedo tropical por los indígenas del río Mirití. *Colombia Amazonica* 3:51–73.

Willey, G. R. (1971): *An introduction to American archaeology*. Vol. 2. Englewood Cliffs: Prentice Hall.

CHAPTER 10

Hunter-Gatherers in Amazonia during the Pleistocene-Holocene Transition

Betty J. Meggers and Eurico Th. Miller

Amazonia is the largest expanse of tropical rain forest on the planet and the least known archeologically. Survey along the major tributaries has identified hundreds of habitation sites with ceramics, but little evidence of earlier occupation. Lithics encountered during survey along power-line transects and riverbanks and in three rock shelters testify to the presence of humans in widely separated parts of the lowlands by at least 13,000 B.P., but establishing whether these locations were within the rain forest and were colonized prior to the adoption of agriculture requires identifying the extent of the forest during the Pleistocene-Holocene transition and the antiquity of plant domestication. In contrast to the acceptance of periods of rain forest reduction in tropical Africa, the existence, magnitude, timing, duration, and character of Amazonian fluctuations remain disputed. Establishing the antiquity of dependence on cultigens is inhibited by the perishable nature of the primary domesticates, manioc and sweet potatoes. On the positive side, the biogeographic interpretation of the distributions of linguistic, genetic, and ethnographic traits and the observations of the subsistence behavior of surviving hunter-gatherers provide independent sets of data that can be compared with the paleoenvironmental and archeological reconstructions.

After a brief summary of the modern ecosystem, discussion will focus on four general topics: (1) the environment between circa 15,000 and 7000 B.P.; (2) the archeological evidence for early human presence; (3) the antiquity of subsistence agriculture, and (4) the ethnographical evidence for the productivity of wild food resources. We conclude that the Amazonian rain forest was colonized by small groups of foragers as early as other parts of South America.

The Present Environment of Amazonia

Tropical rain forest is now the dominant vegetation east of the Andes below about 1000 m elevation, where average monthly temperature varies less than 2.5°C, annual precipitation exceeds 1500 mm, the dry season lasts less than three months, and relative humidity normally exceeds 80% (Meggers, 1996). Bisection by the equator produces alternating rainy seasons north and south

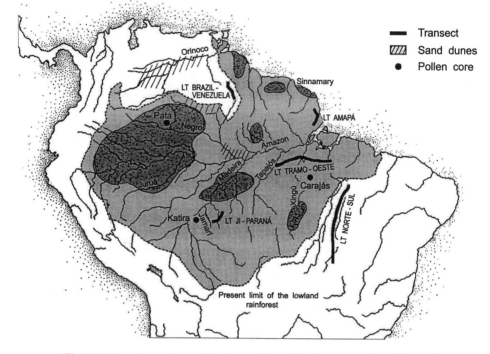

Fig. 10.1. Locations of power-line transects and riverine surveys in relation to the present limit of the rain forest; the remnants assuming a 40% decrease in precipitation with a similar distribution; and the locations of dune fields (after Van der Hammen and Absy, 1994; Meggers, 1996).

of the east-flowing Amazon, reducing the normal crest during annual flooding to less than 10 m. The canopy makes a major contribution to the present climate by moderating solar radiation and by generating 50% or more of the rainfall by evapo-transpiration. Enclaves of open vegetation and savanna of various magnitudes occur especially across the northeast, where local precipitation is under 1500 mm, the dry season exceeds three months, and edaphic conditions are unsuitable for forest.

The relatively uniform appearance of the forest vegetation throughout the lowlands conceals high local and regional diversity. Individuals of most species are dispersed, palms and bamboos being the principal exceptions. Flowering and fruiting are typically seasonal. Birds, reptiles, amphibians, small mammals, monkeys, and invertebrates populate the canopy, whereas most terrestrial vertebrates are rare, small, and scattered. Grassland fauna is relatively impoverished, with only 6 species of ungulates compared with 68 species in tropical Africa (Bourlière, 1973: 281).

Ecologists recognize two principal types of ecosystems: the *varzea* and the *terra firme*. Although each incorporates considerable local diversity, they differ significantly in their potential for human exploitation (Sioli, 1984; Barrow, 1985; Campbell et al., 1986; Junk, 1997; Herrera et al., 1978; Padoch et al., 1999). The varzea, which constitutes only some 2% of Amazonia, is restricted to the flood plains of the "white-water" rivers that drain from the Andes. Annual deposition of nutrient-rich sediment renews the soil, and dissolved nutrients support an aquatic food chain extending from microorganisms to mammals. The flora is adapted to seasonal inundation. The remaining 98%, known as the terra firme, consists of unflooded land drained by the nutrient-poor clear- and black-water tributaries originating in the Guayana and Brazilian shields. Fish resources are relatively impoverished and aquatic reptiles and mammals are rare or absent. The vegetation is adapted to nutrient-poor soil and seasonal rainfall.

The size of the indigenous population in A.D. 1500 is unknown, and recent estimates range from less than 1 million to more than 5 million. Early European explorers reported large settlements along the varzea, but their existence has been questioned by historians and is unsupported thus far by archeology (Meggers, 1995). Autonomous communities of hunter-gatherers have persisted until the past few decades in the interior of the Guianas, in southwestern Brazil, and along the eastern base of the Andes in Peru, Ecuador, and Colombia. Their subsistence and settlement behavior reveal in-depth knowledge of the rain forest flora and fauna and their interactions, as well as sophisticated measures for maintaining their long-term sustainable exploitation.

The Environment between circa 15,000 and 7000 B.P.

The Amazonian rain forest is characterized by exceptional biodiversity. It has been argued that isolation conducive to speciation could be provided by river-channel migration (Salo, 1987), sea-level rise (Nores, 1999), lowered temperature (Bush, 1994) or could even have developed in the context of forest stability (Colinvaux and Oliveira, 2001). By contrast, the existence of episodes of rain forest fragmentation of varying extents, durations, and intensities during and since the Pleistocene is increasingly supported by palynological, geological, paleoclimatological, sedimentological, and biogeographical evidence (Behling and Hooghiemstra, 2001; Latrubesse and Ramonell, 1994; Prance, 1982; see Hooghiemstra and Van der Hammen, 1998, for an extensive bibliography).

Abrupt declines in sediment deposition in pollen cores and associated gaps in carbon-14 dates in seven widely separated locations imply that the late glacial maximum "was represented probably by a hiatus of several thousand years, indicative of drier climates than before or after" (Ledru et al., 1998: 233). Five of the sequences are in central Brazil, but two are from

Amazonia. At Carajás in the east, a hiatus in sedimentation, implying a long drought between circa 22,000 and 13,000 B.P., correlates with low lake levels in other parts of South America (Sifeddine et al., 2001; all dates are uncalibrated). At Pata in the northwest, where "some 80% of total pollen grains can be assigned to trees of the tropical lowland forest" (Colinvaux, 1997: 5), no deposition occurred between circa 30,000 and 18,000 B.P. and between circa 14,000 and 6000 B.P. The existence of similar hiatuses in widely separated regions makes it "plausible that these discontinuities of the sedimentary record can be attributed to climatic periods with reduced precipitation regimes" (Oliveira, 1996: 132; also Ledru et al., 1998). This assessment is supported by analysis of the oxygen isotopic composition of planktonic foraminifera, from a marine sediment core west of the Amazon fan, which indicates a reduction in discharge of at least 60% during the Younger Dryas (Maslin and Burns, 2000).

Eolian sand extends for some 400 km along the left side of the Amazon east of the Negro and covers several thousand square kilometers between the Branco and Negro to the north (fig. 10.1; Iriondo and Latrubesse, 1994; Van der Hammen and Hooghiemstra, 2000: 731). Although the principal dunes may have formed during or prior to the late glacial maximum, a more recent origin for the small associated sand fields near the lower Negro is suggested by a layer of eolian sand deposited in the nearby Caverna da Pedra Pintada between circa 10,000 and 7000 B.P. (Santos et al., 1993; Roosevelt et al., 1996). Declines in the diversity of forest species and explosions of pioneer plants between circa 10,000 and 8000 B.P. and between 6000 and 4000 B.P. in French Guiana, where annual precipitation now exceeds 3000 mm, also imply significant lowering of the water table during these times (Charles-Dominique et al., 1998). Sand dunes and a strong regression of the forest during a dry Holocene interval have also been reported in lowland Bolivia (Servant et al., 1981).

Recent hydrology research in central Amazonia has shown that "despite a five-fold reduction in infiltration rates following forest clearing by heavy machinery . . . this decrease was insufficient to produce overland flow and erosion" (Bonell, 1998: 240). Similar lack of runoff, plus the removal of "massive amounts of material" during low sea level and the redeposition of older sediments eroded from the lower terrace, can account for the absence of an expected increase in the frequency of grass pollen on the Amazon shelf during periods of forest fragmentation (Nittrouer et al., 1995: 189; Van der Hammen and Hooghiemstra, 2000: 736; cf. Haberle and Maslen, 1999).

The combination of lowered rainfall and a 100 m drop in sea level during the late glacial maximum produced a 20–25 m decline in the mean water level of the lower Solimões (Tricart, 1985; Müller et al., 1995). The Amazon and its principal eastern tributaries cut deep channels during this time, and sedimentation occurred only after the sea level began to rise. Since the modern configuration of the varzea (flood plain) is estimated to have an antiquity no greater than circa 5000 years (Irion et al., 1997: 27), its present abundance of subsistence resources would not have been available to early Holocene colonists. The tributaries draining the Guayana and Brazilian shields depend on local rainfall, and the substantial declines recorded during recent brief El Niño episodes imply that longer droughts would have inflicted significant moisture stress on both terrestrial and aquatic biota (Carvalho, 1952; Molion and Moraes, 1987; Nobre and Renno, 1985).

Correlation of the various forms of botanical and climatic evidence suggests a decline of about 40% in annual precipitation between circa 29,000 and 13,000 B.P. (Van der Hammen and Hooghiemstra, 2000). Maintaining the present pattern of distribution and decreasing the 2500 mm isohyet by 40% preserves the rain forest in the west and restricts it to isolated patches in central and eastern Amazonia (fig. 10.1). The disjunct occurrences of various species of savanna plants north and south of the Amazon (Giulietti and Pirani, 1988; Harley, 1988) and the existence of relict savanna enclaves in northeastern Amazonia (Eden, 1974) testify to past continuity of open vegetation across the eastern lowlands. Its extent, distribution, duration, and composition undoubtedly fluctuated in time and space, and rain forest probably persisted along watercourses as well as in topographically favorable locations (Meave et al., 1991; Kellman et al., 1994; Servant et al., 1993; Behling and Hooghiemstra, 2001). This scenario is compatible with the palynological evidence for persistence of rain forest along the eastern margin of the Andes (Colinvaux et al., 1997) and for episodes of forest replacement at Carajás and Katira, as well as presence of gaps in the sediment cores (fig. 10.1). Even a 25% reduction in rainfall would place these locations in a diagonal corridor across eastern Amazonia linking the llanos of the Orinoco with the uplands of eastern Brazil. The boundaries of these regions are hypothetical and would have been subject to temporal fluctuations (Hooghiemstra and Van der Hammen, 1998: fig. 3c; Turcq et al., 1998; this reconstruction is rejected by Colinvaux et al., 2000).

Evidence for Early Human Presence

Two principal sources of evidence are available for assessing the antiquity and context of human colonization of Amazonia. Archeological remains provide direct evidence in the form of artifacts and the locations and dates of sites. Linguistic, genetic, and cultural distributions provide indirect evidence of past population movements and replacements.

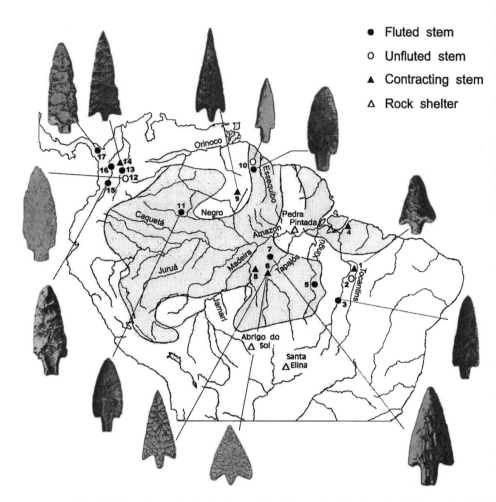

- ● Fluted stem
- ○ Unfluted stem
- ▲ Contracting stem
- △ Rock shelter

Fig. 10.2. Locations of rock shelters and isolated projectile points in relation to the limit of the rain forest, assuming a 25% reduction in precipitation with a similar distribution (after Van der Hammen and Absy, 1994): *1,* Itaguatins, TO, length 4.3 cm; *2,* Darcinópolis, TO; *3,* Upper Araguaia, TO/MT, length 15.0 cm; *4,* Ilha Cotijuba, PA, length 6.1 cm (Hilbert, 1998: fig. 2); *5,* Middle Xingu, PA, length 17.4 cm (Hilbert, 1998: fig. 4); *6,* Middle Tapajós, PA, length 8.5 and 6.4 cm (Simões, 1976: figs. 1 and 2); *7,* Cara Preta, AM, length 10.5 cm (Museu do Homem do Norte, Manaus); *8,* Apuí, AM (private collection); *9,* Igarapé Murupu, RR, length 9.2 cm; *10,* Mazaruni District, Guyana, length 8.0 and 10.6 cm (Evans and Meggers, 1960); *11,* Upper Negro, AM, length 13.0 cm (private collection); *12,* Sabana de Bogotá (Reichel-Dolmatoff, 1986: fig. 10); *13,* Middle Magdalena (López, 1998, fig. 1.2j); *14,* Puerto Berrío (Ardilla and Politis, 1989, fig. 3.3); *15,* Restrepo, Valle, length ca. 8.5 cm (Ardilla, 1991: fig. 4.4); *16,* Niquía, Medellín, length ca. 11.0 cm (Ardilla Calderón, 1991: fig. 4.3); *17,* Golfo de Urubá (Ardilla Calderón and Politis, 1989: fig. 2.2).

Archeological Evidence

ISOLATED PROJECTILE POINTS

Large, symmetrical, bifacial, stemmed projectile points have been encountered in widely separated parts of Amazonia (fig. 10.2). Length ranges from circa 4 to 17 cm. One variety has narrow horizontal shoulders and a parallel-sided, unfluted or fluted stem. Another has barbed shoulders and a tapered stem with a rounded base. A third has ephemeral shoulders and a rounded stem (Evans and Meggers, 1960: pl. 8b). All are undated and, except for the unfluted, stemmed example from the LT (Linha de Transmissáo, or transmission line) Norte-Sul transect, their archeological context is unknown.

The first two types also occur in forested parts of the Cauca and Magdalena Valleys of Colombia (fig. 10.2). Obsidian hydration dates from artifacts associated with the parallel-sided, stemmed Restrepo type extend from circa 11,000 to 9000 B.P. (Ardilla Calderón, 1991; Lleras and Gnecco, 1986). A carbon-14 date of 10,350 ± 90 B.P. was obtained from a site producing an unfluted example (López, 1998). As in Amazonia, the bulk of the associated artifacts are expedient tools. Climatic reconstructions indicate wetter conditions than exist today and mixed open- and closed-tropical vegetation. It is noteworthy that projectile points of the lanceolate El Jobo and the fishtail-fluted types, which are characteristic in Panama and Venezuela to the north and down the Andes to southern cone, appear to be absent in Amazonia and rare in Colombia.

ROCK SHELTERS

Stratigraphic excavations have been conducted in three rock shelters: Caverna da Pedra Pintada, on the left bank of the lower Amazon, and Abrigo do Sol and Santa Elina, both on the southern margin of the present rain forest (fig. 10.2). None has produced bifacial stone projectile points of the kind described above.

At Caverna da Pedra Pintada, a Paleoindian level some 30 cm thick was separated from more recent occupation by about 30 cm of sterile, wind-blown sand (Roosevelt et al., 1996). Among more than 30,000 lithics collected, only 24 were shaped tools, the majority unifacial. Percussion is the dominant technique, but pressure flaking also occurs. Subsistence remains consist of fruits, seeds, and nuts from wild plants and a wide range of aquatic, terrestrial, and aerial fauna. The 56 carbon-14 dates obtained from this occupation range from circa 11,000 to 10,000 B.P. Although the authors argue for a rain forest context, the similarity of the dates, lithic complex, subsistence remains, hearths, and rock art to those in the earliest levels at Boquete rock shelter in Minas Gerais suggests similar adaptation to a patchy

Fig. 10.3. Projectile point of the Restrepo type from TO-TO-01, Darcinópolis, and scrapers and cores from various sites in the LT Norte-Sul transect

environment (Prous, 1991; Kipnis, 1998), consistent with its location in the arid corridor.

At Abrigo do Sol in the Chapada dos Parecís, percussion-flaked cores, flakes, and expedient tools also predominate, some with lateral retouch or evidence of use. Charred palm fruits were the only surviving subsistence remains. Thirty-two carbon-14 dates extending from 14,700 to 5760 B.P. tend to become more recent toward the interior of the shelter, reflecting recession of the drip line. An isolated carbon-14 date of 19,400 ± 195 B.P. is problematical, but more recent than the earliest date from Santa Elina not far to the east (Miller, 1987).

Santa Elina, in the Serra das Araras, is now in a mixed *cerrado* and seasonal tropical forest environment. The earliest assemblage, consisting of unifacial flakes associated with Glossotherium remains, has a Uranium-Thorium date of 27,000 ± 2000 B.P. The second assemblage, characterized by modern fauna, lithics, and hematite fragments, is carbon-14 dated between circa 13,000 and 7000 B.P. This and some 40 other shelters in the region contain rock art of uncertain antiquity (Vilhena Vialou et al., 1999).

OPEN SITES

Lithic sites encountered by Miller during salvage investigations along power-line transects document human presence in several parts of Amazonia by at least 13,000 B.P. Surface indications are rare and most sites were revealed by test pits in the foundations of towers and in excavations of footings, which consisted of four large pits totaling up to 464 m^3.

Lithics were encountered at 299 of 933 tower sites investigated in the LT Norte-Sul transect, which extends along the Tocantins for 516 km in the states of Maranhão and Tocantins, as well as at three substations and in 24 locations around and between towers (fig. 10.1). Surface materials oc-cur only where exposed by erosion; occupation refuse typically begins at a depth of about 40 cm and occasionally continues to 7.0 m. Charcoal is abundant, and the 21 carbon-14 dates obtained extend from 11,300 ± 90 to 1150 ± 60 B.P., with a problematic outlier of 23,280 ± B.P.

Lithics were also encountered at 162 of 330 tower sites along the LT Brazil–Venezuela transect in northern Roraima (fig. 10.1). An additional 68 sites were identified between towers, many of which also had ceramic oc-cupations. Here too, refuse began approximately 40 cm below the surface. Charcoal was again abundant and 13 carbon-14 dates were obtained, ex-tending from 13,660 ± 430 to 1170 ± 60 B.P.

Excavations at a substation near the coast in Amapá produced carbon-14 dates of 12,840 ± 210 and 6650 ± 60 B.P. Excavations along the LT Ji-Paraná–Rolim de Moura transect in Rondônia produced quartz flakes and a single carbon-14 date of 13,720 ± 160 B.P. (fig. 10.1; Eurico Th. Miller, manuscript). Preliminary reconnaissance along the LT Tramo-Oeste tran-sect, extending across terra firme between the Tocantins and the Tapajós, also identified lithic sites.

Contemporary occupation of riverbanks is attested by salvage investiga-tions in two widely separated regions prior to construction of hydroelectric dams. Two sequential complexes underlay up to 6 m of later refuse along the Jamarí, a right bank tributary of the upper Madeira in southwestern

Amazonia (fig. 10.1; Miller et al., 1992). Sites of the initial Itapipoca phase were on sterile soil 10 to 20 m above high-water level and 3.6 to 6.0 m below the present surface. At two sites, artifacts were concentrated in a layer 15 cm thick extending 32 and 56 m along the bank. At another, they were dispersed in a layer up to 90 cm thick. The lithic inventory is characterized by large, irregular, percussion-flaked bifaces and scrapers, hammer stones, and flakes with and without retouch, as well as exhausted cores. Four carbon-14 dates extend from 8230 ± 100 to 6970 ± 60 B.P., but the earliest levels are undated. With one exception, the sites of the succeeding Pacatuba phase are in different locations; thickness of the deposit varied between 40 and 310 cm. The lithic complex consists of smaller expedient tools as well as cores, flakes, and microflakes, some showing wear. Four carbon-14 dates range from 6090 ± 130 to 5210 ± 70 B.P.

Survey along the lower Sinnamary, which flows through rain forest on the central coast of French Guiana, identified 273 loci of human activity (fig. 10.1). A detailed relative sequence extending from the initial occupation of the region into the colonial period is complemented by 125 carbon-14 dates. The two earliest are 14,990 and 12,190 B.P., and seven others fall between 9910 and 8390 B.P. Although no distinct preceramic levels were identified, the vast majority of the flakes and expedient tools were encountered at a site with three dates in the 9900–8000 B.P. range (Vacher et al., 1998).

The lithic assemblages from all these sites consist almost exclusively of expedient tools and debitage produced from flakes and cores of quartzite, quartz, chalcedony, sandstone, and chert. Some exhibit use damage on one or more margins; more rarely, an edge has been retouched. The single exception is an unfluted, stemmed projectile point excavated at Darcinópolis, Tocantins (figs. 10.2 and 10.3). The transects follow high land between watersheds, and the sites appear to be temporary camps for tool manufacture. Where multiple carbon-14 dates exist, they indicate that some were reoccupied during millennia. No significant temporal or spatial differences in the lithics are evident from preliminary inspection, although the time span of the carbon-14 dates suggests some might be expected from more detailed analysis. By contrast, a decline in size was evident in the Jamarí sequence.

Comparing the locations of the sites with the climatic reconstructions shows that all the rock shelters, most of the power-line transects, and the Jamarí riverbank sites are in regions likely to have had more open vegetation during the Pleistocene-Holocene transition, even with only a 25% reduction in precipitation (fig. 10.2). The presence of lithic sites along the Sinnamary and Caquetá, however, as well as the existence of isolated projectile points in other locations likely to have remained forested even with a 40%

Fig. 10.4. Scrapers and flakes with evidence of use from various sites in the LT Norte-Sul transect

Fig. 10.5. Lithics with and without evidence of use from sites at the southern end of the LT Brazil-Venezuela transect with carbon-14 dates between 9100 ± 80 and 5460 ± 70 B.P.

Fig. 10.6. Quartz hammer stones and exhausted cores from sites in the LT Brazil-Venezuela transect with carbon-14 dates between 13,660 ± 430 and 10,470 ± 230 B.P.

reduction imply occupation of the rain forest by hunter-gatherers during this period.

Biogeographical Evidence

Biogeographers interpret the disjunct occurrences of Amazonian plants and animals of the same genus or species as evidence of the disruption of formerly contiguous communities by fragmentation of their rain forest habitat. Several linguistic, genetic, and cultural-element distributions are also compatible with the existence of more open vegetation across the central lowlands during the Pleistocene-Holocene transition (Meggers, 1975, 1982, 1994).

LINGUISTIC DISTRIBUTIONS

Whereas other parts of South America are dominated by a single linguistic stock, Amazonia is characterized by exceptional heterogeneity. The most recent classification identifies two widespread stocks, Ge-Pano-Carib and

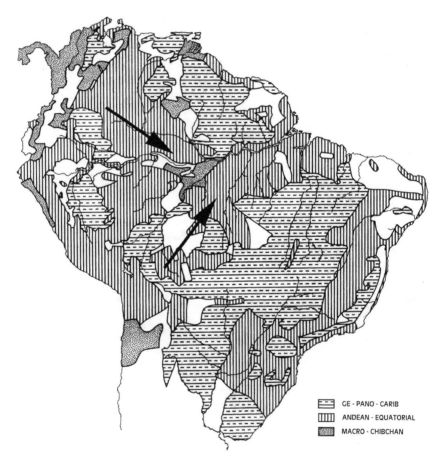

GE - PANO - CARIB
ANDEAN - EQUATORIAL
MACRO - CHIBCHAN

Fig. 10.7. Distributions of Ge-Pano-Carib and Equatorial-Tucanoan languages. The disjunct distributions of Ge, Pano, and Carib speakers east, west, and north of the central Amazon and their association with more open vegetation suggest that they were replaced by Equatorial speakers as they expanded from their homelands in the west when the rain forest coalesced during the early Holocene (after Greenberg, 1987).

Equatorial-Tucanoan, several smaller stocks, and numerous isolated languages (Greenberg, 1987). Families of Ge-Pano-Carib have disjunct distributions. Ge speakers extend across eastern Brazil, where scrub and cerrado vegetation prevail; Carib speakers are concentrated in the Guianas and adjacent eastern Venezuela, where rain forest is interrupted by enclaves of savanna, and Pano speakers occupy similar mixed habitats along the eastern margin of the central Andes. Their shared linguistic antecedents and preference for open vegetation suggest that the ancestral Proto-Ge-Pano-Carib speakers colonized the central lowlands during the Pleistocene-Holocene transition, when these conditions prevailed. Expansion of the rain forest would have isolated the populations in peripheral locations, favoring their differentiation by evolutionary drift. The concentration of Equatorial-Tucanoan speakers in the western lowlands, where rain forest persisted, suggests that adaptation to this environment facilitated their eastward expansion and replacement of the Ge-Pano-Carib speakers. Lexicostatistical estimates for completion of the differentiation of Proto-Ge, Proto-Carib, Proto-Arawak, and Proto-Tupí by circa 5000 B.P. are compatible with the carbon-14 dates for coalescence of the rain forest (Migliazza, 1982).

CULTURAL DISTRIBUTIONS

A few cultural traits not subject to environmental constraints that would inhibit their retention have distributions implying displacement of an earlier by a later variant. For example, among two kinds of racks used for roasting meat, the platform raised on four vertical posts occurs south, east, and north of the central lowlands, whereas the tripod is restricted to the intervening region. Explanations for the shadows on the surface of the moon also fall into two categories. Groups that interpret the dark patches as the face of a man who committed incest with his sister and fled to escape revenge are centrally located, whereas those with different explanations occupy the periphery. Other traits with concentric distributions include pole snares and simple nooses for capturing animals and finger positions used to hold arrows prior to release (Meggers, 1982, 1994).

GENETIC DISTRIBUTIONS

The frequencies of 13 marker-gene systems among 17 indigenous populations from different parts of the lowlands were subjected to principal-component analysis. The plots of the first three components, which explain 93% of the original variance, suggest dispersion from the central Amazon toward the north, south, east, and west (Rothhammer and Silva, 1992: fig. 2a,c). Examining the linguistic affiliations of the groups included reveals that the Ge-Pano-Carib speakers are distributed peripherally to the Equatorial-Tucanoan speakers, in keeping with other evidence for expansion of the latter into rather than out of the central lowlands during the Pleistocene-Holocene transition.

The distributional patterns of these linguistic, genetic, and cultural traits imply replacement of an earlier population associated with mixed vegetation by a later one associated with rain forest, providing independent support for the environmental fluctuations inferred from various kinds of biological, geological, sedimentological, and climatological evidence.

Evidence for Plant Domestication

The antiquity of agriculture in Amazonia is a matter of speculation. The rare archeobotanical remains from habitation sites are restricted to charred seeds, fruits, and wood from wild species. No cultigens were included among the "very abundant" plant remains in Teso dos Bichos on Marajó, occupied between circa 1200 and 930 B.P. (Roosevelt, 1991: 375–376, 313–314). No phytoliths of maize, squash, arrowroot, or other domesticates were identified in a pollen core taken 90 km north of Manaus in central Amazonia with dates from circa 7700 and 3000 B.P. (Piperno and Becker, 1996). Although the appearance of particulate carbon and pollen from secondary vegetation at 8000 B.P. or earlier is often considered to be proxy evidence of forest clearing by humans (Athens and Ward, 1999; Gnecco, 1998; Piperno, 1990: 672, 1998: 420–421; Bush et al., 1989; Behling, 1996), this assessment is disputed by ecologists, who point out that ignition is possible only after an extended dry season and that lightning is an alternative agent (Servant et al., 1981; Martin et al., 1995; Soubiès, 1979–80; Charles-Dominique et al., 1998; Lucas et al., 1993; Middleton et al., 1997; Hansen and Rodbell, 1995). They also question the anthropogenic origin of the vine forests in eastern and southwestern Amazonia, noting their correlation with special climatic conditions (Gentry, 1993: 511) and extensive recent deforestation (Pérez-Salicrup et al., 2001). Similarly, a natural origin for the large single stands of babassu *(Orbignya martiana)* is implied in its unique adaption to survive fire by sending the apical shoot downward for several years before turning it toward the surface (Balée, 1989: 97).

The only tropical cultigen identifiable from pollen or phytoliths is maize (Piperno, 1985: 196). Dates as early as 7000 B.P. have been obtained from pollen cores in eastern Colombia and Ecuador (Bush et al., 1989; Gnecco and Mora, 1997), and similar antiquity has been reported in Panama, coastal Ecuador, and Peru (Cooke and Ranere, 1992; Pearsall, 1995: 183; Bonavia and Grobman, 1998). The questionable nature of the contexts and dating (Fritz, 1994; Smith, 1995), the absence of macrobotanical remains even on the arid coast of Peru prior to circa 3000 B.P. (Van der Merwe et al., 1993; Pearsall, 1999), stable isotope studies indicating significant consumption only after that time (Schoeninger, 1996; Burger and Van der Merwe, 1990), and especially the failure of maize to become an important component of the diet in Mexico until about 5000 B.P. argue against the reliability of the early Amazonian dates (Blake et al., 1992; Minnis, 1992; Smith, 1995; see Fritz, 1994, for details).

Suggestive indirect evidence for plant domestication is provided by the archeological sequence on the Jamarí, a tributary of the upper Madeira in southwestern Amazonia (fig. 10.1; Miller et al., 1992: 37–38). Stratigraphic excavations at three sites revealed horizons of black soil 45 to 80 cm thick containing lithics between the late preceramic Pacutuba–phase occupation, which exhibits no soil discoloration, and the pottery-making Jamarí phase. The lithic complex of this Massangana phase expands the inventory of the preceding Pacatuba phase by the addition of small mortars and pestles, anvil stones, axes with flaked and pecked surfaces, grinding stones impregnated with hematite, and fragments of hematite. Nineteen carbon-14 dates from six sites place its inception circa 4800 B.P. and the introduction of pottery circa 2500 B.P. Although the genesis of black soil is unclear, its omnipresence in habitation sites of pottery-making groups throughout the lowlands suggests that the associated subsistence and settlement behavior had been adopted in southwestern Amazonia by 4800 B.P. (Woods, 1995). This hypothesis is compatible with phylogeographic evidence for the domestication of manioc in this region (Olsen and Schaal, 1999, 2001).

Ethnographic Evidence

Although surviving Amazonian hunter-gatherers are generally considered degraded agriculturalists, several share elements of folklore, games, puberty rites, and other cultural traits not only with one another but also with North American hunter-gatherers, suggesting that they perpetuate an ancient way of life (Meggers, 1964). The production of palm starch by groups now isolated north and south of the rain forest suggests that it may have been a widespread subsistence resource prior to the adoption of manioc. It remains the primary staple of the Warao at the delta of the Orinoco (Meggers, 2001; Wilbert, 1976). *Mauritia flexuosa*, the principal source of the starch, is ubiquitous in the lowlands, occurring not only in old oxbows, on river margins, in seasonal swamps, and in gallery forests, but also in savannas and forest understory, and it often forms dense pure stands (Moore, 1973). The average yield of 60 kg per trunk is comparable to that of Indonesian species of sago palms (Ruddle et al., 1978). The combination of vitamins and minerals obtained from fruits and seeds, protein and fat provided by insect larvae developed in rotting trunks, and starch would have constituted a balanced and reliable diet in a variety of habitats (Moore, 1973; Clement, 1993).

The viability of a wild-food diet has been documented by comprehensive investigations of the behavior of the Nukak, who inhabit the rain forest in

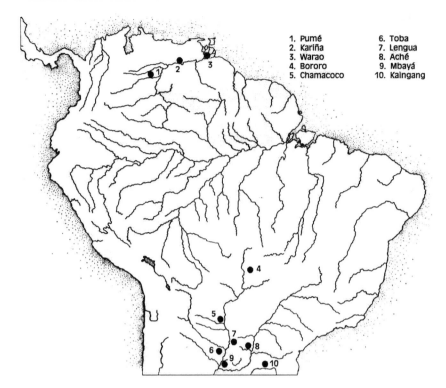

1. Pumé   6. Toba
2. Kariña   7. Lengua
3. Warao   8. Aché
4. Bororo   9. Mbayá
5. Chamacoco 10. Kaingang

Fig. 10.8. The locations of surviving hunter-gatherers who extract starch from the trunks of palms. Their disjunct distributions north and south of the rain forest suggest that this subsistence resource was a major source of carbohydrates throughout the intervening lowlands before the domestication of manioc (after Meggers, 2001).

southeastern Colombia. The Nukak live in bands of 10 to 30 individuals and move camp every three to five days as local subsistence resources decline. Their diet consists primarily of plants (fruits, seeds, tubers), small game, fish, insects, turtles, and honey. Nutritional analyses testify to a balanced intake of oils, acids, fats, proteins, and carbohydrates. They enhance the carrying capacity of their environment by several kinds of "involuntary management" that were probably widespread during earlier times, especially selective felling of trees to increase the abundance of useful species, bringing fruits to camp, where preservation of the canopy favors sprouting of seeds, and perpetuating the resulting concentrations by occupying adjacent areas during revisits (Politis and Rodríguez, 1994; Politis et al., 1997). Evidence that similar potential exists elsewhere in the Amazonian rain forest was provided by 100 members of a Yanomami community in southern Venezuela; they were able to maintain good health when forced to subsist for a year on wild foods following destruction of their gardens by fire during the exceptionally dry summer of 1972 (Lizot, 1974).

Conclusions

Palynological, geological, sedimentological, hydrological, paleoclimatological, and biogeographical data support the existence of episodes of reduction of the rain forest across the central lowlands to enclaves separated by more open vegetation of varying composition, extent, location, and duration, including mixed-gallery forest, parkland, savanna, and cerrado, prior to circa 7000 B.P.

This fractionation is reflected in the concentric distributions of linguistic, genetic, and cultural traits, implying displacement of an earlier human population by a later one across the central lowlands. The linguistic classification identifies the initial colonists as Ge-Pano-Carib speakers who were isolated on the periphery when Equatorial speakers expanded from homelands in the west as the rain forest coalesced.

Although archeological evidence for human occupation by circa 13,000 B.P. has been encountered mainly in regions likely to have been dominated by open vegetation, the isolated occurrences of bifacial stone projectile points and the existence of a few campsites in locations that would have remained forested under the most extreme scenario indicate that this distribution reflects sampling bias rather than reality. The predominance of similar expedient tools in open sites and rock shelters throughout the lowlands implies the existence of generalized foraging strategies focusing on similar combinations of wild resources. Both the archeological record and the ethnographic documentation of the subsistence behavior of surviving rain forest hunter-gatherers testify to the sustainability and nutritional adequacy of a wild-food diet.

All the early Holocene archeobotanical remains reported thus far in Amazonia represent wild plants, except in eastern Colombia and Ecuador, where maize pollen and phytoliths have been dated circa 7000 B.P. Since this antiquity precedes the domestication of maize in Mexico, it is disputed. The anthropogenic origin of charcoal and pollen of secondary vegetation is also dubious in the context of periodic drought. By contrast, the hypothesis that the black soil in preceramic sites along the Jamarí in southwestern Amazonia is proxy evidence for the adoption of shifting cultivation by 4800 B.P. is compatible with phylogeographical research identifying this region as the locus of manioc domestication.

Our reconstruction of human adaptation to mixed rain forest and open environments in Amazonia during the Pleistocene-Holocene transition draws on paleoenvironmental, climatic, archeological, ethnographic, and linguistic evidence and evaluates the information in each category against the others. The result is a generalized overview that incorporates what seems to us to be the best fit of each type of data. We hope that specialists in the various disciplines will find it sufficiently intriguing to examine it objectively rather than dismiss it out of hand.

Acknowledgments

The unpublished data from the power-line transects were obtained by Eurico Th. Miller, who conducted the fieldwork and analyses under the auspices of the Centrais Elétricas do Norte do Brasil S.A. (ELETRONORTE), Brasília. Funds for carbon-14 dating were provided by the Neotropical Lowland Research Program, National Museum of Natural History, Washington, D.C. We are grateful to both of these organizations for their long-term support.

References

Ardilla Calderón, G. I. (1991): The peopling of northern South America. In *Clovis: Origins and adaptations,* edited by R. Bonnichsen and K. Turnmire, 261–282. Eugene: Oregon State University, Center for the Study of the First Americans.

Ardilla Calderón, G. I., and G. G. Politis (1989): Nuevos datos para un viejo problema: Investigación y discusiones en torno del poblamiento de América del Sur. *Boletín del Museo del Oro* 23:3–45.

Athens, J. S., and J. V. Ward (1999): The late Quaternary of the western Amazon: Climate, vegetation, and humans. *Antiquity* 73:287–302.

Balée, W. (1989): The culture of Amazonia forests. *Advances in Economic Botany* 7: 1–21.

Barrow, C. J. (1985): The development of the *várzeas* (floodlands) of Brazilian Amazonia. In *Change in the Amazon Basin,* edited by J. Hemming, vol. 1, 108–128. Manchester University Press.

Behling, H. (1996): First report on new evidence for the occurrence of *Podocarpus* and possible human presence at the mouth of the Amazon during the late-glacial. *Vegetation History and Archaeobotany* 5(3):241–246.

Behling, H., and H. Hooghiemstra (2001): Neotropical savanna environments in space and time: Late Quaternary interhemispheric comparisons. In *Interhemispheric climate linkages,* edited by V. Markgraf, 307–323. San Diego: Academic Press.

Blake, M., B. S. Chisholm, J. E. Clark, and K. Mudar (1992): Non-agricultural staples and agricultural supplements: Early Formative subsistence in the Soconusco region, Mexico. In *Transitions to agriculture in prehistory,* edited by A. B. Gebauer and T. D. Price, 133–151. Madison: Prehistory Press.

Bonavia, D., and A. Grobman (1998): Review of evidence on preceramic maize in the central Andean region. *Proceedings of the XIII International Congress on Prehistoric and Protohistoric Sciences* 5:403–406.

Bonell, M. (1998): Possible impacts of climate variability and change on tropical forest hydrology. *Climate Change* 39:215–272.

Bourlière, F. (1973): The comparative ecology of rain forest mammals in Africa and tropical America: Some introductory remarks. In *Tropical forest ecosystems* in *Africa and South America: A comparative review*, edited by B. J. Meggers, E. S. Ayensu, and W. D. Duckworth, 279–292. Washington, D.C.: Smithsonian Institution Press.

Burger, R. L., and N. J. Van der Merwe (1990): Maize and the origin of highland Chavín civilization: An isotopic perspective. *American Anthropologist* 92:85–95.

Bush, M. B. (1994): Amazonian speciation: A necessarily complex model. *Journal of Biogeography* 21:5–17.

Bush, M. B., D. R. Piperno, and P. A. Colinvaux (1989): A 6000-year history of Amazonian maize cultivation. *Nature* 340:303–305.

Campbell, D. G., D. Daly, G. Prance, and V. Maciel (1986): Quantitative ecological inventory of terra firme and várzea tropical forest on the Rio Xingu, Brazilian Amazon. *Brittonia* 38:369–393.

Carvalho, J.C.M. (1952): *Notas de viagem ao Rio Negro*. Rio de Janeiro: Museu Nacional, Publicações Avulsas 9.

Charles-Dominique, P., P. Blanc, D. Larpin, M.-P. Ledru, B. Riéra, C. Sarthou, M. Servant, and C. Tardy (1998): Forest perturbations and biodiversity during the last ten thousand years in French Guiana. *Acta Oecologica* 19:295–302.

Clement, C. R. (1993): Native Amazonian fruits and nuts: Composition, production, and potential use for sustainable development. In *Tropical forests, people, and food*, edited by C. M. Hladik et al., 139–152. Paris: UNESCO.

Colinvaux, P. A. (1997): The Ice-Age Amazon and the problem of diversity. *Review of Archaeology* 19:1–10.

Colinvaux, P. A., and P. E. de Oliveira (2001): Amazon plant diversity and climate through the Cenozoic. *Palaeogeography, Palaeoclimatology, Palaeoecology* 166:51–63.

Colinvaux, P. A., M. B. Bush, M. Steinitz-Kannan, and M. C. Miller (1997): Glacial and postglacial pollen records from the Ecuadorian Andes and Amazon. *Quaternary Research* 48:69–78.

Colinvaux, P. A., P. E. de Oliveira, and M. B. Bush (2000): Amazonian and Neotropical plant communities on glacial time-scales: The failure of the aridity and refuge hypotheses. *Quaternary Science Reviews* 19:141–169.

Cooke, R. G., and A. J. Ranere (1992): Precolumbian influences on the zoogeography of Panama: An update based on archaeofaunal and documentary data. In *Biogeography of Mesoamerica*, edited by S. Darwin and A. Weldon, 21–58. New Orleans: Tulane University.

Eden, M. J. (1974): Paleoclimatic influences and the development of savanna in southern Venezuela. *Journal of Biogeography* 1:95–109.

Evans, C., and B. J. Meggers (1960): *Archeological investigations in British Guiana*. Bureau of American Ethnology Bulletin 177. Washington, D.C.: Smithsonian Institution.

Fritz, G. L. (1994): Reply. *Current Anthropology* 35:639–643.

Gentry, A. H. (1993): Diversity and floristic composition of lowland tropical forest in Africa and South America. In *Biological relationships between Arctica and South America*, edited by P. Goldblatt, 500–547. New Haven: Yale University Press.

Giulietti, A. M., and J. R. Pirani (1988): Patterns of geographic distribution of some plant species from the Espinhaço range, Minas Gerais and Bahia, Brazil. In *Proceedings of a workshop on Neotropical distribution patterns,* edited by P. E. Vanzolini and W. R. Heyer, 39–69. Rio de Janeiro: Academia Brasileira de Ciência.

Gnecco, C. (1998): Paisajes antropogénicos en el pleistoceno final y holoceno temprano en Colombia. *Revista de Antropología y Arqueología* 10(1):45–61.

Gnecco, C., and S. Mora (1997): Late Pleistocene/early Holocene tropical forest occupations at San Isidro and Peña Roja, Colombia. *Antiquity* 71:683–690.

Greenberg, J. (1987): *Language in the Americas.* Stanford: Stanford University Press.

Haberle, S. G., and M. A. Maslin (1999): Late Quaternary vegetation and climate change in the Amazon Basin based on a 50,000-year pollen record from the Amazon fan, ODP Site 932. *Quaternary Research* 51:27–38.

Hansen, B.C.S., and D. T. Rodbell (1995): A late-glacial/Holocene pollen record from the eastern Andes of northern Peru. *Quaternary Research* 44:216–227.

Harley, R. M. (1988): Evolution and distribution of Eriope (Labiatae) and its relatives in Brazil. In *Proceedings of a workshop on Neotropical distribution patterns,* edited by P. E. Vanzolini and W. R. Heyer, 71–120. Rio de Janeiro: Academia Brasileira de Ciências.

Herrera, R., C. F. Jordan, H. Klinge, and E. Medina (1978): Amazon ecosystems: Their structure and function with particular emphasis on nutrients. *Interciencia* 3:223–232.

Hilbert, K. (1998): Nota sobre algumas pontas-de-projétil da Amazônia. *Estudos Ibero-Americanos* 24:291–310.

Hooghiemstra, H., and T. Van der Hammen (1998): Neogene and Quaternary development of the Neotropical rain forest: The forest refugia hypothesis, and a literature review. *Earth-Science Reviews* 44:147–183.

Irion, G., W. J. Junk, and J.A.S.N. de Melo (1997): The large central Amazonian River floodplains near Manaus: Geological, climatological, hydrological, and geomorphological aspects. In *The central Amazonian floodplain: Ecology of a pulsing system,* edited by W. J. Junk, 23–46. Berlin: Springer-Verlag.

Iriondo, M., and E. M. Latrubesse (1994): A probable scenario for a dry climate in central Amazonia during the late Quaternary. *Quaternary International* 21:121–128.

Junk, W. J., ed. (1997): *The central Amazon floodplain: Ecology of a pulsing system.* Berlin: Springer-Verlag.

Kellman, M., R. Tackaberry, N. Brokaw, and J. Meave (1994): Tropical gallery forests. *National Geographic Research and Exploration* 10:92–103.

Kipnis, R. (1998): Early hunter-gatherers in the Americas: Perspectives from central Brazil. *Antiquity* 72:581–592.

Latrubesse, E. M., and C. G. Ramonell (1994): A climatic model for southwestern Amazonia in last glacial times. *Quaternary International* 21:163–169.

Ledru, M.-P., J. Bertaux, and A. Sifeddine (1998): Absence of last glacial maximum records in lowland tropical forests. *Quaternary Research* 49:233–237.

Lizot, J. (1974): El río de los Periquitos: Breve relato de un viaje entre los Yanomami del Alto Siapa. *Antropológica* 37:3–23.

Lleras, H., and C. Gnecco (1986): Puntas de proyectil en el Valle de Popayán. *Boletín del Museo del Oro* 17:44–57.

López, C. E. (1998): Evidence of late Pleistocene/early Holocene occupations in the tropical lowlands of the middle Magdalena Valley. In *Recent advances in the archaeology of the northern Andes*, edited by A. Oyuela-Caycedo and J. S. Raymond, 1–9. Los Angeles: UCLA Institute of Archaeology.

Lucas, Y., F. Soubiès, A. Chauvel, and T. Desjardins (1993): Estudos do solo revelam alterações climáticas da Amazônia. *Ciência Hoje* 16(93):36–39.

Martin, L., J. Bertaux, M.-P. Ledru, P. Mourguiart, A. Sifeddine, F. Soubiès, and B. Turcq (1995): Perturbaciones del régimen de las lluvias y condiciones de tipo El Niño en América del Sur tropical desde hace 7000 años. *Bulletin de l'Institut Frances d'études Andines* 24:595–605.

Maslin, M. A., and S. J. Burns (2000): Reconstruction of the Amazon Basin effective moisture availability over the past 14,000 years. *Science* 290:2285–2287.

Meave, J., M. Kellman, A. MacDougall, and J. Rosales (1991): Riparian habitats as tropical forest refugia. *Global Ecology and Biogeography Letters* 1:69–76.

Meggers, B. J. (1964): North and South American cultural connections and convergences. In *Prehistoric man in the New World*, edited by J. D. Jennings and E. Norbeck, 511–526. Chicago: University of Chicago Press.

———. (1975): Application of the biological model of diversification to cultural distributions in tropical lowland South America. *Biotropica* 7:141–161.

———. (1982): Archeological and ethnographic evidence compatible with the model of forest fragmentation. In *Biological diversification in the tropics*, edited by G. T. Prance, 483–496. New York: Columbia University Press.

———. (1994): Biogeographical approaches to reconstructing the prehistory of Amazonia. *Biogeographica* 70:97–110.

———. (1995): Amazonia on the eve of European contact: Ethnohistorical, ecological, and anthropological perspectives. *Revista de Arqueología Americana* 8:91–115.

———. (1996): *Amazonia: Man and culture in a counterfeit paradise.* 2nd ed. Washington, D.C.: Smithsonian Institution Press.

———. (2001): The mystery of the Marajoara. *Amazoniana* 16:421–440.

Middleton, B. A., E. Sanchez-Rojas, B. Suedmeyer, and A. Michels (1997): Fire in a tropical dry forest of Central America: A natural part of the disturbance regime? *Biotropica* 29:515–517.

Migliazza, E. C. (1982): Linguistic prehistory and the refuge model in Amazonia. In *Biological diversification in the tropics*, edited by G. T. Prance, 497–519. New York: Columbia University Press.

Miller, E. Th. (1987): Pesquisas arqueológicas paleoindígenas no Brasil Ocidental. *Estudios Atacameños* 8:37–61.

Miller, E. Th., et al. (1992): *Arqueologia nos empreendimentos hidrelétricos da Eletronorte; Resultados preliminares.* Brasília: Centrais Elétricas do Norte do Brasil S.A.

Minnis, P. E. (1992): Earliest plant cultivation in the desert borderlands of North America. In *The origins of agriculture*, edited by C. W. Cowan and P. J. Watson, 121–141. Washington, D.C.: Smithsonian Institution Press.

Molion, L.C.B., and J. C. de Moraes (1987): Oscilação Sul e descarga de rios en America do Sul tropical. *Revista Brasileira de Engeneria, Caderno de Hidrologia* 5: 53–63.

Moore, H. E., Jr. (1973): Palms in the tropical forest ecosystems of Africa and South America. In *Tropical forest ecosystems in Africa and South America: A comparative review*, edited by B. J. Meggers, E. S. Ayensu, and W. D. Duckworth, 63–88. Washington, D.C.: Smithsonian Institution Press.

Müller, J., G. Irion, J. Nunes de Mello, and W. Junk (1995): Hydrological changes of the Amazon during the last glacial-interglacial cycle in central Amazonia (Brazil). *Naturwissenschaften* 82:232–235.

Nittrouer, C. A., S. A. Kuehl, R. W. Sternberg, A. G. Figueiredo Jr., and L.E.C. Faria Jr. (1995): An introduction to the geological significance of sediment transport and accumulation on the Amazon continental shelf. *Marine Geology* 125:177–192.

Nobre, C. A., and N. O. Renno (1985): Droughts and floods in South America due to the 1982–1983 ENSO episode. Houston, American Meteorological Society. *Proceedings of the 16th Conference on Hurricanes and Tropical Meteorology:* 131–133.

Nores, M. (1999): An alternative hypothesis for the origin of Amazonian bird diversity. *Journal of Biogeography* 26:475–485.

Oliveira, P. E. de (1996): Glacial cooling and forest disequilibrium in western Amazonia. *Academia Brasileira de Ciência, Anales* 68(supp. 1):129–138.

Olsen, K. M., and B. A. Schaal (1999): Evidence on the origin of cassava: Phylogeography of *Manihot esculenta*. *Proceedings of the National Academy of Science* 96: 5586–5591.

———. (2001): Microsatellite variation in cassava (*Manihot esculenta*, Euphorbiaceae) and its wild relatives: Further evidence for a southern Amazonian origin of domestication. *American Journal of Botany* 88:131–142.

Padoch, D., J. M. Ayres, M. Pinedo-Vasquez, and A. Henderson, eds. (1999): *Várzea: Diversity, development, and conservation of Amazonia's whitewater floodplains.* New York: Botanical Garden Press.

Pearsall, D. M. (1995): Domestication and agriculture in the New World tropics. In *Last hunters–first farmers*, edited by T. G. Price and A. B. Gebauer, 157–192. Santa Fe: School of American Research.

———. (1999): Agricultural evolution and the emergence of Formative societies in Ecuador. In *Pacific Latin America in prehistory*, edited by M. Blake, 161–170. Pullman: Washington State University Press.

Pérez-Salicrup, D. R., V. L. Sork, and F. E. Putz (2001): Lianas and trees in liana forest of Amazonian Bolivia. *Biotropica* 33:34–47.

Piperno, D. R. (1985): Phytolith analysis and tropical paleo-ecology: Production and taxonomic significance of siliceous forms in New World plant domesticates and wild species. *Review of Palaeobotany and Palynology* 45:185–228.

———. (1990): Aboriginal agriculture and land usage in the Amazon basin, Ecuador. *Journal of Archaeological Science* 17:665–677.

———. (1998): Paleoethnobotany in the Neotropics from microfossils: New insights into ancient plant use and agricultural origins in the tropical forest. *Journal of World Prehistory* 12:393–449.

Piperno, D. R., and P. Becker (1996): Vegetational history of a site in the central Amazon basin derived from phytolith and charcoal records from natural soils. *Quaternary Research* 45:202–209.

Politis, G. G., and J. Rodríguez (1994): Algunos aspectos de subsistencia de los Nukak de la Amazonia Colombiana. *Colombia Amazónica* 7:169–207.

Politis, G. G., G. A. Martínez, and J. Rodríguez (1997): Caza, recolección y pesca como estrategia de explotación de recusos en florestas tropicales lluviosas: Los Nukak de la amazonía colombiana. *Revista Española de Antropología Americana* 27:167–197.

Prance, G. T., ed. (1982): *Biological diversification in the tropics*. New York: Columbia University Press.

Prous, A. (1991): Fouilles de l'Abri du Boquete, Minas Gerais, Brésil. *Journal de la Société des Américanistes* 77:77–109.

Reichel-Dolmatoff, G. (1986): *Arqueología de Colombia: Un texto introductorio*. Bogotá: FUNBOTANICA.

Roosevelt, A. C. (1991): *Moundbuilders of the Amazon*. New York: Academic Press.

Roosevelt, A. C., M. L. da Costa, C. L. Machado, M. Michab, N. Mercier, H. Valladas, J. Feathers, W. Barnett, M. I. da Silveira, A Henderson, J. Silva, B. Chernoff, D. S. Reese, J. A. Holman, N. Toth, and K. Schick (1996): Paleoindian cave dwellers in the Amazon: The peopling of the Americas. *Science* 272:373–384.

Rothhammer, F., and C. Silva (1992): Gene geography of South America: Testing models of population displacement based on archeological evidence. *American Journal of Physical Anthropology* 89:441–446.

Ruddle, K., D. Johnson, P. K. Townsend, and J. D. Rees (1978): *Palm sago: A tropical starch from marginal lands*. Honolulu: University Press of Hawaii.

Salo, J. (1987): Pleistocene forest refuges in the Amazon: Evaluation of the biostratigraphical, lithostratigraphical, and geomorphological data. *Annali Zoologica Fennici* 24:203–211.

Santos, J., O. Schneider, B. W. Nelson, and C. A. Giovannini (1993): Campos de dunas; corpos de areia sob leitos abandonados de grandes rios. *Ciência Hoje* 16(93):22–25.

Schoeninger, M. J. (1996): Stable isotope studies in human evolution. *Evolutionary Anthropology* 4:83–98.

Servant, M., J.-C. Fontes, M. Rieu, and J.-F. Saliège (1981): Phase climatiques arides holocènes dans le sud-ouest de l'Amazonie (Bolivie). *Comptes Rendus de l'Academi des Sciences, Paris*, 2nd ser., 292:1295–1297.

Servant, M., J. Maley, B. Turcq, M.-L. Absy, P. Brenac, M. Fournier, and M.-P. Ledru (1993): Tropical forest changes during the late Quaternary in African and South American lowlands. *Global and Planetary Change* 7:25–40.

Sifeddine, A., L. Martin, B. Turcq, C. Volkmer-Ribeiro, R. Soubiès, and K. Suguio (2001): Variations of the Amazonian rain forest environment: A sedimentological record covering 30,000 years. *Palaeogeography, Palaeoclimatology, Palaeoecology* 108:221–235.

Simões, M. F. (1976): Nota sobre duas pontas de projetil da Bacia Tapajós (Pará). *Belém: Boletim do Museu Paraense Emílio Goeldi* 62.

Sioli, H., ed. (1984): *The Amazon: Limnology and landscape ecology of a mighty river and its basin.* Dordrecht: Dr. W. Junk Publishers.

Smith, B. (1995): The origins of agriculture in the Americas. *Evolutionary Anthropology* 3:174–184.

Soubiès, F. (1979–80): Existence d'une phase sèche en Amazonie brésilienne datée par la présence de charbons dans les sols (6.000–3.000 ans B.P.). *Cahiers O.R.S.T.O.M.,* Geology ser., 11:133–148.

Tricart, J. (1985): Evidence of upper Pleistocene dry climates in northern South America. In *Environmental change and tropical geomorphology,* edited by I. Douglas and T. Spencer, 197–217. London: Allen and Unwin.

Turcq, B., A. Sifeddine, L. Martin, M. L. Absy, F. Soubiès, K. Suguio, and C. Volkmer-Ribeiro (1998): Amazonia rainforest fires: A lacustrine record of 7000 years. *Ambio* 27:139–142.

Vacher, S., S. Jérémie, and J. Briand, eds. (1998): *Amérindiens du Sinnamary (Guyane): Archéologie en forêt équatoriale.* Documents d'Archéologie Française 70. Paris: Editions de la Maison des Sciences de l'Homme.

Van der Hammen, T., and H. Hooghiemstra (2000): Neogene and Quaternary history of vegetation, climate, and plant diversity in Amazonia. *Quaternary Science Reviews* 19:725–742.

Van der Hammen, T., and M. L. Absy (1994): Amazonia during the last glacial. *Palaeogeography, Palaeoclimatology, Palaeoecology* 109:247–261.

Van der Merwe, N. J., J. A. Lee-Thorp, and J. S. Raymond (1993): Light, stable isotopes and the subsistence base of Formative cultures at Valdivia, Ecuador. In *Prehistoric human bone: Archaeology at the molecular level,* edited by J. B. Lambert, 63–97. Berlin: Springer-Verlag.

Vilhena Vialou, A., P. D. de Blasis, L. Figuti, P. Paillet, and D. Vialou (1999): Art rupestre et habitats préhistoriques au Mato Grosso (Brésil). *BAR International Series* 746:9–22.

Wilbert, J. (1976): *Manicaria saccifera* and its cultural significance among the Warao Indians of Venezuela. *Botanical Museum Leaflets* 24(10):275–332. Cambridge: Harvard University.

Woods, W. I. (1995): Comments on the black earths of Amazonia. *Papers and Proceedings of Applied Geography Conferences* 18:159–165.

Contributors

Brit Asmussen's doctoral research in the School of Archaeology and Anthropology at the Australian National University, Canberra, Australia, concerns the analysis of several middle to late Holocene occupation deposits in rock shelters of Central Queensland. Her research interests include social and economic change, human ecology, plant use, taphonomy, and methodology.

William P. Barse is a senior archeologist at URS Corporation and a research associate in the Department of Anthropology at the Smithsonian Institution, Washington, D.C. His work in the Orinoco River, Venezuela, provides a framework for understanding the Archaic cultures, climatic conditions in the Holocene, and early Formative village life. He has published in *Science* and *Interciencia*.

F. David Bulbeck is a specialist in Indonesian historical archaeology. He works at the School of Archaeology and Anthropology, in the Australian National University, Canberra, Australia. He has published articles in the *Bulletin of the Indo-Pacific Prehistory Association, Perspectives in Human Biology,* and *Australian Historical Archaeology.*

Joanna Casey is a professor of anthropology at the University of South Carolina. She is interested in the Middle Stone Age, origins of food production, ethnoarchaeology, and gender in northern Ghana. She has published in the series "British Archaeological Reports International" and *Antiquity.*

Richard G. Cooke is a staff scientist at the Smithsonian Tropical Research Institute, Panama. His interests are the archaeology of the New World, Central American land bridge, archaeozoology, and ideology: Recent publications can be found in *Quaternary International,* the *Journal of World Prehistory,* and *World Archaeology.*

Cristóbal Gnecco is a professor of anthropology with Universidad del Cauca, Colombia. His research interests include the early settlement of the Americas and the political economy of ancient South American societies. He has published articles in *Antiquity* and *Quaternary International.* He is also the author of several books published in Colombia.

Raquel Martí is with the Universidad Nacional de Educación a Distancia, Madrid, Spain. She is interested in the Middle and Later Stone Age of Central Africa and has worked in several African countries. Her publications include articles in *Quaternary International,* the *Journal of Archaeological Science,* and *Nyame Akuma.*

Betty J. Meggers is a research associate in the National Museum of Natural History, Smithsonian Institution. Her interests include cultural evolution and diffusion, ecology, Amazonian prehistory, coastal Ecuador, and pre-Columbian transpacific contact. Her awards include decorations and honorary doctorates from several Latin American countries. Recent publications include articles in *Latin American Antiquity, Amazoniana,* and *Revista de Arqueología Americana.*

Julio Mercader is a professor of anthropology with George Washington University, Washington, D.C. His interest is in Paleolithic adaptations to tropical forest environments, and he has excavated in Central and West Africa. Recent publications include articles in the *American Journal of Physical Anthropology, Azania, Quaternary International, Quaternary Research,* the *Journal of Archaeological Science,* the *Journal of Anthropological Research, Evolutionary Anthropology,* and *Science.*

Eurico Th. Miller is an environmental analyst for cultural patrimony on the staff of Eletronorte and has conducted surveys and excavations in Brazil.

Santiago Mora is a professor of anthropology at St. Thomas University in New Brunswick, Canada. His primary research interests include native resource management, savanna and tropical forest adaptive systems, and Latin American archaeology and history. He has published in *Antiquity* and *Latin American Archaeology Reports.*

Anthony J. Ranere is professor of anthropology at Temple University, Philadelphia, Pennsylvania, and research associate at the Smithsonian Tropical Research Institute, Panama. He has carried out research in Panama on the human use of Central American tropical forests. Recent publications include articles in *Nature, World Archaeology,* and *American Anthropologist.*

Anne-Marie Sémah is a senior staff scientist at the Institut de Recherche pour le Développement, in France. She has worked on the paleoenvironment of hominid sites in Java and Lapita sites in New Caledonia. Her recent publications include articles in *Comptes Rendus de l'Academie des Sciences, Paris* and *Journal de la Société des Océanistes.*

François Sémah is a professor at the Muséum National d'Histoire Naturelle, Paris, France. His honors include the Silver Medal from the Centre National Recherche Scientifique, and he was a senior staff scientist at C.N.R.S. until 1999. Sémah has worked in Southeast France, Polynesia, and Indonesia. Recent publications include articles in the *Journal of Human Evolution* and the *Journal of Archaeological Science.*

Truman Simanjuntak is a senior staff scientist at the Center for Archaeological Research, Puslit Arkeologi, Jakarta, Indonesia. His research interests include the Paleolithic and the Neolithic. He has published in the *Bulletin of the Indo-Pacific Prehistory Association.*

Index

Note: Page numbers in italics indicate illustrations.